Spectral Synthesis

Pure and Applied Mathematics

A Series of Monographs and Textbooks

RECENT TITLES

GLEN E. BREDON. Introduction to Compact Transformation Groups

WERNER GREUB, STEPHEN HALPERIN, AND RAY VANSTONE. Connections, Curvature, and Cohomology: Volume I, De Rham Cohomology of Manifolds and Vector Bundles. Volume II, Lie Groups, Principal Bundles, and Characteristic Classes. *In preparation:* Volume III, Cohomology of Principal Bundles and Homogeneous Spaces

XIA DAO-XING. Measure and Integration Theory of Infinite-Dimensional Spaces: Abstract Harmonic Analysis

RONALD G. DOUGLAS. Banach Algebra Techniques in Operator Theory

WILLARD MILLER, JR. Symmetry Groups and Their Applications

ARTHUR A. SAGLE AND RALPH E. WALDE. Introduction to Lie Groups and Algebras

T. BENNY RUSHING. Topological Embeddings

JAMES W. VICK. Homology Theory: An Introduction to Algebraic Topology

E. R. KOLCHIN. Differential Algebra and Algebraic Groups

GERALD J. JANUSZ. Algebraic Number Fields

A. S. B. HOLLAND. Introduction to the Theory of Entire Functions

WAYNE ROBERTS AND DALE VARBERG, Convex Functions

A. M. OSTROWSKI. Solution of Equations in Euclidean and Banach Spaces, Third Edition of Solution of Equations and Systems of Equations

H. M. EDWARDS. Riemann's Zeta Function

SAMUEL EILENBERG. Automata, Languages, and Machines: Volume A. *In preparation:* Volume B

MORRIS HIRSCH AND STEPHEN SMALE. Differential Equations, Dynamical Systems, and Linear Algebra

WILHELM MAGNUS. Noneuclidean Tesselations and Their Groups

J. DIEUDONNÉ. Treatise on Analysis, Volume IV

FRANÇOIS TREVES. Basic Linear Partial Differential Equations

WILLIAM M. BOOTHBY. An Introduction to Differentiable Manifolds and Riemannian Geometry

BRAYTON GRAY. Homotopy Theory: An Introduction to Algebraic Topology

ROBERT A. ADAMS. Sobolev Spaces

JOHN J. BENEDETTO. Spectral Synthesis

In preparation

D. V. WIDDER. The Heat Equation

IRVING E. SEGAL. Mathematical Cosmology and Extragalactic Astronomy

J. DIEUDONNÉ. Treatise on Analysis, Volume II, enlarged and corrected printing

Spectral Synthesis

JOHN J. BENEDETTO
Department of Mathematics
University of Maryland

ACADEMIC PRESS, INC. New York London San Francisco 1975
A Subsidiary of Harcourt Brace Jovanovich, Publishers

Prof. John J. Benedetto

Born 1939 in Boston. Received B.A. from Boston College in 1960, M.A. from Harvard University in 1962, and Ph.D. from University of Toronto in 1964. Assistant professor at New York University from 1964 to 1965; and research associate at University of Liège and the Institute for Fluid Dynamics and Applied Mathematics from 1965 to 1966; employed by RCA and IBM from 1960 to 1965. At University of Maryland, assistant professor from 1966 to 1967, associate professor from 1967 to 1973, and professor beginning in 1973. Visiting positions include the Scuola Normale Superiore at Pisa from 1970 to 1971 and 1974 (spring) and MIT in 1973 (fall); also Senior Fulbright-Hays Scholar from 1973 to 1974.

Licensed edition for the Western Hemisphere
Academic Press, Inc., New York/London/San Francisco,
A Subsidiary of Harcourt Brace Jovanovich, Publishers.

Originally published in the series "Mathematische Leitfäden" edited by G. Köthe and G. Trautmann.

Library of Congress Cataloging in Publication Data

Benedetto, John.
Spectral synthesis.

(Pure and applied mathematics, a series of monographs and textbooks ; 66)
Bibliography: p.
Includes indexes.
1. Spectral synthesis (Mathematics)
2. Locally compact Abelian groups. 3. Tauberian theorems. I. Title. II. Series.
QA3.P8 vol. 66 [QA403] 510'.8s [515'.78] 75-13765
ISBN 0-12-087050-9

AMS (MOS) 1970 Subject Classifications: 20H10, 50C05

Printed in Germany

Setting: William Clowes & Sons Ltd., London.
Printer: Johannes Illig, Göppingen

Introduction

Purpose

The major topic in this book is spectral synthesis. The purpose of the book can be described as follows:

A. To trace the development of spectral synthesis from its origins in the study of Tauberian theorems;

B. To draw attention to other mathematical areas which are related to spectral synthesis;

C. To give a thorough (although not encyclopedic) treatment of spectral synthesis for the case of $L^1(G)$;

D. To introduce the "integration" and "structure" problems that have emerged because of the study of spectral synthesis.

A and **B.** The first two points are discussed in Chapters 1 and 2, and are the major reasons that such a large bibliography has evolved. By the end of Chapter 2, the significant relationship between Tauberian theorems and spectral synthesis is not only firmly established, but the extent to which this relationship is still undetermined is emphasized by the open "*C*-set-*S*-set" problem. Also, Chapters 1 and 2 view spectral synthesis amidst other problems and influences.

C. By contrast, Chapter 3 is (or at least is meant to be) more business-like, and synthesis is essentially the only topic under discussion.

It is fashionable and important to exposit spectral synthesis results in the setting of regular Banach algebras (or even more generally). One of the reasons that I have chosen the "special" $L^1(G)$ case is because of the two problems mentioned in the fourth point; neither problem has reached the stage where the algebraic structure has been successfully exploited, and a presentation of the analytic techniques available for $L^1(G)$ (and not always available more generally) seemed the reasonable thing to do.

Of course, L^1 is still the right setting for applications and also the proper setting to exposit best the range of topics I have treated.

D. The "integration" problem (e.g. Sections 3.2.1, 3.2.2, 3.2.4, 3.2.8, 3.2.9) is to find the relationship between spectral synthesis and integration theories. The problem was essentially posed by Beurling, and at this time very little is known except that such a relationship exists at a fundamental level.

The "structure" problem (e.g. Sections 3.2.15–3.2.19) is one with a more secular flavor. Basically, one would like to know the intrinsic properties of a distribution, such as its support, when we have a knowledge of its Fourier transform, and vice-versa. This is precisely the sort of information that one must have at hand in order to solve spectral synthesis problems. Essentially this is the problem of determining the finer structure of Schwartz distributions by using Fourier analysis. Schwartz, of course, studied the structure of distributions and obtained representations (of distributions) in terms of derivatives of ordinary functions.

Notation

G will always be a locally compact abelian group with dual group Γ. The problems that we've considered are quite classical, and my intent is not altered if one takes $G = \mathbf{R}$, the real numbers, or $G = \mathbf{Z}$, the integers; in these cases $\Gamma = \mathbf{R}$, the real numbers again, and $\Gamma = \mathbf{T} = \mathbf{R}/2\pi\mathbf{Z}$, respectively. For these two examples, "(γ,x)" is "$e^{i\gamma x}$".

Also, notationally, $\mathbf{Q}$ denotes the rational numbers, $\mathbf{C}$ the complex numbers, $\mathbf{R}^+ = \{r \in \mathbf{R} : r \geqslant 0\}$, $\mathbf{Z}^+$ is $\{0, 1, 2, \ldots\}$ or $\{1, 2, \ldots\}$, ∂X is the boundary of X, int X is the interior of X, $X^\sim$ is the complement of X, and supp φ is the support of the function φ.

Depending on the situation, $\mathbf{T}$ will be considered as $\{z \in \mathbf{C} : |z| = 1\}$ or as an interval of length 2π on $\mathbf{R}$.

Bibliography

I have already mentioned that the large bibliography is due to the historical notes and the fact that many other topics related to synthesis are mentioned and referenced.

I have not provided as large a bibliography for current topics in synthesis, and several issues, in which there is presently a good deal of activity, are not adequately discussed. A sampling of these issues is: complex methods in synthesis; operational calculus problems; extensions to the non-abelian case; isomorphism problems; the relation between arithmetic and synthesis; tensor algebra in harmonic analysis; the theory of multipliers and p-spaces; and probabilistic methods in synthesis. Consequently, some of the most active and competent workers in spectral synthesis are not duly listed in the bibliography. On the other hand, we do not feel it is too big a step from this book to their work.

Exercises

The exercises contain both problems and remarks. Many of the problems are easy although some of the accompanying remarks, usually referenced, may contain more difficult material. In any case, a reading of the problems will provide added perspective.

Acknowledgements

My first thanks go to C. R. Warner who introduced me to the spectral synthesis problem. There has also been the generous and often ingenious assistance of G. Helzer, R. Johnson, and G. Salmons who were always willing to discuss mathematics with me, and whose explicit influence is to be found in some of the exercises. Further, I have benefited mathematically from conversations and correspondence (in ways they may not remember) with S. Antman, A. Atzmon, L. Gårding, C. Graham, J.-P. Kahane, Y. Katznelson, R. Kaufman, C. McGehee, D. Niebur, J. Osborn, F. Ricci, D. Sweet, and P. Wolfe. I wrote Chapter 2 while I was a guest at MIT and Chapter 3 while I was a guest at the Scuola Normale Superiore; I would like to thank K. Hoffman and E. Vesentini of these institutions for the hospitality extended to me. My final thanks go to G. Maltese for being a vital source of encouragement to me and to G. Köthe for asking me to write the book.

Boston and Pisa, Summer 1974 J. J. BENEDETTO

Table of contents

Index of notation

The first list of notation includes sets and spaces; the second includes specific elements, operations, and the remaining symbols. Symbols that are used only in the section where they are introduced are not generally included in either list.

I.

1 The spectral synthesis problem

1.1 The Fourier transform of $L^1(G)$

1.1.1 Prerequisites. We suppose as known the basic theory of harmonic analysis as found in [Rudin, 5, Chapter 1] and the basic theory of commutative Banach algebras as found in [Loomis, 1, Chapter 4]. In this Paragraph 1.1, we quickly restate some of the necessary results from harmonic analysis in order to establish notation; and proceed far enough along to give a proof of one form of Wiener's Tauberian theorem (Theorem 1.1.3).

1.1.2 $L^1(G)$ and $L^\infty(G)$. Let G be a Hausdorff locally compact abelian group (LCAG), written additively, with Haar measure and dual group Γ. Haar measure is translation invariant on the Borel sets of G and such measures are unique up to multiplicative constants. We write $\hat{G} = \Gamma$ so that $\hat{\Gamma} = G$. Haar measure on G or Γ is denoted by m.

$L^1(G)$ (resp. $L^\infty(G)$) is the space of $\mathbf{C}$-valued integrable (resp. measurable and essentially bounded) functions with respect to m, where two functions are identified if they are equal a.e. (resp. equal locally a.e.). (To be more precise about the definition of $L^\infty(G)$ we make this parenthetical remark. An m-measurable set A is *locally null* if $m(A \cap K) = 0$ for each compact subset $K \subseteq G$. A property holds *locally* a.e. if it is true outside of a locally null set. $\mathscr{L}^\infty(G)$ is the space of all m-measurable functions Φ for which a constant $a_\Phi > 0$ exists such that $\{x \in G : |\Phi(x)| > a_\Phi\}$ is locally null. We write $\Phi \sim \Psi$ if $\{x : \Phi(x) \neq \Psi(x)\}$ is locally null, and so $L^\infty(G) = \mathscr{L}^\infty(G)/\sim$. Any non-discrete non-$\sigma$-compact LCAG contains a locally null set A with $mA = \infty$.) We designate the usual norms on $L^1(G)$ and $L^\infty(G)$ by $\|\ \|_1$ and $\|\ \|_\infty$, respectively.

As is well-known $L^\infty(G)$ is the Banach space dual of $L^1(G)$ and we define the duality between $L^1(G)$ and $L^\infty(G)$ by

$$\forall f \in L^1(G) \quad \text{and} \quad \forall \Phi \in L^\infty(G), \quad \langle \Phi, f \rangle = \int_G \Phi(x) f(x)\, dx.$$

$L^1(G)$ is a commutative Banach algebra with convolution

$$\forall f, g \in L^1(G), \quad f * g(x) = \int_G f(y) g(x - y)\, dy$$

as the multiplicative operation. $L^1(G)$ has a unit if and only if G is discrete.

1.1.3 Banach algebras. Let X be a commutative Banach algebra. An ideal $I \subseteq X$ is *regular* if X/I has a multiplicative identity. The *maximal ideal space*, X^m, of X is the collection of all the regular maximal ideals in X. $\gamma \in X'$ is a *multiplicative element* if $\langle \gamma, xy \rangle = \langle \gamma, x \rangle \langle \gamma, y \rangle$ for each $x, y \in X$. The following is a basic theorem in Banach algebras: *the set of non-zero multiplicative elements of the dual space, X', is identified with X^m by the mapping $\gamma \mapsto$ (kernel of γ) $\in X^m$*, where $\gamma \in X' \setminus \{0\}$ is multiplicative. As such X^m is given the induced weak $*$ topology from X'; and the *Gelfand transform* of $x \in X$ is the continuous function $\hat{x}: X^m \to \mathbf{C}$ defined by

$$\hat{x}(\gamma) = \langle \gamma, x \rangle.$$

X is *semi-simple* if the intersection of all of the regular maximal ideals in X is $\{0\}$. If $\hat{X}$ is the set of Gelfand transforms of X then X is semi-simple if and only if

$$x \in X \setminus \{0\} \quad \Rightarrow \quad \hat{x} \in \hat{X} \setminus \{0\}.$$

X is a regular Banach algebra if for each $\gamma \in X^m$ and each closed set $E \subseteq X^m \setminus \{\gamma\}$, there is $\hat{x} \in \hat{X}$ for which $\hat{x}(\gamma) = 1$ and $\hat{x}(E) = 0$. $L^1(G)$ is a commutative regular semi-simple Banach algebra whose maximal ideal space is Γ.

1.1.4 The theory of Fourier transforms for $L^1(G)$, $L^2(G)$, and $M(G)$. The *Fourier transform of $f \in L^1(G)$* is a function $\varphi: \Gamma \to \mathbf{C}$ defined as

$$\hat{f}(\gamma) = \varphi(\gamma) = \int_G f(x)\,(\gamma, x)\,dx, \qquad \gamma \in \Gamma;$$

and the space of Fourier transforms of $L^1(G)$ elements is denoted by $A(\Gamma)$. The Gelfand transform of $L^1(G)$ is the Fourier transform [Rudin, 5, Theorem 1.2.2]. The Fourier transform map, $L^1(G) \to A(\Gamma)$ is bijective; $A(\Gamma)$ is contained in the space $C_0(\Gamma)$ of continuous functions which vanish at infinity and it is a Banach algebra with pointwise multiplication and norm

$$\forall \hat{f} = \varphi \in A(\Gamma), \qquad \|\varphi\|_A = \|f\|_1.$$

A feature of Fourier transforms (and Laplace transforms) which is the key for their importance in applications is the fact that

$$\forall f, g \in L^1(G), \qquad (f * g)\hat{}\,(\gamma) = \hat{f}(\gamma)\hat{g}(\gamma).$$

This fact and some of the above remarks are expressed as

Proposition 1.1.1. *The Fourier transform bijective map*

$$L^1(G) \quad \to \quad A(\Gamma)$$

is an isometry and an algebraic isomorphism.

$M(G)$, the space of *bounded Radon measures*, is the Banach space dual of $C_0(G)$, taken with the sup norm $\| \ \|_\infty$. $L^1(G)$ is imbedded in $M(G)$ because of the Radon–Nikodym theorem, and the *Fourier transform of $\mu \in M(G)$* is a function $\varphi: \Gamma \to \mathbf{C}$ well-defined by

$$\hat{\mu}(\gamma) = \varphi(\gamma) = \int_G (\gamma, x)\,d\mu(x), \qquad \gamma \in \Gamma.$$

The operation of convolution in $L^1(G)$ extends naturally to $M(G)$ and is given by

$$\forall \mu, \nu \in M(G) \quad \text{and} \quad \forall f \in C_0(G), \qquad \langle \mu * \nu, f \rangle = \langle \mu_x, \langle \nu_y, f(x+y) \rangle \rangle.$$

As such, $M(G)$ is a commutative Banach algebra with unit, δ, defined by $\langle \delta, f \rangle = f(0)$. The space of Fourier transforms of $M(G)$ elements is denoted by $B(\Gamma)$; and the Fourier transform map, $M(G) \to B(\Gamma)$, is bijective. $B(\Gamma)$ is characterized by *Bochner's theorem* as the space of finite linear combinations of continuous positive definite functions ($\varphi : \Gamma \to \mathbf{C}$ is *positive definite* if

$$\forall \gamma_1, \ldots, \gamma_n \in \Gamma \quad \text{and} \quad \forall c_1, \ldots, c_n \in \mathbf{C}, \qquad \sum_{j,k} c_j \bar{c}_k \varphi(\gamma_j - \gamma_k) \geqslant 0).$$

It is easy to check that the elements of $B(\Gamma)$ are bounded and uniformly continuous. The *inversion theorem* is

Theorem 1.1.1. *Let $f \in L^1(G) \cap B(G)$. Then $\hat{f} \in L^1(\Gamma)$ and*

$$(1.1.1) \qquad f(x) = \int_\Gamma \hat{f}(\gamma) \overline{(\gamma, x)} \, d\gamma.$$

Remark 1. The Haar measure on Γ in (1.1.1) is uniquely determined by the given Haar measure on G.

2. Note that if $f \notin C_0(G)$ then $f \notin L^1(G) \cap B(G)$.

3. Compare Theorem 1.1.1 with Exercise 1.3.5.

The *Parseval–Plancherel theorem* is

Theorem 1.1.2. *Let $L^1(G) \cap L^2(G)$ and $L^2(\Gamma)$ be taken with their respective L^2-norms. With these norms the Fourier transform is an isometry,*

$$(1.1.2) \qquad L^1(G) \cap L^2(G) \quad \to \quad L^2(\Gamma),$$

onto a dense subspace of $L^2(\Gamma)$. As such there is a unique extension of (1.1.2) *to a bijective isometry*

$$\begin{aligned} L^2(G) &\to L^2(\Gamma) \\ f &\mapsto \hat{f}. \end{aligned}$$

Further,

$$\forall f, g \in L^2(G), \qquad \int_G f(x) \overline{g(x)} \, dx = \int_\Gamma \hat{f}(\gamma) \overline{\hat{g}(\gamma)} \, d\gamma.$$

We denote the usual L^2-norm by $\| \;\; \|_2$.

Corollary 1.1.2.1. (M. Riesz) *$A(\Gamma) = L^2(\Gamma) * L^2(\Gamma)$ and*

$$\forall \varphi, \psi \in L^2(\Gamma), \qquad \|\varphi * \psi\|_A \leqslant \|\varphi\|_2 \|\psi\|_2.$$

1.1.5 $\forall\ \varphi \in A(\Gamma)$, supp φ is σ-compact. Let $C_c(G)$ be the space of continuous functions with compact support. Also note that if $f, g \in L^1(G)$ are equal outside of a locally null set then $f = g$ a.e.

We say that $U \subseteq \Gamma$ is a *compact neighborhood* if U is compact and int $U \neq \emptyset$.

Proposition 1.1.2. a) *If $f \in L^1(G)$ then there is $g = f$ a.e. (i.e., f and g really define the same element in $L^1(G)$) such that* supp g *is σ-compact.*

b) *For each $\varphi \in A(\Gamma)$,* supp φ *is σ-compact.*

Proof. a) Since $C_c(G)$ is dense in $L^1(G)$ let $\lim \|f_n - f\|_1 = 0$, where $f_n \in C_c(G)$, and set $K_n = \overline{\{x : f_n(x) \neq 0\}} = \operatorname{supp} f_n$.

Therefore, for any measurable set $B \subseteq A = \{x : f(x) \neq 0\} \setminus (\cup K_n)$, we compute that $\int_B f(x)\,dx = 0$; and so $mA = 0$.

By the definition of A and K_n,

$$\{x : f(x) \neq 0\} \subseteq A \cup (\cup K_n).$$

Define

$$g(x) = \begin{cases} f(x) & \text{for } x \in \cup K_n \\ 0 & \text{for } x \notin (\cup K_n)^{\sim}. \end{cases}$$

b.i) For $\varphi \in A(\Gamma)$ note that

$$\{\gamma : \varphi(\gamma) \neq 0\} = \bigcup_{n=1}^{\infty} \{\gamma \in \Gamma : |\varphi(\gamma)| \geqslant 1/n\} = \bigcup_{n=1}^{\infty} E_n,$$

where E_n is compact. The compactness follows since φ vanishes at infinity.

b.ii) Thus it is sufficient to prove that in a LCAG X the closure of a σ-compact set $H = \cup H_n$, H_n compact, is σ-compact.

Let U be a compact symmetric neighborhood of 0 and set

$$Y = \bigcup_{n=1}^{\infty} (U + \ldots + U),$$

where the sum on the right-hand side contains n terms.

Hence, Y is a σ-compact open (and therefore closed) subgroup of X [Hewitt and Ross, 1, I, pp. 33–34]. Consequently, we can choose $A \subseteq X$ for which

$$X = \bigcup_{\gamma \in A} Y_\gamma, \quad Y_\gamma = \gamma + Y,$$

is a disjoint union.

Each H_n is covered by a finite number of Y_γ and so there is a sequence $B \subseteq A$ with the property that

$$H \subseteq \bigcup_{\gamma \in B} Y_\gamma.$$

$\bigcup_{\gamma \in B} Y_\gamma$ is σ-compact and closed, the latter fact following since Y_γ is open and

$$\bigcup_{\gamma \in B} Y_\gamma = \Big(\bigcup_{\gamma \in A \setminus B} Y_\gamma\Big)^\sim.$$

q.e.d.

1.1.6 The point at infinity is an S-set. Let $A_c(\Gamma) = A(\Gamma) \cap C_c(\Gamma)$.

Proposition 1.1.3. $\overline{A_c(\Gamma)} = A(\Gamma)$.

Proof. In light of Cor. 1.1.2, the fact that $C_c(\Gamma)$ is dense in $L^2(\Gamma)$, and the continuity of the map,

$$\begin{array}{ccc} L^2(\Gamma) \times L^2(\Gamma) & \to & A(\Gamma) \\ (\varphi, \psi) & \mapsto & \varphi * \psi, \end{array}$$

we need only observe that $\operatorname{supp} \varphi * \psi$ is compact for $\varphi, \psi \in C_c(\Gamma)$.

q.e.d.

Proposition 1.1.3 is precisely the statement that the point at infinity (of Γ) is a spectral synthesis set (cf. Theorem 1.2.1 and Paragraph 1.4). This result is used in the proof of Wiener's Tauberian theorem.

1.1.7 An approximate identity technique. The following result yields the fact that for any two disjoint closed sets $E_1, E_2 \subseteq \Gamma$, where E_1 is compact, there is $\varphi \in A(\Gamma)$ such that $\varphi = 1$ on E_1 and $\varphi = 0$ on E_2. In particular, $A(\Gamma)$ is a regular Banach algebra. We shall have more to say about such matters in Paragraph 1.2 and shall develop the cryptic proof below more fully there.

Proposition 1.1.4. *Let $E \subseteq \Gamma$ be compact and let $V \subseteq \Gamma$ have positive measure and compact closure. There is $\varphi \in A(\Gamma)$ such that $0 \leqslant \varphi \leqslant 1$,*

$$\|\varphi\|_A \leqslant (m(E - V)/mV)^{1/2},$$

and

$$\varphi(\gamma) = \begin{cases} 1 & \text{for } \gamma \in E \\ 0 & \text{for } \gamma \in (E + V - V)^\sim. \end{cases}$$

Proof. Define

$$\varphi(\gamma) = \frac{1}{mV} \chi_V * \chi_{E-V}(\gamma).$$

q.e.d.

This result should also be compared with Exercise 2.5.1c.

1.1.8 Wiener's theorem on the inversion of Fourier series. For each $\varphi \in A(\Gamma)$, define the *zero set of* φ to be

$$Z\varphi = \{\gamma \in \Gamma : \varphi(\gamma) = 0\}.$$

Thus, if $X \subseteq A(\Gamma)$ then the *zero set of* X is

$$Z(X) = \{\gamma \in \Gamma : \forall \varphi \in X, \varphi(\gamma) = 0\} = \cap \{Z\varphi : \varphi \in X\}.$$

Further, if $E \subseteq \Gamma$ is closed then we define

$$k(E) = \{\varphi \in A(\Gamma) : E \subseteq Z\varphi\}$$

and

$$j(E) = \{\varphi \in A(\Gamma) : E \cap \operatorname{supp} \varphi = \emptyset\}.$$

$Z\varphi$ is closed, $k(E)$ is a closed ideal, $j(E)$ is an ideal, and $j(E) \subseteq k(E)$. Let $A(E) = A(\Gamma)/k(E)$ and $A_j(E) = A(\Gamma)/\overline{j(E)}$ be taken with their quotient topologies. It is then an elementary fact from the theory of Banach algebras that $A(E)$ and $A_j(E)$ are commutative Banach algebras. Clearly, if E is not compact then $A(E)$ and $A_j(E)$ do not have units. Also, E is compact if and only if $A(E)$ has a unit; and if E is compact, then $A_j(E)$ has a unit if and only if $\overline{j(E)} = k(E)$ (cf. Theorem 1.4.1b). $A(E)$ can be considered as the set of restrictions of elements of $A(\Gamma)$ to E.

Proposition 1.1.5b is *Wiener's theorem on the inversion of Fourier series.*

Proposition 1.1.5. *Let* $E \subseteq \Gamma$ *be closed.*

a) *The set of maximal ideals of both* $A(E)$ *and* $A_j(E)$ *is identified with* E. *If* E *is compact then* E *is the maximal ideal space of* $A(E)$.

b) *If* E *is compact and* $Z\varphi \cap E = \emptyset$ *for* $\varphi \in A(\Gamma)$ *then there is* $\psi \in A(\Gamma)$ *such that*

$$\forall \gamma \in E, \qquad \psi(\gamma) = 1/\varphi(\gamma).$$

Proof. a.i) We shall show that the set of maximal ideals in $A_j(E)$ can be identified with E; the same proof works for $A(E)$.

We know (e.g. Exercise 1.1.6) that $M_\gamma = \{\varphi \in A(\Gamma) : \varphi(\gamma) = 0\}$ is a regular maximal ideal in $A(\Gamma)$. Let $c: A(\Gamma) \to A_j(E)$ be the canonical map. The canonical image, cM_γ, of M_γ in $A_j(E)$ is clearly an ideal.

If $\gamma \in E$ then $\overline{j(E)} \subseteq M_\gamma$. For this γ let $cJ \supseteq cM_\gamma$, $J \subseteq A(\Gamma)$, be a proper ideal. (J is defined as the set of all elements $\psi \in A(\Gamma)$ for which $\psi + \overline{j(E)} \in cJ$.) Thus $M_\gamma + \overline{j(E)} \subseteq J$ and so $M_\gamma \subseteq J$ since $\gamma \in E$.

By the definition of the canonical multiplication in the quotient we see that J is an ideal. Consequently, $M_\gamma = J$ since M_γ is maximal; and thus $cM_\gamma \subseteq A_j(E)$ *is a maximal ideal if* $\gamma \in E$.

Conversely, let $cM \subseteq A_j(E)$ be a maximal ideal. As before, M is an ideal.

If $\varphi \in \overline{j(E)}$ we see that $\varphi \in M$ since $c\varphi$ is the zero element of cM.

Therefore $ZM \subseteq E$.

Let $J \subseteq A(\Gamma)$ be an ideal. Then $cM \subseteq cJ$ if $M \subseteq J$. Since cM is maximal we have that $cM = cJ$.

If $\varphi \in J$ then $c\varphi \in cM$ and so $\varphi \in M$. Hence, $J \subseteq M$, from which we conclude that M is maximal.

By Exercise 1.1.6, $M = M_\gamma$; and $\gamma \in E$ since $ZM \subseteq E$.

Consequently, *if* $cM \subseteq A_j(E)$ *is a maximal ideal then* $M = M_\gamma$ *for some* $\gamma \in E$.

a.ii) If E is compact then $A(E)$ has a unit and so every ideal is regular.

b) Using part a) and the basic characterization of invertible elements in Banach algebras with unit, we have that $c\varphi \in A(E)$ is invertible if and only if the Gelfand transform $\widehat{c\varphi}$ never vanishes, i.e.,

$$Z\varphi \cap E = \emptyset.$$

Thus, there is $\psi \in A(\Gamma)$ such that

$$\forall \gamma \in E, \qquad \psi(\gamma)\,\varphi(\gamma) = 1.$$

q.e.d.

Wiener proved Proposition 1.1.5b for the case $\Gamma = \mathbf{T}$ and $E = \mathbf{T}$. His proof used a technique of collecting local information to make a global statement; we shall develop this method fully in Paragraph 2.4. In Proposition 1.1.6 we present a different proof to Wiener's original result (trivially generalized to any compact group).

Take $\varphi \in A(\Gamma)$ and define

$$L_\varphi : A(\Gamma) \to A(\Gamma)$$
$$\psi \mapsto \varphi\psi$$

(cf. Exercise 2.4.8 and Exercise 3.1.3). L_φ is a well-defined continuous linear map. Obviously, L_φ can be defined for $\varphi \in B(\Gamma)$. The *resolvent set*, $\rho(L_\varphi)$, of L_φ is the set of $z \in \mathbf{C}$ such that

$$(L_\varphi - zI)^{-1}: \quad A(\Gamma) \to A(\Gamma)$$

is a continuous linear map. $\rho(L_\varphi) \subseteq \mathbf{C}$ is an open set.

Proposition 1.1.6. *Let* Γ *be a compact group.*

a) *For each* $\varphi \in A(\Gamma)$, $\rho(L_\varphi)^\sim = \{z \in \mathbf{C} : \varphi(\gamma) = z, \gamma \in \Gamma\} = C_\varphi$.

b) *Assume* $\varphi \in A(\Gamma)$ *never vanishes. Then*

(1.1.3) $\qquad \forall \psi \in A(\Gamma)\ \exists\, \theta \in A(\Gamma)$ such that $\psi = \theta\varphi$.

c) *Assume $\varphi \in A(\Gamma)$ never vanishes. Then $1/\varphi \in A(\Gamma)$.*

Proof. c) is immediate from b).

To prove b) note that because of part a) we have $0 \in \rho(L_\varphi)$; and so $L_\varphi^{-1}: A(\Gamma) \rightarrow A(\Gamma)$ exists and is continuous.

It is then sufficient to define $\theta = L_\varphi^{-1}(\psi)$.

a) In proving part b) we only used the inclusion

$$C_\varphi^\sim \subseteq \rho(L_\varphi)$$

which we now verify. The opposite inclusion is trivial.

Let $w \in C_\varphi^\sim$ and define $\delta_w = d(w, C_\varphi)$, "d" being Euclidean distance (note that C_φ is compact).

We set

$$\Delta_w = \sup_{\gamma \in \Gamma} |\varphi(\gamma) - w|$$

and

(1.1.4) $$\psi(\gamma) = (\overline{\varphi(\gamma)} - \bar{w})(\varphi(\gamma) - w) - \tfrac{1}{2}(\Delta_w^2 + \delta_w^2),$$

noting that $\Delta_w \geqslant \delta_w$, $\psi \in A(\Gamma)$, and

$$\|\psi\|_\infty = \tfrac{1}{2}(\Delta_w^2 - \delta_w^2).$$

Thus $\zeta = \frac{1}{2}(\Delta_w^2 + \delta_w^2) \notin \{z : |z| \leqslant \|\psi\|_\infty\}$.

By Beurling's formula [Loomis, 1, p. 75] for the spectral radius of L_ψ we conclude that $\zeta \in \rho(L_\psi)$ (cf. the Remark below).

Consequently,

$$L_\psi - \zeta I = (L_{\bar{\varphi}} - \bar{w}I)(L_\varphi - wI)$$

has a continuous inverse; and so

$$(L_\varphi - wI)^{-1}: \quad A(\Gamma) \rightarrow A(\Gamma)$$

exists and is continuous. Therefore $w \in \rho(L_\varphi)$.

q.e.d.

Remark. Because of (1.1.3) we mention the *Cohen factorization theorem* (1959):

(1.1.5) $$A(\Gamma)\,A(\Gamma) = A(\Gamma).$$

Actually (1.1.5) is true for quite general algebras and modules (cf. Exercise 1.4.1). Salem proved (1.1.5) for $\Gamma = \mathbf{Z}$ and Rudin proved the $\Gamma = \mathbb{R}$ case. The reason that the proof of Proposition 1.1.6b does not prove the factorization theorem for $A(\Gamma)$, Γ σ-compact (in which case we can trivially define functions $\varphi \in A(\Gamma)$ which never vanish), is that, although $\psi \in B(\Gamma)$ in (1.1.4) and L_ψ is well-defined, we can not apply the spectral radius formula. Note that the Cohen factorization theorem is trivial for $A(\Gamma)$, Γ compact, since $1 \in A(\Gamma)$ (cf. (1.1.7)). We refer to [Koosis, 1; Pták, 1] for elegant proofs of (1.1.5).

1.1.9 Wiener's Tauberian theorem. With Proposition 1.1.3 and Proposition 1.1.5b we shall now prove the *Wiener Tauberian theorem* (Theorem 1.1.3). We use the notation

$$\tau_y f(x) = f(x-y).$$

Theorem 1.1.3. *Given Γ and $\varphi \in A(\Gamma)$. φ never vanishes if and only if Γ is σ-compact and*

$$\forall \psi \in A(\Gamma) \quad \text{and} \quad \forall \varepsilon > 0, \qquad \exists\, c_1, \ldots, c_n \in \mathbf{C} \quad \text{and} \quad \exists\, x_1, \ldots, x_n \in G$$

such that

$$\text{(1.1.6)} \qquad \left\| \psi(\cdot) - \sum_1^n c_j\, \varphi(\cdot)(\cdot, x_j) \right\|_A < \varepsilon.$$

Proof. i) Given the approximation property (1.1.6), we prove that φ never vanishes. Let $\varphi(\gamma) = 0$, and choose $\varepsilon > 0$ and $\psi \in A(\Gamma)$ such that $\psi(\gamma) > \varepsilon$. This contradicts (1.1.6).

ii) The opposite direction provides more of a challenge.
If $Z\varphi = \emptyset$ then Γ is σ-compact by Proposition 1.1.2.
Observe that

$$\text{(1.1.7)} \qquad \forall \theta \in A_c(\Gamma)\ \exists\, \theta_1 \in A_c(\Gamma) \quad \text{such that } \theta = \varphi\theta_1.$$

This follows from Proposition 1.1.5b by setting $E = \operatorname{supp}\theta$, choosing $\theta_2 \in A(\Gamma)$ for which $\theta_2 = 1/\varphi$ on E, and noting that $\varphi\theta_1 = \theta$, where $\theta_1 = \theta\theta_2$.
Take $\varepsilon > 0$ and $\psi \in A(\Gamma)$. We can choose $\theta_1 \in A_c(\Gamma)$ such that

$$\|\psi - \theta_1\|_A < \varepsilon/3$$

by Proposition 1.1.3, and then can choose $\theta_2 \in A_c(\Gamma)$ such that

$$\theta_1 = \varphi\theta_2$$

by (1.1.7).
Next, pick $\theta = \hat{h} \in A(\Gamma)$ with the properties that $\operatorname{supp} h = K$ is compact and

$$\|\theta_2 - \theta\|_A < \varepsilon/(3\|\varphi\|_A).$$

The left-hand side of (1.1.6) is less than or equal to

$$\|\psi - \theta_1\|_A + \|\varphi\theta_2 - \varphi\theta\|_A + \|\varphi\theta - \textstyle\sum c_j\, \varphi(\cdot)(\cdot, x_j)\|_A.$$

Letting $\hat{f} = \varphi$ it is sufficient to prove that

$$\left\| h * f - \sum_1^n c_j \tau_{x_j} f \right\|_1 < \varepsilon/3 \tag{1.1.8}$$

for some $c_1, \ldots, c_n \in \mathbf{C}$ and $x_1, \ldots, x_n \in G$.

Since the map $G \to L^1(G)$, $x \mapsto \tau_x f$, is uniformly continuous there is a symmetric neighborhood U of $0 \in G$ such that

$$\forall x \in U, \qquad \|\tau_x f - f\|_1 < \varepsilon/(3\|h\|_1).$$

Because K is compact there is a finite collection $U_j = x_j + U$, $x_j \in K$, $j = 1, \ldots, n$, which covers K.

We set $K_1 = K \cap U_1$,

$$K_j = \left(K \setminus \bigcup_1^{j-1} K_i \right) \cap U_j, \qquad j = 2, \ldots, n,$$

and $c_j = \int_{K_j} h(x)\,\mathrm{d}x$.

Consequently,

$$h * f(x) - \sum_{j=1}^n c_j \tau_{x_j} f(x) = \sum_{j=1}^n \int_{K_j} h(y)[\tau_y f(x) - \tau_{x_j} f(x)]\,\mathrm{d}y,$$

noting that for $y \in K_j$, $x_j - y \in x_j - U_j = U$.

Thus

$$\left\| h * f - \sum_1^n c_j \tau_{x_j} f \right\|_1 \leqslant \sum_1^n \int_{K_j} |h(y)| \|\tau_y f(\cdot) - \tau_{x_j} f(\cdot)\|_1\,\mathrm{d}y < \varepsilon/3.$$

q.e.d.

We do not explicitly use the fact that Γ is σ-compact to prove the sufficient conditions that $Z\varphi = \emptyset$. Compare the above proof with Theorem 2.1.4, where the factorization is also explicit but without the use of Proposition 1.1.5b.

Exercises 1.1.

1.1.1 Non-vanishing Fourier transforms

a) Let Haar measure on $\mathbf{R}$ be chosen so that $m(0, 1) = 1$ and thus Haar measure on $\hat{\mathbf{R}}$ is determined by $m(0, 1) = 1/2\pi$. Compute

i) $\qquad \hat{f}(\gamma) = \Gamma(1 + i\gamma) \qquad$ for $f(x) = \mathrm{e}^{-\mathrm{e}^x}\mathrm{e}^x$.

ii) $$\hat{f}(\gamma) = \frac{\sqrt{\pi}}{c} \exp(-\gamma^2/4c^2) \qquad \text{for } f(x) = e^{-c^2x^2}, \qquad c > 0.$$

Thus in both cases $\hat{f} \in A(\mathbf{R})$ never vanishes.

b) Let Γ be σ-compact. Construct a non-vanishing element $\varphi \in A(\Gamma)$.

1.1.2 A special Banach algebra (cf. Exercise 3.1.1)

Let X be the vector space of Fourier transforms $\hat{f} = \varphi$ of functions $f \in L^2(\mathbf{T})$ which are continuous on $\left[-\frac{\pi}{2}, \frac{\pi}{2}\right]$. Define the norm

$$\forall \hat{f} = \varphi \in X, \qquad \|\varphi\| = \|\varphi\|_2 + \sup\left\{|f(x)| : x \in \left[-\frac{\pi}{2}, \frac{\pi}{2}\right]\right\}.$$

a) With this norm and the operation of pointwise multiplication, prove that X is a commutative semi-simple Banach algebra with unit and that its maximal ideal space X^m is identified with $\mathbf{Z}$.

b) Prove that $\{\varphi \in X : \text{card supp}\, \varphi < \infty\}$ is dense in X (cf. Proposition 1.1.3).

1.1.3 $\overline{A(\Gamma)} = C_0(\Gamma)$

It is trivial to prove that $\overline{A(\Gamma)} = C_0(\Gamma)$ by the Stone–Weierstrass theorem (cf. Proposition 1.3.2). Obtain the same result by assuming that $\overline{A(\Gamma)}$ is contained properly in $C_0(\Gamma)$. (*Hint*: Let $\mu \in M(\Gamma) \setminus \{0\}$ annihilate $\overline{A(\Gamma)}$. By Fubini's theorem, $\hat{\mu} \in B(G) \subseteq L^\infty(G)$ annihilates $L^1(G)$, and thus $\hat{\mu}$ is a locally null continuous function. Consequently $\hat{\mu} = 0$ so that $\mu = 0$ (the desired contradiction) since the Fourier transform $M(\Gamma) \to B(G)$ is bijective).

Note that if card $\Gamma \geqslant \aleph_0$ then $A(\Gamma)$ is a set of first category in $C_0(\Gamma)$ even though $\overline{A(\Gamma)} = C_0(\Gamma)$. If $A(\Gamma) = C_0(\Gamma)$ then card $\Gamma < \aleph_0$ [Segal, 4]; the simplest possible proof of this fact is found in [Graham, 4].

1.1.4 Riesz products and the Fourier coefficients of Cantor-Lebesgue measures

A *Riesz product* is an infinite product

(E1.1.1) $$R(r) = \prod_{j=1}^{\infty} (1 + a_j \cos(r_j r + \alpha_j)), \qquad 0 < |a_j| \leqslant 1, a_j, r_j, r, \alpha_{1j} \in \mathbf{R}.$$

They were introduced by [F. Riesz, 1] to compute the Fourier coefficients of the Cantor–Lebesgue measure supported by the triadic Cantor set $C \subseteq [0, 2\pi]$ (e.g. part c) below). A readable sketch of their properties, particularly those of which we have the least interest in this book, is given in [Keogh, 1]. The partial products, $R_k(r)$, of $R(r)$ are non-negative since $0 < |a_j| \leqslant 1$. A sequence $\{r_j : j = 1, \ldots\} \subseteq \mathbf{Z}^+$ is *lacunary* if $\inf_j r_{j+1}/r_j > 1$. Riesz products are used to prove *Sidon's theorem* [Zygmund, 2, I, pp. 247–248] : if $\sum_1 c_j e^{ir_j\gamma}$ is the Fourier series of $\varphi \in L^\infty(\mathbf{T})$ and $\{r_j : j = 1, \ldots\}$ is lacunary, then $\varphi \in A(\mathbf{T})$.

a) Compute that if $a_j = a$ then

(E1.1.2)

$$R_k(r) = \sum_{\varepsilon_j = 0, \pm 1} \left(\frac{a}{2}\right)^{|\varepsilon_1| + \ldots + |\varepsilon_k|} \exp i(\varepsilon_1 \alpha_1 + \ldots + \varepsilon_k \alpha_k) \exp ir(\varepsilon_1 r_1 + \ldots + \varepsilon_k r_k).$$

b) The *perfect symmetric set* $E \subseteq \mathbf{T}$ *determined by* $\{\xi_k : k = 1, \ldots\} \subseteq (0, 1/2)$ is formed geometrically as follows: set $E = \bigcap E_k$, and write

$$E_k \bigcup_{j=1}^{2^k} E_j^k,$$

where

$$E_1^1 = [0, 2\pi\xi_1] \quad \text{and} \quad E_2^1 = [2\pi(1-\xi_1), 2\pi],$$
$$E_1^2 = [0, 2\pi\xi_1\xi_2], \qquad E_2^2 = [2\pi\xi_1(1-\xi_2), 2\pi\xi_1],$$
$$E_3^2 = [2\pi(1-\xi_1), 2\pi(1-\xi_1) + 2\pi\xi_1\xi_2], \quad \text{and} \quad E_4^2 = [2\pi(1-\xi_1\xi_2), 2\pi],$$

etc. Thus, E_k is the union of 2^k closed intervals E_j^k, each of length $2\pi\xi_1 \ldots \xi_k$, and to form E_{k+1} we construct

$$E_{2j-1}^{k+1}, E_{2j}^{k+1} \subseteq E_j^k, \qquad j = 1, \ldots, 2^k.$$

Prove that E is a compact totally disconnected set without isolated points, that each $\gamma \in E \subseteq [0, 2\pi]$ can be written as

(E1.1.3) $\quad \gamma = 2\pi \sum_1^\infty \varepsilon_j r_j,$

where $\varepsilon_j = 0, 1$, $r_1 = 1 - \xi_1$, and $r_k = \xi_1 \ldots \xi_{k-1}(1 - \xi_k)$ for $k \geqslant 2$, and that $\mathrm{m}E = \lim_j 2^{j+1}\pi\xi_1 \ldots \xi_j$.

Let $C^\infty(\mathbf{T})$ be the space of infinitely differentiable functions $\varphi : [0, 2\pi] \to \mathbf{C}$ such that $\varphi^{(k)}(0) = \varphi^{(k)}(2\pi)$ for each $k \geqslant 0$. The *distributional derivative* H' of $H \in L^1(\mathbf{T})$ is defined by

$$\forall \varphi \in C^\infty(\mathbf{T}), \qquad \langle H', \varphi \rangle = -\frac{1}{2\pi} \int_0^{2\pi} H(\gamma)\, \varphi'(\gamma)\, d\gamma,$$

where $\int_0^{2\pi} d\gamma = 2\pi$.

The *Cantor–Lebesgue function for E* is the continuous increasing function F on $[0, 2\pi]$ defined as

$$F(\gamma) = 2\pi \sum_1^\infty \varepsilon_j/2^j \qquad \text{on } E,$$

where γ and $\{\varepsilon_j\}$ are related by (E1.1.3), and extended continuously with line segments. The *Cantor–Lebesgue* measure $\mu \in M(\mathbf{T})$ *corresponding to F* with $\operatorname{supp} \mu \subseteq E$ is defined by the Riesz representation theorem as

$$\mu = \delta + F',$$

where F' is the distributional derivative of F. Supp μ is defined rigorously in Paragraph 1.3.6. It is easy to check that $\|\mu\|_1 = 1$ and that

$$\text{(E1.1.4)} \qquad \forall\, n \in \mathbf{Z}, \qquad \hat{\mu}(n) = \frac{1}{2\pi} \int_0^{2\pi} e^{in\gamma}\, dF(\gamma).$$

c) Let μ be the Cantor–Lebesgue measure corresponding to a given F and E. Compute

$$\forall n \in \mathbf{Z}, \qquad \hat{\mu}(n) = \prod_{j=1}^{\infty} \cos \pi n r_j,$$

where r_j is defined in (E1.1.3). (*Hint*: An approximation for the Stieltjes integral in (E1.1.4) is

$$\text{(E1.1.5)} \qquad \frac{1}{2^k} \sum_{\varepsilon_j = 0,1} \exp 2\pi i n(\varepsilon_1 r_1 + \ldots + \varepsilon_k r_k),$$

noting that F increases by $1/2^k$ between E_j^k and E_{j+1}^k. Calculate that

$$\prod_{j=1}^{k} \cos \pi n r_j$$

is equal to the expression in (E1.1.5)).

If E is determined by $\xi_j = 1/3, j = 1, \ldots$, then $E = C$ is the (triadic) *Cantor set*. In this case we have $\hat{\mu}(n) = \prod_{j=1}^{\infty} \cos(2\pi n/3^j)$. Thus $\hat{\mu}(3^m) = 1$ for each $m \in \mathbf{Z}^+$ and hence $\hat{F}(n) \neq o(1/|n|)$, as $|n| \to \infty$, even though, as is well-known, $\hat{F}(n) = o(1/|n|)$, as $|n| \to \infty$, since F is a function of bounded variation.

1.1.5 Regular maximal ideals in Banach algebras

Let X be a commutative Banach algebra.

a) If $I \subseteq X$ is a (proper) regular ideal prove that I is contained in a regular maximal ideal.

b) If $I \subseteq X$ is a regular maximal ideal prove that I is closed. (*Hint*: Since I is regular there is an element $u \in X$ such that $x - ux \in \bar{I}$ for each $x \in X$. If $I \neq \bar{I}$ then $u \in \bar{I} = X$ since I is maximal. To obtain a contradiction first note that

$$\forall x \in X, \quad x = \left(\sum_0^\infty (u-y)^j x - u\sum_0^\infty (u-y)^j x\right) + y\sum_0^\infty (u-y)^j x$$

for $\|u - y\| < 1$. Thus X is the ideal generated by I and y, and so $y \notin I$ if $\|u - y\| < 1$. Hence $u \notin \bar{I}$).

c) Prove that every maximal ideal in X is regular if and only if $XX = X$.

Thus, by the Cohen factorization theorem, *every maximal ideal in* $A(\Gamma)$ *is regular* (*and therefore closed*). Closely related to this matter, and at one point using a lemma by Cohen, [Varopoulos, 1] investigates the continuity of positive linear forms on Banach algebras X with an involution.

1.1.6 Regular maximal ideals in $A(\Gamma)$

For each $\gamma \in \Gamma$ define

$$M_\gamma = \{\varphi \in A(\Gamma) : \varphi(\gamma) = 0\}.$$

Prove that $I \subseteq A(\Gamma)$ is a closed regular maximal ideal if and only if $I = M_\gamma$ for some $\gamma \in \Gamma$. It is interesting that Theorem 1.2.5, which is a form of the Tauberian theorem, provides an immediate proof of the fact that closed maximal ideals in $A(\Gamma)$ are regular (cf. Exercise 1.1.5).

1.2 Approximate identities in $A(\Gamma)$

1.2.1 $\emptyset$ is a strong Ditkin set. The construction of approximate identities, which are norm bounded and contained in specific subsets of $A(\Gamma)$, constitutes an essential technique in spectral synthesis.

The first result involves a standard technique used to approximate δ, as a measure on G, in various distribution spaces and for various topologies. It is our simplest approximation procedure for $A(\Gamma)$, and the convolution trick that is employed is an essential tool for much of this section.

Notationally, set $\tilde{f}(x) = \overline{f(-x)}$. Recall that G is metric if and only if Γ is σ-compact [Hewitt and Ross, 1, I, p. 397].

Proposition 1.2.1. *There is a directed system* $\{\varphi_U\} \subseteq A(\Gamma)$, *where* $\hat{f}_U = \varphi_U$, *such that* $f_U, \varphi_U \geqslant 0, f_U \in A_c(G), \|\varphi_U\|_A = 1$,

(1.2.1) $\quad \forall\, \varphi \in A(\Gamma), \quad \lim_U \|\varphi - \varphi\varphi_U\|_A = 0,$

and for all compact subsets $E \subseteq \Gamma$

$$\lim_{U} \varphi_U = 1, \qquad \text{uniformly on } E. \tag{1.2.2}$$

If G is metric, or, equivalently, if Γ is σ-compact, $\{\varphi_U\}$ can be chosen as a sequence.

Proof. i) For each compact neighborhood U of $0 \in G$ let V be a compact symmetric neighborhood of $0 \in G$ satisfying $V + V \subseteq U$; and choose a non-negative element $g_V \in L^2(G)$ such that $\operatorname{supp} g_V \subseteq V$ and $\int_G g_V(x)\,dx = 1$.

Define

$$f_U = g_V * \tilde{g}_V \geqslant 0 \quad \text{and} \quad \hat{f}_U = \varphi_U. \tag{1.2.3}$$

ii) Clearly, f_U is a continuous positive element of $L^1(G)$.
By the Plancherel or inversion theorem, $\varphi_U \in L^1(\Gamma)$; and so by the inversion result $f_U \in A(G)$. The fact that $f_U \in A_c(G)$ is then clear from (1.2.3); in fact, supp $f_U \subseteq U$.

iii) Now

$$\int_G f_U(x)(\gamma, x)\,dx = \int_G\int_G g_V(y) g_V(x-y)(\gamma, x)\,dx\,dy =$$

$$\int_G\int_G \overline{g_V(y)} g_V(u)(\gamma, y)(\gamma, u)\,du\,dy = \hat{g}_V(\gamma)\overline{\int_G g_V(y)(-\gamma, y)\,dy} = \tag{1.2.4}$$

$$\tilde{\hat{g}}_V(\gamma)\,\hat{g}_V(\gamma) = (\hat{g}_V(\gamma))^2.$$

iv) Since V is symmetric we can take $g_V(x) = g_V(-x)$ (e.g. $g_V = (1/mV)\chi_V$) so that because g_V is real, we have

$$\hat{g}_V(\gamma) = \int_G g_V(-x)(\gamma, x)\,dx = \int_G g_V(x)\overline{(\gamma, x)}\,dx = \overline{\hat{g}_V(\gamma)};$$

thus $\hat{g}_V$ is real.
Consequently, in this case, $\varphi_U \geqslant 0$ by (1.2.4). Also from (1.2.4), we compute

$$\|\varphi_U\|_A = \int_G f_U(x)\,dx = \left(\int_G g_V(x)\,dx\right)^2 = 1.$$

v) Now, given $\varepsilon > 0$ and $f \in L^1(G)$, choose U as above, so that by the uniform continuity of the function $y \mapsto \tau_y f$ (cf. the proof of Theorem 1.1.3), we have that

$$\forall\, y \in U, \quad \|\tau_y f - f\|_1 < \varepsilon.$$

Consequently, since $f_U \geqslant 0$ and $\int_G f_U(x)\,dx = 1$,

$$\|f * f_U - f\|_1 \leqslant \int_G f_U(y)\,\|\tau_y f - f\|_1\,dy < \varepsilon;$$

and so (1.2.1) is true.

For (1.2.2) let $\varphi \in A(\Gamma)$ equal 1 on E and apply (1.2.1).

vi) To define precisely the directed system, set $U \geqslant W$ if $U \subseteq W$, where U and W are compact basis elements at 0.

q.e.d.

A subset $X \subseteq A(\Gamma)$ contains an *approximate identity* $\{\varphi_\alpha\}$ if $\{\varphi_\alpha\} \subseteq X$ is a directed system and, for each $\varphi \in X$, $\lim_\alpha \|\varphi - \varphi\varphi_\alpha\|_A = 0$; an approximate identity $\{\varphi_\alpha\}$ contained in X is *bounded* if $\{\varphi_\alpha\}$ is $\|\ \|_A$-norm bounded.

1.2.2 The point at infinity is a strong Ditkin set. Our next two results were first stated generally by [Godement, 2, p. 126]. The major difference between them and Proposition 1.2.1 is that the functions are chosen in $A_c(\Gamma)$.

Proposition 1.2.2. *There is a directed system* $\{\varphi_\alpha\} \subseteq A_c(\Gamma)$, *where* $\hat{f}_\alpha = \varphi_\alpha$, *such that* $f_\alpha, \varphi_\alpha \geqslant 0, f_\alpha \in A(G)$, $\|\varphi_\alpha\|_A = 1$, *and*

$$\forall\, \gamma \in \Gamma, \qquad \lim_\alpha \varphi_\alpha(\gamma) = 1.$$

If G is metric, or, equivalently, if Γ is σ-compact, $\{\varphi_\alpha\}$ can be chosen as a sequence.

Proof. i) For each n and each compact neighborhood V of $0 \in \Gamma$ we can find (e.g. [Hewitt and Ross, 1, I, pp. 254–255]) a compact neighborhood $U \subseteq \Gamma$ such that $V \subseteq U$ and

$$\frac{m(U^{\sim} \cap (U+V))}{mU} < \frac{1}{n}. \tag{1.2.5}$$

This condition is obviously satisfied for $\Gamma = \mathbb{R}$. Generally, (1.2.5) is a reasonable statement; its proof for arbitrary Γ, although not requiring any structure theory or other powerful results, is however quite technical. We shall assume (1.2.5) as known.

ii) Consider the triplet $\alpha = (n, V, U)$, with the notation from (1.2.5), and set

$$\varphi_\alpha = \frac{1}{mU}\chi_U * \tilde{\chi}_U \geqslant 0.$$

Define $\alpha_1 \geqslant \alpha_2$ if $\frac{1}{n_1} \leqslant \frac{1}{n_2}$, $V_1 \subseteq V_2$, and $U_1 \subseteq U_2$. Clearly, if we are given α_1 and α_2 we can choose α_3 such that $n_3 > \max(n_1, n_2)$ and $V_3 = U_1 \cup U_2$. Thus, $\{\varphi_\alpha\}$ is a directed system.

iii) $\varphi_\alpha \in A_c(\Gamma)$ by Corollary 1.1.2 and $f_\alpha \in A(G)$ by the inversion theorem. Also, by the computation in Proposition 1.2.1,

$$f_\alpha(x) = \frac{1}{mU} \int_\Gamma \chi_U * \tilde{\chi}_U(\gamma)(-\gamma, x)\,d\gamma = \frac{1}{mU} \hat{\chi}_U(-x)\overline{\hat{\chi}_U(-x)} \geqslant 0.$$

Thus, $\|\varphi_\alpha\|_A = \int_G f_\alpha(x)\,dx = \hat{f}_\alpha(0)$, and so

$$\|\varphi_\alpha\|_A = \frac{1}{mU} \int_\Gamma \chi_U(\gamma)\tilde{\chi}_U(-\gamma)\,d\gamma = \frac{1}{mU} \int_\Gamma \chi_U^2(\gamma)\,d\gamma = 1.$$

iv) Given $\gamma \in \Gamma$. It remains to verify that $\lim_\alpha \varphi_\alpha(\gamma) = 1$. If $\alpha = (n, V, U)$ and $\gamma \in V$ then, using (1.2.5) and the fact that $m(\gamma + U) = mU$, we compute

$$1 \geqslant \varphi_\alpha(\gamma) = \frac{m(U \cap (\gamma + U))}{mU} \geqslant \frac{m(\gamma + U) - m(U^\sim \cap (V + U))}{mU} \geqslant 1 - \frac{1}{n}.$$

This yields the result.

q.e.d.

Another approach to deal with (1.2.5) is given in [Hewitt and Ross, 1, II, pp. 299–301].

We are now in a position to make a refinement of Proposition 1.1.3. Let $E \subseteq \Gamma$ be closed. For $\varphi \in \overline{j(E)}$ we define the "multiplier norm"

$$|||\varphi||| = \sup\{\|\varphi\psi\|_A/\|\psi\|_A : \psi \in k(E) \setminus \{0\}\}.$$

Clearly, $|||\varphi||| \leqslant \|\varphi\|_A$. $E \subseteq \Gamma$ is a *strong Ditkin set* if there is a directed system $\{\varphi_\alpha\} \subseteq j(E)$ such that

$$\forall\, \varphi \in k(E), \qquad \lim_\alpha \|\varphi - \varphi\varphi_\alpha\|_A = 0$$

and

$$\sup_\alpha |||\varphi_\alpha||| < \infty$$

(cf. [Saeki, 7, Definition 2]). If Γ is σ-compact and metrizable then $E \subseteq \Gamma$ is a strong Ditkin set if and only if there is a sequence $\{\varphi_n : n = 1, \ldots\} \subseteq j(E)$ such that

$$\forall\, \varphi \in k(E), \qquad \lim_{n \to \infty} \|\varphi - \varphi\varphi_n\|_A = 0.$$

It turns out that intervals $[a, b] \subseteq \mathbb{R}$ are strong Ditkin sets (e.g. Exercise 1.2.6c), and so, by drawing the appropriate picture, we see that $\{\varphi_n : n = 1, \ldots\}$ can not generally be $\| \ \|_A$-norm bounded.

Using Proposition 2.5.3 and the fact that closed sets $E \subseteq \mathbf{T}$ for which $mE = 0$ are disjoint from some translates of roots of unity, the following is easy to check: *given a closed set $E \subseteq \mathbf{T}$ for which $mE = 0$; then*

$$\forall\, \varphi \in j(E), \qquad \|\varphi\|_A = \|\!|\varphi|\!\|$$

(cf. the remark on sets of measure 0 in Section 1.2.4).

The following result implies that the point at infinity (of Γ) is a strong Ditkin set.

Theorem 1.2.1. *There is a directed system $\{\varphi_\alpha\} \subseteq A_c(\Gamma)$, with each $\|\varphi_\alpha\|_A = 1$, such that for all $\varphi \in A(\Gamma)$, $\lim_\alpha \|\varphi - \varphi\varphi_\alpha\|_A = 0$. In particular, if $E \subseteq \Gamma$ is compact then*

$$\lim_\alpha \varphi_\alpha = 1, \qquad \text{uniformly on } E.$$

If G is metric, or, equivalently, if Γ is σ-compact, then $\{\varphi_\alpha\}$ can be chosen as a sequence.

Proof. i) Take $\{\varphi_\alpha\}$ as in Proposition 1.2.2.

We shall prove that if V is a compact neighborhood of $0 \in G$ then

$$(1.2.6) \qquad \lim_\alpha \int_V f_\alpha(x)\,\mathrm{d}x = 1,$$

where $\hat{f}_\alpha = \varphi_\alpha$.

The idea of proof for (1.2.6) is straightforward. Note that $f_\alpha \in L^2(G)$ since $f_\alpha^2 \leqslant f_\alpha \sup_x f_\alpha(x)$ and $f_\alpha \in A(G)$. Thus, by Parseval's formula,

$$\int_G \chi_V(x) f_\alpha(x)\,\mathrm{d}x = \int_\Gamma \varphi_\alpha(\gamma)\,\overline{\hat{\chi}_V(\gamma)}\,\mathrm{d}\gamma;$$

so that if $\bar{\hat{\chi}}_V \in A(\Gamma) \cap L^1(\Gamma)$ (instead of just being in $A(\Gamma) \cap L^2(\Gamma)$) then

$$\lim_\alpha \int_V f_\alpha(x)\,\mathrm{d}x = \lim_\alpha \int_\Gamma \varphi_\alpha(\gamma)\,\overline{\hat{\chi}_V(\gamma)}\,\mathrm{d}\gamma = \overline{\int_\Gamma \hat{\chi}_V(\gamma)\,\mathrm{d}\gamma} = \chi_V(0) = 1,$$

by Proposition 1.2.2 and the inversion theorem.

We omit the technical detail to handle the case that $\hat{\chi}_V \in L^2(\Gamma) \setminus L^1(\Gamma)$; it is simply a matter of constructing functions g and h for which $g \leqslant \chi_V \leqslant h$, $g(0) = h(0) = 1$, and $\hat{g}, \hat{h} \in A(\Gamma) \cap L^1(\Gamma)$, by the convolution technique we've been using.

ii) Take $\hat{f} = \varphi \in A(\Gamma)$.

For any $\varepsilon > 0$ choose $\alpha_{f,\varepsilon}$ and a compact neighborhood U of $0 \in G$ such that

$$\forall\, y \in U, \qquad \|\tau_y f - f\|_1 < \varepsilon/3$$

and

$$\forall\, \alpha \geqslant \alpha_{f,\varepsilon}, \qquad 1 - \int_U f_\alpha(x)\,dx < \varepsilon/(3\|f\|_1)$$

(by (1.2.6)).

Then, for $\alpha \geqslant \alpha_{f,\varepsilon}$,

$$\begin{aligned}\|\varphi - \varphi\varphi_\alpha\|_A &\leqslant \int_G \left| f(x)\left(1 - \int_U f_\alpha(y)\,dy\right)\right| dx \\ &+ \int_G \left| f(x)\int_U f_\alpha(y)\,dy - \int_U f_\alpha(y) f(x-y)\,dy \right| dx \\ &+ \int_G \left| \int_{G\setminus U} f_\alpha(y) f(x-y)\,dy \right| dx \\ &\leqslant \frac{\varepsilon}{3} + \int_U f_\alpha(y)\|\tau_y f - f\|_1\,dy \\ &+ \int_{G\setminus U} f_\alpha(y)\left(\int_G |f(x-y)|\,dx\right) dy < \varepsilon.\end{aligned}$$

q.e.d.

1.2.3 Points in Γ are strong Ditkin sets. Strong Ditkin sets were first defined (as such) and systematically studied by [Wik, 1], and a complete classification of strong Ditkin sets for quite general groups has been made in [Meyer, 2; Meyer and Rosenthal, 1; Rosenthal, 1; Gilbert, 2; B. Schreiber, 1; Saeki, 7] (Saeki's paper was submitted for publication in the spring of 1969). Theorem 1.2.2b, which implies that each one-point set in Γ is a strong Ditkin set, goes back much further and we shall discuss its history in Chapter 2. The following result specializes Proposition 1.1.4 in a useful way; the additional conclusion, (1.2.7), is necessary to prove Theorem 1.2.2. The relevant parts of Proposition 1.2.3 and Theorem 1.2.2 should be compared with Exercise 2.5.1c.

Proposition 1.2.3. *Given $r > 1$, $\varepsilon > 0$, a compact subset $K \subseteq G$, $\lambda \in \Gamma$, and a neighborhood $W \subseteq \Gamma$ of λ. There is $\varphi \in A_c(\Gamma)$, such that $\varphi = 1$ on a neighborhood of λ, $\varphi = 0$ on $W^\sim$, $0 \leqslant \varphi \leqslant 1$, $\|\varphi\|_A < r$, and*

$$(1.2.7) \qquad \forall\, x \in K, \qquad \|\varphi(\cdot) - (\cdot, x)\varphi(\cdot)\|_A < \varepsilon.$$

Proof. Without loss of generality take $\lambda = 0$. By the definition of the topology on Γ, the subset $U \subseteq \Gamma$, for which

$$(1.2.8) \qquad \forall\, \gamma \in U \quad \text{and} \quad \forall\, x \in K, \qquad |1 - (\gamma, x)| < \varepsilon/2r,$$

is an open neighborhood of $0 \in \Gamma$.

Let V be any compact symmetric neighborhood of $0 \in \Gamma$ such that $V - V \subseteq U \cap W$.

By the regularity of Haar measure, there is a compact neighborhood E of $0 \in \Gamma$ satisfying

$$m(E - V) < r^2 mV;$$

and E can be chosen so that $E + V - V \subseteq U \cap W$.

Setting $\varphi = (1/mV)\chi_V * \chi_{E-V}$ (as in Proposition 1.1.4) we obtain all the conclusions except (1.2.7); in particular, $\varphi = 1$ on E and $\varphi = 0$ off of $E + V - V$.

We have

$$(1.2.9) \qquad \begin{aligned}(mV)(\varphi(\cdot)(\cdot, x) - \varphi(\cdot)) &= \chi_V * ((\cdot, x)\chi_{E-V} - \chi_{E-V}) \\ &\quad + ((\cdot, x)\chi_{E-V}) * ((\cdot, x)\chi_V - \chi_V).\end{aligned}$$

Using Cor. 1.1.2, (1.2.8), and (1.2.9), we compute

$$\forall\, x \in K, \qquad (mV)\|\varphi(\cdot)(\cdot, x) - \varphi(\cdot)\|_A \leqslant \frac{\varepsilon}{r}[mV\, m(E - V)]^{1/2} < \varepsilon(mV).$$

q.e.d.

Theorem 1.2.2. *Given $\lambda \in \Gamma$ and $r > 1$.*

a) *There is a directed system $\{\psi_\alpha\} \subseteq A_c(\Gamma)$ such that $\psi_\alpha = 1$ on a neighborhood of λ, $0 \leqslant \psi_\alpha \leqslant 1$, $\|\psi_\alpha\|_A < r$, and*

$$(1.2.10) \qquad \forall\, \varphi \in k(\{\lambda\}), \qquad \lim_\alpha \|\varphi\psi_\alpha\|_A = 0.$$

b) *There is a directed system $\{\theta_\alpha\} \subseteq A_c(\Gamma) \cap j(\{\lambda\})$, norm bounded by*

$$\forall\, \alpha, \qquad \|\theta_\alpha\|_A < 1 + r,$$

such that

$$(1.2.11) \qquad \forall\, \varphi \in k(\{\lambda\}), \qquad \lim_\alpha \|\varphi - \varphi\theta_\alpha\|_A = 0.$$

c) *If G is σ-compact, or, equivalently, if Γ is metric, the directed system in* a) *can be chosen as a sequence. If G and Γ are both σ-compact the directed system in* b) *can be chosen as a sequence.*

Proof. Without loss of generality, let $\lambda = 0$.

a) The triplets $\alpha = (n, K, W)$, where $K \subseteq G$ is compact and W is a compact neighborhood of $0 \in \Gamma$, form a directed set with an order defined by

$$\alpha \geqslant \beta\colon\ n_\alpha \geqslant n_\beta,\ \ K_\alpha \supseteq K_\beta,\ \ W_\alpha \subseteq W_\beta.$$

For each α we choose $\hat{g}_\alpha = \psi_\alpha$, corresponding to φ in Proposition 1.2.3, such that

$$\forall\, x \in K_\alpha, \qquad \|\psi_\alpha(\cdot) - (\cdot, x)\,\psi_\alpha(\cdot)\|_A < 1/n_\alpha.$$

Now, for $\hat{f} = \varphi \in k(\{0\})$ and $\varepsilon > 0$, take a compact set $K \subseteq G$ for which

$$\int_{K\sim} |f(x)|\,dx < \varepsilon/(4r).$$

Since $\int_G f(x)\,dx = \varphi(0) = 0$, we have

$$f * g_\alpha(x) = \int_G f(y)(g_\alpha(x-y) - g_\alpha(x))dy,$$

and so, if $\alpha = (n, K, W)$ satisfies $\frac{1}{n} < \varepsilon/(2\|\varphi\|_A)$, then

$$\|\varphi\psi_\alpha\|_A \leqslant \int_K |f(y)|\,\|\tau_y g_\alpha - g_\alpha\|_1\,dy + \int_{K\sim} |f(y)|\,\|\tau_y g_\alpha - g_\alpha\|_1\,dy \leqslant \frac{1}{n}\|\varphi\|_A + 2r\frac{\varepsilon}{4r} < \varepsilon.$$

Thus,

$$\forall\, \beta \geqslant \alpha, \qquad \|\varphi\psi_\beta\|_A < \varepsilon.$$

b) Choose $\{\varphi_\beta\}$ as in Theorem 1.2.1 and $\{\psi_n\}$ as in part a).
Set $\theta_\alpha = \varphi_\beta - \varphi_\beta\psi_n$ and define $\alpha_1 \geqslant \alpha_2$ if $\beta_1 \geqslant \beta_2$ and $\eta_1 \geqslant \eta_2$.
Then, if $\varphi \in k(\{0\})$, we observe that

$$\|\varphi - \varphi\theta_\alpha\|_A \leqslant \|\varphi - \varphi\varphi_\beta\|_A + \|\varphi\psi_n\|_A\|\varphi_\beta\|_A.$$

The result follows from Theorem 1.2.1 and part a).

q.e.d.

1.2.4 Idempotent measures and strong Ditkin sets. One of Wik's major tools in studying strong Ditkin sets, and one used and refined by various of the subsequent investigators mentioned earlier, centered about P. J. Cohen's classification of *idempotent measures* μ (i.e., those for which $\mu * \mu = \mu$). We refer to [Rudin, 5, Chapter 3] and [Amemiya and Itô, 1] for a full discussion and elegant proof, respectively, of Cohen's result. In order to state Cohen's theorem (in the context developed by Wik and Rosenthal) we define $\mathscr{R} \subseteq \mathscr{P}(\Gamma)$ to be the smallest algebra of subsets of Γ (i.e., closed under finite unions and complements) which contains all of the cosets of arbitrary subgroups of Γ. The result is

Assume $X \subseteq \Gamma$ is not in $\mathscr{R}$ and that $\{\varphi_n : n = 1, \ldots\} \subseteq B(\Gamma)$ satisfies the following property:

$$(1.2.12) \qquad \lim_{n\to\infty} \varphi_n(\gamma) = \begin{cases} 0 & \text{for } \gamma \in X, \\ 1 & \text{for } \gamma \notin X; \end{cases}$$

then $\lim_n \|\mu_n\|_1 = \infty$, *where* $\hat{\mu}_n = \varphi_n$.

Using this fact it has been shown, for example, that the only strong Ditkin sets in $\Gamma = \mathbf{T}$ with Haar (i.e., Lebesgue) measure 0 are finite sets and that those with positive Haar measure contain an interval. If $\mu \in M(G)$ is idempotent and $\|\mu\|_1 > 1$ then $\|\mu\|_1 \geqslant (1 + \sqrt{2})/2$ and this constant is attained by the norm of some such μ [Saeki, 2; 3].

1.2.5 A characterization of $\overline{j(E)}$ and $k(E)$. The point for us at this moment is that although $A(\Gamma)$ is normal (cf. Proposition 1.1.4 and the remark preceding it) (1.2.12) tells us that it has definite possibilities for being rather "abnormal". In order to determine just how regular $A(\Gamma)$ can be, we shall investigate the regularity properties of the ideals, i.e., we shall investigate its "local regularity" properties.

Proposition 1.2.4. *Let $I \subseteq A(\Gamma)$ be an ideal.*

a) *$A_c(\Gamma) \cap I$ is norm dense in I.*

b) *If $E \subseteq \Gamma$ is compact and $ZI \cap E = \emptyset$ then there is $\varphi \in I$ such that $\varphi = 1$ on E.*

Proof. a) Consider $\{\varphi_\alpha\} \subseteq A_c(\Gamma)$ as in Theorem 1.2.1. For each $\varphi \in I$, $\lim_\alpha \|\varphi - \varphi\varphi_\alpha\|_A = 0$ and $\varphi\varphi_\alpha \in I$.

b) By hypothesis, for each $\gamma \in E$, there is $\varphi_\gamma \in I$ such that $\varphi_\gamma(\gamma) = 1$.

Since I is an ideal, $\psi_\gamma = \varphi_\gamma \bar{\varphi}_\gamma \in I$. Thus $\psi_\gamma \geqslant 0$, and there is a neighborhood V_γ of γ such that $\psi_\gamma \geqslant 1/2$.

By the compactness we can choose a finite set, $\{V_{\gamma_1}, \ldots, V_{\gamma_n}\}$, of such neighborhoods which covers E. Define

$$\psi = \sum_1^n \psi_{\gamma_j}.$$

We apply Proposition 1.1.5b (noting that $Z\psi \cap E = \emptyset$) and have the existence of $\theta \in A(\Gamma)$ for which $\varphi = \theta\psi = 1$ on E.

q.e.d.

Proposition 1.2.5. *Let $I \subseteq A(\Gamma)$ be a closed ideal and set $E = ZI$.*

a) *If $\varphi \in A_c(\Gamma)$ and $ZI \subseteq \operatorname{int} Z\varphi$ (i.e., $ZI \cap \operatorname{supp} \varphi = \emptyset$) then $\varphi \in I$.*

b) *$\overline{j(E)} \subseteq I \subseteq k(E)$, and so $k(E)$ is the largest closed ideal with zero set E and $\overline{j(E)}$ is the smallest closed ideal with zero set E.*

Proof. a) By Proposition 1.2.4b there is $\psi \in I$ such that $\psi = 1$ on $\operatorname{supp} \varphi$. Therefore $\varphi = \varphi\psi \in I$.

b) The statement about $k(E)$ is obvious.

From the Proposition 1.2.4a) and the fact that I is closed we need only to prove that $A_c(\Gamma) \cap j(E) \subseteq I$.

Let $\varphi \in A_c(\Gamma) \cap j(E)$ have the property that $ZI \cap \operatorname{supp} \varphi = \emptyset$ (since $ZI = E$). Apply part a).

q.e.d.

1.2.6 A generalization of Wiener's Tauberian theorem. We have the following generalization of *Wiener's Tauberian theorem.*

Theorem 1.2.3. *Let $I \subseteq A(\Gamma)$ be a closed ideal and assume that*

$$ZI \subseteq \operatorname{int} Z\varphi,$$

where $\varphi \in A(\Gamma)$. Then $\varphi \in I$.

Proof. By Theorem 1.2.1 or Exercise 1.2.1 let $\lim_\alpha \|\varphi - \varphi\varphi_\alpha\|_A = 0$, where $\varphi_\alpha \in A_c(\Gamma)$. Then $\operatorname{supp} \varphi\varphi_\alpha \subseteq \operatorname{supp} \varphi$ and $\operatorname{supp} \varphi \cap ZI = \emptyset$, so that, from Proposition 1.2.5a), $\varphi\varphi_\alpha \in I$; and hence $\varphi \in I$ since I is closed.

q.e.d.

Note that both Theorem 1.1.3 and Theorem 1.2.3 depend essentially on Proposition 1.1.3 and Proposition 1.1.5b. We do not use the full strength of Theorem 1.2.1 in Theorem 1.2.3 but only employ Exercise 1.2.1 which depends heavily on Proposition 1.1.3. To see that Theorem 1.2.3 actually generalizes Theorem 1.1.3 we first need the following result recorded by [Godement, 2, p. 127].

Proposition 1.2.6. a) *If $I \subseteq L^1(G)$ is a closed ideal then it is a closed translation invariant subpsace.*

b) *If $I \subseteq L^1(G)$ is a closed translation invariant subspace then it is a closed ideal.*

Proof. a) For any $f \in L^1(G)$ choose $\{g_n : n = 1, \ldots\} \subseteq L^1(G)$ such that $\lim_n \|f - f * g_n\|_1 = 0$ (cf. the *Hint* of Exercise 1.2.1). Also, $\tau_x(f * g_n) = (\tau_x g_n) * f$, and so

$$\lim_n \|(\tau_x g_n) * f - \tau_x f\|_1 = 0.$$

Thus, if I is a closed ideal and $f \in I$ we have $\tau_x f \in I$ since $(\tau_x g_n) * f \in I$.

b) The proof of part b) is the proof of (1.1.8) which in fact did not involve the hypothesis that $\hat{f}$ never vanished.

q.e.d.

An Hahn–Banach argument is also often used to verify Proposition 1.2.6 [Rudin, 5, pp. 157–158].

Theorem 1.2.4. *Let $I \subseteq A(\Gamma)$ be a closed ideal. If for all $\gamma \in \Gamma$ there is an element $\varphi \in I$ for which $\varphi(\gamma) \neq 0$ then $I = A(\Gamma)$.*

Proof. Take $\varphi \in A(\Gamma)$ and note that $ZI = \emptyset$ by hypothesis.
Thus, we apply Theorem 1.2.3 directly.

q.e.d.

As a corollary we obviously obtain Theorem 1.1.3 because of Proposition 1.2.6. Further, in the context of Banach algebras, *Wiener's Tauberian theorem* has the form

Theorem 1.2.5. *Let $I \subseteq A(\Gamma)$ be a proper closed ideal. Then I is contained in a closed regular maximal ideal* (*cf. Exercise* 1.1.5a).

Proof. $E = ZI \neq \emptyset$, for otherwise we contradict Theorem 1.2.4.
Thus, $I \subseteq k(E) \subseteq M_\gamma$, where $\gamma \in E$ and M_γ is the maximal ideal corresponding to γ (cf. Exercise 1.1.6).

q.e.d.

1.2.7 A remark about Drury's theorem. The discussion prior to Proposition 1.2.4 about the separation properties of $A(\Gamma)$ can be rounded out for the time being by mention of some recent theorems due to Drury et al. Essentially these results determine a property of $A(\Gamma)$ which allows us to separate Helson sets (e.g. Section 1.3.13) from closed sets while maintaining a norm boundedness (cf. (1.2.12)). We are necessarily vague at this point but will be more explicit in Section 3.1.18.

Exercises 1.2

1.2.1 The point at infinity is a C-set

Prove that for each $\varphi \in A(\Gamma)$ there is a sequence $\{\varphi_n : n = 1, \ldots\} \subseteq A_c(\Gamma)$ such that $\lim \|\varphi - \varphi\varphi_n\|_A = 0$. This is equivalent to the statement that the point at infinity of Γ is a C-set (cf. (E1.2.3) and Section 1.4.9). The proof is much simpler than the proof of Theorem 1.2.1. (*Hint*: Given $\hat{f} = \varphi \in A(\Gamma)$ and $\varepsilon > 0$. Using the uniform continuity of the function $G \to L^1(G)$, $x \mapsto \tau_x f$, find $\psi \in A(\Gamma)$ such that $\|\varphi - \varphi\psi\|_A < \varepsilon/2$ (cf. Proposition 1.2.1); from Proposition 1.1.3 there is $\theta \in A_c(\Gamma)$ such that $\|\psi - \theta\|_A < \varepsilon/(2\|\varphi\|_A)$. Then $\|\varphi - \varphi\theta\|_A < \varepsilon$).

1.2.2 The Tauberian theorem for compact G

A statement of Theorem 1.2.3 for the case of G compact (and Γ discrete) is: if $I \subseteq A(\Gamma)$ is a closed ideal and $\varphi \in A(\Gamma)$ satisfies the condition,

(E1.2.1) $\quad ZI \subseteq Z\varphi,$

then $\varphi \in I$. Prove this directly and briefly from first definitions and Exercise 1.2.1. (*Hint*: $\psi \in A_c(\Gamma)$ has the form $\sum_{\gamma \in F} a_\gamma \chi_\gamma$, where card $F < \infty$, and $\varphi\psi = \sum_{\gamma \notin ZI} b_\gamma \chi_\gamma \in I$; apply Exercise 1.2.1).

1.2.3 The Tauberian theorem and spectral synthesis properties of points

a) For arbitrary G and Γ we would like to find conditions which, when coupled with (E1.2.1), imply that $\varphi \in I$. Prove that if a closed ideal $I \subseteq A(\Gamma)$ and $\varphi \in A(\Gamma)$ satisfy (E1.2.1) and if $ZI = \{\gamma\}$, a set consisting of a single point, then $\varphi \in I$. (*Hint*: From

Theorem 1.2.2, $\lim_{\alpha} \|\varphi - \varphi\varphi_\alpha\|_A = 0$, and note that $ZI \subseteq \operatorname{int} Z(\varphi\varphi_\alpha)$; apply Proposition 1.2.5).

Let us remark that this theorem uses the Tauberian result, Proposition 1.2.5, and Theorem 1.2.2. Actually the full strength of Theorem 1.2.2 is not necessary but only the fact that $\{\gamma\} \subseteq \Gamma$ is an S-set, i.e.,

(E1.2.2) $\quad \forall\, \varepsilon > 0 \quad \text{and} \quad \forall\, \varphi \in k(\{\gamma\}), \qquad \exists\, \psi \in j(\{\gamma\}) \quad \text{such that } \|\varphi - \psi\|_A < \varepsilon.$

(E1.2.2) or the stronger statement,

(E1.2.3) $\quad \forall\, \varepsilon > 0 \quad \text{and} \quad \forall\, \varphi \in k(\{\gamma\}), \qquad \exists\, \psi \in j(\{\gamma\}) \quad \text{such that } \|\varphi - \varphi\psi\|_A < \varepsilon,$

can be proven directly in an easier way than Theorem 1.2.2 (the analogy is the same as that between Exercise 1.2.1 and Theorem 1.2.1).

b) Prove (E1.2.3) without resorting to the complications of Theorem 1.2.2. (E1.2.3) is the statement that $\{\gamma\}$ is a C-set (cf. Exercise 1.2.1 and Section 1.4.9).

1.2.4 Primary ideals and spectral synthesis properties of points

We continue to discuss the subject of Exercise 1.2.3 but from an algebraic point of view. A *primary ideal* $I \subseteq A(\Gamma)$ is an ideal that is contained in precisely one closed regular maximal ideal (cf. Exercise 1.1.5b, c). Prove that the following statements are equivalent (and true by Exercise 1.2.3):

a) *Every closed primary ideal in $A(\Gamma)$ is a (closed regular) maximal ideal.*

b) *Every one point set $\{\gamma\} \subseteq \Gamma$ is an S-set.*

(*Hint*: To prove that a) implies b) note that $j(\{\gamma\})$ is primary taking into account that each closed regular maximal ideal in $A(\Gamma)$ is of the form $k(\{\gamma\})$ (e.g. Exercise 1.1.6). Conversely, if I is primary we have $j(\{\gamma\}) \subseteq \bar{I} \subseteq k(\{\gamma\})$ by Proposition 1.2.5, and so a) follows immediately from b).

1.2.5 The Nullstellensatz and spectral synthesis

Let X be a commutative ring. An ideal $I \subseteq X$ is *prime* if

(E1.2.4) $\quad xy \in I \quad \text{and} \quad x \notin I \quad \Rightarrow \quad y \in I;$

and an ideal $I \subseteq X$ is *algebraically primary* if

(E1.2.5) $\quad xy \in I \quad \text{and} \quad x \notin I \quad \Rightarrow \quad \exists\, n \in \mathbf{Z}^+ \text{ such that } y^n \in I.$

a) Prove that each closed primary ideal in $A(\Gamma)$ is prime and hence algebraically primary.

b) Prove that each closed algebraically primary ideal I, and hence each closed prime ideal, in $A(\Gamma)$ is primary (and thus maximal). (*Hint*: I is contained in a maximal ideal by Theorem 1.2.5. Let $I \subseteq M_{\gamma_1} \cap M_{\gamma_2}$, where M_γ is defined in Exercise 1.1.6, and take $\varphi_i \in j(\{\gamma_i\})$ such that $\varphi_i(\gamma_j) \neq 0$ for $i \neq j$ and $\psi = \varphi_1 \varphi_2 = 0$. Now $\varphi_i \notin M_{\gamma_j}$ for $i \neq j$ and hence $\varphi_i \notin I$ does not imply $\varphi_j^n \in I$).

Thus the closed primary, closed prime, closed algebraically primary, and maximal ideals are equivalent in $A(\Gamma)$.

Remark 1. The fundamental problems of both the number theoretic and algebraic geometric aspects of commutative ring theory have analogues in the algebraic view to spectral synthesis that we have hinted at in Exercise 1.2.4, and which we shall develop. For this reason we recall that a commutative ring X is *Noetherian* (named after E. Noether because of her fundamental paper in 1921) if every ideal in X is finitely generated. An ideal in a Noetherian ring is the finite intersection of algebraic primary ideals.

We first make a remark on the number theoretic situation. The Noetherian rings which are integrally closed and in which proper prime ideals are maximal are called *Dedekind domains*, and these are characterized by the property that each ideal is a unique finite product of prime ideals. This result follows from the above general intersection property of ideals in Noetherian rings. The ring of integers in an algebraic number field is a Dedekind domain; and, consequently, we have Kummer's very important unique factorization property. (In his attempt to prove Fermat's conjecture, Kummer introduced the notion of an ideal and investigated the unique factorization of integers in certain algebraic number fields.)

A basic example from classical algebraic geometry of a Noetherian ring is the polynomial ring $X = \mathbf{C}[z, w]$, $z, w \in \mathbf{C}$. The fact that the maximal ideals in $\mathbf{C}[z, w]$ are all generated by two elements, $P_a(z) = z - a$ and $P_b(w) = w - b$, where $a, b \in \mathbf{C}$, is the first version of the *Nullstellensatz* [Kaplansky, 2, p. 19]. The ideal J generated by the polynomial $P(z) = z$ is prime but not maximal. It turns out that if $S \subseteq \mathbf{C} \times \mathbf{C}$ and $J \subseteq \mathbf{C}[z, w]$ is the largest possible ideal such that

$$\forall\, P \in J, \qquad P = 0 \text{ on } S,$$

then J is the intersection of a finite number of prime ideals. Thus the zero set of the ideals J and I generated by $P(z) = z$ and $P(z) = z^2$, respectively, is the w-"axis"; clearly $I \subseteq J$ properly and I is an algebraic primary ideal. If $I \subseteq \mathbf{C}[z, w]$ is an ideal we define $\sqrt{I} = \{P : \exists\, n > 0 \text{ such that } P^n \in I\}$. Then the full *Hilbert Nullstellensatz* [Kaplansky, 2, p. 19] depends essentially on the theorem that $\sqrt{I}$ is the intersection of the maximal ideals containing it.

2. We return to the situation of $A(\Gamma)$. In light of Exercise 1.2.5 and our remark on Noetherian rings we ask if each closed ideal in $A(\Gamma)$ is the intersection of maximal ideals. This problem of "unique factorization in $A(\Gamma)$" is the problem of spectral synthesis (cf. Paragraph 1.4).

1.2.6 Approximate identities and strong Ditkin sets

a) Recalling the convention from Exercise 1.1.1 for Haar measure on **R** and **Я**, we define

$$\forall\, \lambda > 0, \qquad \Delta_\lambda(\gamma) = \begin{cases} 1 - \dfrac{|\gamma|}{\lambda} & \text{for } |\gamma| \leqslant \lambda, \\ 0 & \text{for } |\gamma| > \lambda. \end{cases}$$

Prove that $\Delta_\lambda = \dfrac{2\pi}{\lambda}\chi_{\lambda/2} * \chi_{\lambda/2}$ where $\chi_{\lambda/2}$ is meant to be the characteristic function of $[-\lambda/2, \lambda/2] \subseteq$ **Я** for this exercise. Compute $f_\lambda \in L^1(\mathbf{R})$, where $\hat{f}_\lambda = \Delta_\lambda$, as

$$f_\lambda(x) = \begin{cases} \lambda/2\pi & \text{for } x = 0, \\ \dfrac{\lambda}{2\pi}\left(\dfrac{\sin \dfrac{x\lambda}{2}}{x\lambda/2}\right)^2 & \text{for } x \neq 0. \end{cases}$$

Thus $\|\Delta_\lambda\|_A = 1$, $f_\lambda \geqslant 0$, $f_\lambda(x) = \lambda f_1(\lambda x)$, and $\Delta_\lambda(\gamma) = \Delta_1\left(\dfrac{\gamma}{\lambda}\right)$. $\{f_\lambda : \lambda > 0\}$ is the *Fejér kernel* and it is an approximate identity for $L^1(\mathbf{R})$ when we let $\lambda \to \infty$.

b) The *de la Vallée–Poussin kernel*, $\{\nabla_\lambda : \lambda > 0\}$, is defined on **Я** as

$$\forall\, \lambda > 0, \qquad \nabla_\lambda(\gamma) = 2\Delta_{2\lambda}(\gamma) - \Delta_\lambda(\gamma),$$

and its graph is obviously a trapezoid. From a), $\nabla_\lambda \geqslant 0$ and $\|\nabla_\lambda\|_A \leqslant 3$. Prove directly that for each $\varphi \in k(\{0\})$,

$$\lim_{\lambda \to 0} \|\nabla_\lambda \varphi\|_A = 0$$

(e.g. [Kahane and Salem, 4, p. 170; Wik, 1, p. 57]). Thus $\{0\} \subseteq$ **Я** is a strong Ditkin set (cf. Theorem 1.2.2); in fact if $\varphi \in A_c(\text{Я}) \cap k(\{0\})$ and $\theta_n \in A_c(\text{Я})$ is 1 on $[-n, n]$ then for large enough n,

$$\|\varphi - (\theta_n - \nabla_{1/n})\varphi\|_A = \|\nabla_{1/n}\varphi\|_A.$$

c) Prove that each closed interval $[a, b] \subseteq$ **Я** is a strong Ditkin set. (*Hint*: Define

$$\varphi_n = \tau_{-a}\nabla_{1/n} + \tau_{-b}\nabla_{1/n} + \psi_n,$$

where $\psi_n \in A_c(\text{Я})$ is 0 on $[a, b]^\sim$ and $\varphi_n = 1$ on $[a, b]$. Choose θ_n as in b) and check that if $\varphi \in k([a, b]) \cap A_c(\text{Я})$ then

$$\lim \|\varphi - (\theta_n - \varphi_n)\|_A = 0).$$

1.2.7 A characterization of strong Ditkin sets

Let G be σ-compact and metric. Prove that $E \subseteq \Gamma$ is a strong Ditkin set if and only if there is a sequence $\{\mu_n : n = 1, \ldots\} \subseteq M(G)$ such that $\hat{\mu}_n = 1$ on a neighborhood of E and such that for each $\hat{f} = \varphi \in k(E)$,

$$\lim_{n\to\infty} \|\mu_n * f\|_1 = 0.$$

(*Hint*: If E is strong Ditkin let $\mu_n = \delta - f_n$ where $\hat{f}_n = \varphi_n$ and φ_n is the "strong Ditkin approximate identity" for E; this does it. For the converse let $\hat{g}_n = \theta_n$ be an approximate identity for $A(\Gamma)$ with $\|\theta_n\|_A \leqslant 1$. Set $\varphi_n = \theta_n\,(1 - \hat{\mu}_n) \in j(E)$. For each $\varphi \in k(E)$,

$$\|\varphi - \varphi\varphi_n\|_A \leqslant \|\varphi - \varphi\theta_n\|_A + \|\varphi\hat{\mu}_n\|_A,$$

since $\|\theta_n\|_A \leqslant 1$).

1.3 Pseudo-measures

1.3.1 The space $A'(\Gamma)$ of pseudo-measures. The Banach space dual of $A(\Gamma)$ is $A'(\Gamma)$, the space of *pseudo-measures*. The transpose of the Fourier transform map, $L^1(G) \to A(\Gamma)$, is the function

$$F: A'(\Gamma) \to L^\infty(G)$$

defined by

(1.3.1) $\quad \forall\, T \in A'(\Gamma)$ and $\forall\, f \in L^1(G), \quad \langle FT, f\rangle = \langle T, \hat{f}\rangle.$

We write $FT = \hat{T}$, and call $\hat{T}$ the *Fourier transform* of T. The canonical norm on $A'(\Gamma)$ is given by $\|T\|_{A'} = \|\hat{T}\|_\infty$ for each $T \in A'(\Gamma)$. We can well-define the convolution map

$$\begin{aligned} A'(\Gamma) \times A'(\Gamma) &\to A'(\Gamma) \\ (T, S) &\mapsto T * S \end{aligned} \tag{1.3.2}$$

as

$$\forall\, f \in L^1(G), \quad \langle T * S, \hat{f}\rangle = \langle \hat{T}\hat{S}, f\rangle.$$

Thus

$$\forall\, T, S \in A'(\Gamma), \quad \widehat{T * S} = \hat{T}\hat{S}.$$

The following is immediate.

Proposition 1.3.1. *$A'(\Gamma)$ is a commutative Banach algebra with multiplication defined by* (1.3.2) *and with unit δ. Further*

$$F: A'(\Gamma) \to L^\infty(G)$$

is a bijective isometry and algebraic isomorphism.

Note that

(1.3.3) $\quad \forall\, T, S \in A'(\Gamma)$ and $\forall\, \hat{f} = \varphi \in A(\Gamma), \quad \langle T * S, \varphi\rangle = \langle T_\gamma, \langle S_\lambda, \varphi(\lambda + \gamma)\rangle\rangle$

since $\hat{S}f \in L^1(G)$. We shall give an intrinsic (in $A'(\Gamma)$) definition of $T * S$ in Section 2.4.13.

1.3.2 $M(\Gamma) \subseteq A'(\Gamma)$.

Proposition 1.3.2. a) *$M(\Gamma)$ is a subalgebra of $A'(\Gamma)$.*

b) *The duality,*

$$(1.3.4) \qquad \forall\, T \in M(\Gamma) \quad \text{and} \quad \forall\, \varphi \in A(\Gamma), \qquad \langle T, \varphi \rangle = \int_\Gamma \varphi \, \mathrm{d}T,$$

is consistently defined so that

$$(1.3.5) \qquad \forall\, T \in M(\Gamma), \qquad \hat{T}(x) = \int_\Gamma (\gamma, x)\, \mathrm{d}T_\gamma,$$

where $\hat{T}$ in (1.3.5) *is defined by* (1.3.1) *and T is considered as an element of $A'(\Gamma)$ in the pairing of* (1.3.4).

Proof. a) If $T \in M(\Gamma)$, T is defined and linear on $A(\Gamma)$ since $A(\Gamma) \subseteq C_0(\Gamma)$.

T is uniquely determined on $A(\Gamma)$; in fact the algebra $A(\Gamma)$ is separating and self-adjoint so that $\overline{A(\Gamma)} = C_0(\Gamma)$ (in the sup norm) by the Stone–Weierstrass theorem (cf. Exercise 1.1.3).

If $\lim_n \|\varphi_n\|_A = 0$, where $\varphi_n \in A(\Gamma)$, then $\lim_n \|\varphi_n\|_\infty = 0$. Thus, $\lim_n \langle T, \varphi_n \rangle = 0$, and we conclude that $T \in A'(\Gamma)$.

b) For (1.3.5) note that $\langle T_\gamma, (\gamma, x) \rangle$ exists when $T \in M(\Gamma)$.

Also, by Fubini's theorem (which we can use since $T \in M(\Gamma)$),

$$\forall\, f \in L^1(G), \qquad \langle T, \hat{f} \rangle = \int_G f(x) \langle T_\gamma, (\gamma, x) \rangle \, \mathrm{d}x.$$

Consequently, since $\hat{T} \in L^\infty(G)$ and $\langle \hat{T}, f \rangle = \langle T, \hat{f} \rangle$, we have $\hat{T}(x) = \langle T_\gamma, (\gamma, x) \rangle$.

q.e.d.

1.3.3 Operations in $A'(\Gamma)$. By the definition of $\tilde{\varphi}$ and the integral representation in Proposition 1.3.2 we set

$$\forall\, T \in A'(\Gamma) \quad \text{and} \quad \forall\, \varphi \in A(\Gamma), \qquad \langle \tilde{T}, \varphi \rangle = \overline{\langle T, \tilde{\varphi} \rangle}.$$

We define the *translate $\tau_\gamma T$ of $T \in A(\Gamma)$ by γ* as

$$\forall\, \varphi \in A(\Gamma), \qquad \langle \tau_\gamma T, \varphi \rangle = \langle T_\lambda, \varphi(\gamma + \lambda) \rangle = \langle T, \tau_{-\gamma} \varphi \rangle.$$

For the case of δ we also write $\tau_\gamma \delta = \delta_\gamma$.

Note that we have the imbedding

$$\begin{aligned} \Gamma &\to A'(\Gamma) \\ \gamma &\mapsto \tau_\gamma \delta. \end{aligned}$$

In this context Γ is the maximal ideal space of the Banach algebra $A(\Gamma)$ and the induced weak $*$ topology on Γ from $A'(\Gamma)$ is precisely the given locally compact topology on Γ taken as the dual group of G.

If $\mu \in M(\Gamma)$ and $\varphi \in A(\Gamma)$ then

$$\mu * \varphi(\gamma) = \langle \mu_\lambda, \varphi(\gamma - \lambda) \rangle \in A(\Gamma).$$

If $T \in A'(\Gamma)$ we still have

$$\langle T_\lambda, \varphi(\gamma - \lambda) \rangle \in A(\Gamma);$$

in fact, with $\hat{f} = \varphi \in A(\Gamma)$,

$$(1.3.6) \quad \langle T_\lambda, \varphi(\gamma - \lambda) \rangle = \langle T_\lambda, (f(\cdot)(\gamma, \cdot))\hat{\ }(-\lambda) \rangle = \int_G \hat{T}(x)(f(-x)(\gamma, -x))\,dx.$$

Thus we define the convolution

$$(1.3.7) \quad \begin{array}{ccc} A'(\Gamma) \times A(\Gamma) & \to & A(\Gamma) \\ (T, \varphi) & \mapsto & \langle T_\lambda, \varphi(\gamma - \lambda) \rangle \end{array}$$

and we write

$$T * \varphi(\gamma) = \langle T_\lambda, \varphi(\gamma - \lambda) \rangle.$$

When we use the notation, $T * \varphi$, we shall mean (1.3.7) and not the convolution (1.3.2) just in case $\varphi \in A(\Gamma) \cap A'(\Gamma)$.

1.3.4 Pseudo-functions. We shall be interested in various subfamilies of $A'(\Gamma)$. Besides $M(\Gamma) \subseteq A'(\Gamma)$ we now define the space $A'_0(\Gamma)$ of *pseudo-functions* as

$$A'_0(\Gamma) = \{T \in A'(\Gamma) : \hat{T} \text{ vanishes at infinity}\}.$$

$A'_0(\Gamma)$ is quite important for the study of Riemann's sets of uniqueness (e.g. [Benedetto, 6, Chapter 3] and Exercise 2.1.1). Notationally we set

$$M_0(\Gamma) = M(\Gamma) \cap A'_0(\Gamma)$$

and note that $M_0(\Gamma)$ *is a subspace of* $M_c(\Gamma)$, the space of continuous bounded Radon measures on Γ (e.g. Exercise 1.3.1 and Exercise 2.1.2).

1.3.5 Radon measures and A'(Γ). Now write

$$C_c(\Gamma) = \cup\, C_K(\Gamma),$$

where $K \subseteq \Gamma$ is compact and

$$C_K(\Gamma) = \{\varphi \in C_c(\Gamma) : \operatorname{supp} \varphi \subseteq K\}.$$

$C_K(\Gamma)$ is a Banach space with the sup-norm; and a linear functional defined on $C_c(\Gamma)$ which is continuous on each $C_K(\Gamma)$ is a *Radon measure*. The topological vector space properties of the space $D^0(\Gamma)$ of Radon measures are found in [Bourbaki, 1, Chapitre 3; Schwartz, 5, Chapitres 1 and 3]; for our purposes we note that $M(\Gamma) \subseteq D^0(\Gamma)$ (cf. Exercise 1.3.2), and that a directed system $\{\varphi_\alpha\} \subseteq C_c(\Gamma)$ converges to 0 if and only if

$$\exists\, K \subseteq \Gamma, \text{ compact, such that } \forall\, \alpha,\ \operatorname{supp} \varphi_\alpha \subseteq K$$

(1.3.8) and

$$\lim_\alpha \varphi_\alpha = 0, \text{ uniformly on } \Gamma.$$

The notation "$D^0(\Gamma)$" is used because of its relation to distribution theory, e.g. Exercise 1.3.6.

Proposition 1.3.3. *Given $T \in A'(\Gamma)$ and assume that $\langle T, \varphi\rangle \geqslant 0$ for each non-negative function $\varphi \in A(\Gamma)$. Then $T \in M(\Gamma)$.*

Proof. i) We first prove that $T \in D^0(\Gamma)$.

Clearly $\overline{A_c(\Gamma)} = C_c(\Gamma)$ (with the topology described above).

Let $\{\varphi_\alpha\} \subseteq A_c(\Gamma)$ satisfy (1.3.8) and choose a non-negative function $\varphi \in A_c(\Gamma)$ equal to 1 on K. From the uniform convergence there is $\{\varepsilon_\alpha\} \subseteq \mathbf{R}$ tending to 0 such that

$$\forall\, \gamma \in \Gamma, \qquad |\varphi_\alpha(\gamma)| < \varepsilon_\alpha \psi(\gamma).$$

A straightforward calculation yields the fact that the real and imaginary parts of each φ_α are in $A_c(\Gamma)$, and so we assume that φ_α is real-valued.

Thus

$$-\varepsilon_\alpha \psi \leqslant \varphi_\alpha \leqslant \varepsilon_\alpha \psi,$$

so that, by the positivity assumption, $\lim_\alpha \langle T, \varphi_\alpha\rangle = 0$, and, hence, $T \in D^0(\Gamma)$.

ii) From Proposition 1.3.2 and the fact that $T \in D^0(\Gamma)$, we have

$$\forall\, \varphi \in A_c(\Gamma), \qquad \langle T, \varphi\rangle = \int_\Gamma \varphi\, \mathrm{d}T. \tag{1.3.9}$$

Since $T \in A'(\Gamma)$, $\sup\{|\langle T, \varphi\rangle| : \varphi \in A_c(\Gamma) \text{ and } \|\varphi\|_A \leqslant 1\} = \|T\|_{A'} < \infty$; consequently, by Proposition 1.2.2, (1.3.9), the positivity of $T \in D^0(\Gamma)$, and the monotone convergence theorem,

$$\int_\Gamma \mathrm{d}T \leqslant \|T\|_{A'} < \infty.$$

Thus, by standard integration theory, e.g. [Bourbaki, 1, pp. 154–155], $\|T\|_1 = \int_\Gamma \mathrm{d}T$ and therefore $T \in M(\Gamma)$ (as well as $\|T\|_1 = \|T\|_{A'}$).

q.e.d.

If $T \in A'(\Gamma)$ and $\langle T, \varphi \rangle \geqslant 0$ for all non-negative $\varphi \in A(\Gamma)$ then we write $T \geqslant 0$. Obviously, $D^0(\Gamma) = M(\Gamma)$ when Γ is compact.

1.3.6 The support of a pseudo-measure. Our next project is to define the support of $T \in A'(\Gamma)$; this is a crucial notion from the point of view of synthesis. For an open set $U \subseteq \Gamma$ we define

$$A(U) = \{\varphi \in A(\Gamma) : \operatorname{supp} \varphi \subseteq U\}.$$

Given $T \in A'(\Gamma)$; $T = 0$ *on* U if

$$\forall\, \varphi \in A(U), \qquad \langle T, \varphi \rangle = 0.$$

Proposition 1.3.4. *Let $U \subseteq \Gamma$ be open and take $T \in A'(\Gamma)$. If for each $\gamma \in U$ there is an open neighborhood $V_\gamma \subseteq U$ of γ such that $T = 0$ on V_γ then $T = 0$ on U.*

Proof. Let $\varphi \in A(U)$ and choose $\{\varphi_n : n = 1, \ldots\} \subseteq A_c(\Gamma)$ such that $\lim_n \|\varphi - \varphi\varphi_n\|_A = 0$, e.g. Exercise 1.2.1. We shall prove that $\langle T, \varphi\varphi_n \rangle = 0$.

For each V_γ we use Proposition 1.1.4 to choose $\psi_\gamma \in A_c(\Gamma)$ and an open set $N_\gamma \subseteq V_\gamma$ for which $\psi_\gamma = 1$ on N_γ and $\psi_\gamma = 0$ off of V_γ.

$\{N_\gamma : \gamma \in U\}$ is an open cover for the compact set supp $\varphi\varphi_n$ and therefore we can choose $N_{\gamma_1}, \ldots, N_{\gamma_m} \in \{N_\gamma : \gamma \in U\}$ as a finite subcover.

Clearly,

$$\varphi\varphi_n = \varphi\varphi_n[1 - (1 - \psi_{\gamma_1})(1 - \psi_{\gamma_2}) \ldots (1 - \psi_{\gamma_m})].$$

When we expand $1 - (1 - \psi_{\gamma_1}) \ldots (1 - \psi_{\gamma_m})$ we obtain a sum of products of the ψ_{γ_j} so that each of the terms has support in one of the V_{γ_j}.

By hypothesis, then, $\langle T, \varphi\varphi_n \rangle = 0$.

q.e.d.

Because of Proposition 1.3.4 we well-define the *support of* $T \in A'(\Gamma)$ (supp T) to be the complement of the union of all open sets $U \subseteq \Gamma$ such that $T = 0$ on U. Thus, equivalently, the support of $T \in A'(\Gamma)$ is the intersection of all closed sets K for which $T = 0$ on $K^\sim$. Clearly,

Proposition 1.3.5. *Assume that the sequence $\{T_n\colon n = 1, \ldots\} \subseteq A'(\Gamma)$ converges to $T \in A'(\Gamma)$ in the weak $*$ topology. If $E \subseteq \Gamma$ is a closed set such that*

$$\forall n, \qquad \operatorname{supp} T_n \subseteq E$$

then supp $T \subseteq E$.

1.3.7 The product $T\phi$ where $T \in A'(\Gamma)$ and $\phi \in A(\Gamma)$. In order to give further basic properties of the support (e.g. Theorem 1.3.1) we must define the following "multiplication":

$$(1.3.10)\qquad \begin{aligned} A'(\Gamma) \times A(\Gamma) &\to A'(\Gamma) \\ (T, \varphi) &\mapsto T\varphi. \end{aligned}$$

$T\varphi$ is well-defined by the formula

$$\forall\, \psi \in A(\Gamma), \qquad \langle T\varphi, \psi \rangle = \langle T, \varphi\psi \rangle.$$

Observe that

$$\forall\, T \in A'(\Gamma) \quad \text{and} \quad \forall\, \varphi \in A(\Gamma), \qquad \operatorname{supp} T\varphi \subseteq \operatorname{supp} T \cap \operatorname{supp} \varphi.$$

Set $f_-(x) = f(-x)$ and note that $\hat{f}_- = \varphi_-$ if $\hat{f} = \varphi$.

Clearly

$$(1.3.11)\qquad \widehat{T\varphi}(x) = \hat{T} * f_-(x), \qquad \text{where } \hat{f} = \varphi;$$

and if $I \subseteq A(\Gamma)$ is a closed ideal,

$$(1.3.12)\qquad \langle T, I \rangle = 0 \quad \Leftrightarrow \quad TI = 0.$$

We shall verify (1.3.12). Since the point at infinity is a C-set, the implication from right to left is true for any subset $I \subseteq A(\Gamma)$. The other direction is clear using the fact that I is an ideal. The following is also immediate (e.g. Exercise 1.3.4).

Proposition 1.3.6. *Given $T \in A'(\Gamma)$. $T = 0$ on an open set $U \subseteq \Gamma$ if and only if for each $\varphi \in A(U)$, $T\varphi = 0$.*

1.3.8 A dual formulation of Wiener's Tauberian theorem. Because of the Hahn–Banach theorem, Theorem 1.3.1c is equivalent to Theorem 1.1.3 when we assume that φ never vanishes (cf. the discussion in Section 2.1.13). Once again, the proof depends essentially on and follows easily from Proposition 1.1.5b.

Theorem 1.3.1. *Given $T \in A'(\Gamma)$ and $\varphi \in A(\Gamma)$.*

a) *If $\varphi = 0$ on an open set $U \subseteq \Gamma$ then $T\varphi = 0$ on U.*

b) *If $T = 0$ on an open set $U \subseteq \Gamma$ then $T\varphi = 0$ on U.*

c) *If $T\varphi = 0$ then $\varphi = 0$ on $\operatorname{supp} T$.*

Proof. a) and b) are clear.

c.i.) Let $U \subseteq \Gamma$ be a relatively compact open set such that φ never vanishes on $\bar{U}$. Letting $\psi \in A(U)$ we shall first prove that $\langle T, \psi \rangle = 0$.

From Proposition 1.1.5b choose $\theta \in A(\Gamma)$ for which $\theta = 1/\varphi$ on $\bar{U}$ so that $\psi_\varphi = \psi\theta \in A(U)$.

Therefore, by hypothesis on $T\varphi$,

(1.3.13) $\langle T, \psi\rangle = \langle T, \varphi\psi_\varphi\rangle = \langle T\varphi, \psi_\varphi\rangle = 0.$

ii) Let $V = \cup\{U_\alpha \subseteq \Gamma : U_\alpha$ is relatively compact and open, and φ never vanishes on $\bar{U}_\alpha\}$. From Proposition 1.3.4 and (1.3.13), $T = 0$ on V; hence

(1.3.14) $\operatorname{supp} T \subseteq V^\sim.$

iii) If $\gamma \in \operatorname{supp} T$ we shall assume that $\varphi(\gamma) \neq 0$ and obtain a contradiction. There is an open relatively compact neighborhood W of γ such that φ is not 0 on $\overline{W}$ and so, as above, $T = 0$ on W.
On the other hand, $W \cap V^\sim \neq \emptyset$, and this contradicts (1.3.14).

q.e.d.

1.3.9 Radon measures are synthesizable. The following is a standard measure-theoretic fact [Bourbaki, 1, pp. 68–71].

Theorem 1.3.2. a) *Given $T \in D^0(\Gamma)$ and assume $\varphi \in C_c(\Gamma)$ vanishes on* supp T. *Then* $\langle T, \varphi\rangle = 0$.

b) *For each $\mu \in M(\Gamma)$ there is a directed system $\{\mu_\alpha\} \subseteq M(\Gamma)$ such that* $\operatorname{card}\operatorname{supp}\mu_\alpha < \infty$, $\operatorname{supp}\mu_\alpha \subseteq \operatorname{supp}\mu$, $\lim_\alpha \mu_\alpha = \mu$ *in the weak $*$ topology $\sigma(M(\Gamma), C_0(\Gamma))$, and* $\|\mu_\alpha\|_1 = \|\mu\|_1$.

Proof. We shall prove a).

i) Set $K = \operatorname{supp}\varphi$ and $E = \operatorname{supp} T$. By (1.3.8) there is $M_K < 0$ such that for each $\theta \in C_c(\Gamma)$ with $\operatorname{supp}\theta \subseteq K$ we have

$$|\langle T, \theta\rangle| < M_K \|\theta\|_\infty.$$

Given $\varepsilon > 0$, we shall prove that $|\langle T, \varphi\rangle| < \varepsilon$.

ii) Let $V = \{\gamma \in \Gamma : |\varphi(\gamma)| < \varepsilon/(2M_K)\}$; then V is open since φ is continuous, and $E \subseteq V$. Clearly, $E^\sim$ is an open neighborhood of the compact set $V^\sim$.

There is a standard procedure to adapt Urysohn's lemma to locally compact spaces, i.e. there is a continuous function $\psi : \Gamma \to [0, 1]$ such that $\psi = 1$ on $V^\sim$ and supp $\psi \subseteq E^\sim$ (we of course have much stronger results from Paragraphs 1.1 and 1.2).

iii) Note that $E \cap \operatorname{supp}\varphi\psi = \emptyset$ and hence $\langle T, \varphi\psi\rangle = 0$.

Further, $\varphi = \varphi\psi$ on $K \cap V^\sim$ and $|\varphi\psi| \leqslant |\varphi|$ on Γ.

Consequently, since $\varphi = 0$ on $K^\sim$,

$$\|\varphi - \varphi\psi\|_\infty = \sup_{\gamma \in V \cap K} |\varphi(\gamma)(1 - \psi(\gamma))|$$
$$\leqslant 2\sup_{\gamma \in V} |\varphi(\gamma)| \leqslant \varepsilon/M_K.$$

Therefore, noting that $\operatorname{supp}(\varphi - \varphi\psi) \subseteq K$, we compute

$$|\langle T, \varphi \rangle| = |\langle T, \varphi - \varphi\psi \rangle| < M_K \|\varphi - \varphi\psi\|_\infty < \varepsilon.$$

q.e.d.

In terms of harmonic analysis, Theorem 1.3.2 says that each Radon measure in $A'(\Gamma)$ is synthesizable (cf. Paragraph 1.4. and Section 3.2.13). Using this fact and Proposition 1.1.3 we have the following result (cf. Theorem 1.3.1a) which provides a converse to Theorem 1.3.1c in case $T \in D^0(\Gamma)$.

Proposition 1.3.7. *Given $T \in A'(\Gamma) \cap D^0(\Gamma)$ and assume $\varphi \in A(\Gamma)$ vanishes on* $\operatorname{supp} T$. *Then $T\varphi = 0$.*

1.3.10 Heuristics for the notion of spectrum. Now is a convenient time to begin discussion on the notion of spectrum. We shall give various examples in Paragraph 1.5 and develop the idea carefully in Chapter 2.

During the plague years of 1665–1666, Isaac Newton made an important contribution to spectral analysis and synthesis. He discovered that sunlight, as a special case of white light, is actually composed of the continuous spectrum of colors from red to violet; he did this by letting beams of sunlight pass through a prism. Different colors of light correspond to different wavelengths of the sinusoidal light waves emitted by a source [Rossi, 1, Chapter 3]; and each color has its own characteristic index of refraction (with regard to a prism) which is dependent on the frequency of the wave.

In the case of a finite spectrum (of frequencies) this means that a complicated wave Φ is synthesized in terms of its spectrum $\{\lambda_k : k = 1, \ldots, n\} \subseteq \mathfrak{R}$ as

$$\Phi(x) = \sum_1^n a_k e^{i\lambda_k x}. \tag{1.3.15}$$

By a standard computation (which we shall develop in Chapter 2) the *means*

$$\frac{1}{2R} \int_{-R}^{R} \Phi(x) e^{-i\gamma x} dx$$

tend to 0 if $\gamma \neq \lambda_k$ and to a_k if $\gamma = \lambda_k$ (for the case of (1.3.15)). The point is that if quite general "waves" $\Phi \in L^\infty(\mathbf{R})$ have "spectra" contained in $\mathfrak{R}$, and if Φ has a formal Fourier expansion similar to (1.3.15) (but infinite), then it is possible to utilize various *means* on Φ and specify the "spectrum" of Φ. Abel [Beurling, 5] and Riemann [Pollard, 2] summability have been used for this purpose; and, in this context, Proposition 1.3.6 and Theorem 1.3.1a, b are proved using the Riemann localization principle, e.g. [Benedetto, 6, pp. 55ff.; Bochner, pp. 10–11; Kahane and Salem, 4, pp. 166–167]. Since (1.3.15) is the Fourier transform of a measure with support $\{\lambda_1, \ldots, \lambda_n\}$, we would like to associate the concept of support with the intuitive notion of a spectrum for even more general phenomena than (1.3.15). [Beurling, 2; 6] and [Godement, 2] defined the spectrum of $\hat{T} = \Phi \in L^\infty(G)$ equivalent to $\operatorname{supp} T$; Beurling's formulation was in terms of the narrow topology on $L^\infty(\mathbf{R})$ (cf. Paragraph 2.2) and Godement's (viz. (1.4.1)) was in terms of the weak $*$ topology on $L^\infty(G)$.

1.3.11 A characterization of the weak $*$ closed submodules of $A'(\Gamma)$. $A'(\Gamma)$ as a group under addition is a module over the ring $A(\Gamma)$.

Proposition 1.3.8. *The Fourier transform,* $A(\Gamma) \quad \to L^\infty(G)$, *is a bijection from the space of weak $*$ closed submodules of $A'(\Gamma)$ onto the space of weak $*$ closed translation invariant subspaces of $L^\infty(G)$.*

Proof. i) If $M \subseteq A'(\Gamma)$ we write $\mathscr{T} = \hat{M} \subseteq L^\infty(G)$. $\mathscr{T}$ is weak $*$ closed (resp. a vector space) if and only if M is weak $*$ closed (resp. a vector space).

ii) Given a weak $*$ closed translation invariant subspace $\mathscr{T} \subseteq L^\infty(G)$, $\hat{f}_- = \varphi_- \in A(\Gamma)$, and $\hat{T} \in \mathscr{T}$; from (1.3.11) M is a submodule if $\hat{T} * f \in \mathscr{T}$.

Choose a directed system $\{\mu_\alpha\} \subseteq M(G)$ as in Theorem 1.3.2b which converges to f in the weak $*$ topology $\sigma(M(G), C_0(G))$. $\hat{T} * \mu_\alpha \in \mathscr{T}$ by the translation invariance. If $g \in L^1(G)$ then

$$\langle \hat{T} * \mu_\alpha, g\rangle = \langle \mu_\alpha, \langle \hat{T}(x), g(x+y)\rangle\rangle \quad \to \quad \langle f, \langle \hat{T}(x), g(x+y)\rangle\rangle = \langle \hat{T} * f, g\rangle,$$

since $\langle \hat{T}(x), g(x+y)\rangle$ is bounded and continuous.

(The fact, $\hat{T} * f \in \mathscr{T}$, can be proved by the Hahn–Banach theorem [Benedetto, 2, p. 168].)

iii) Let $M \subseteq A'(\Gamma)$ be a weak $*$ closed submodule.

It is sufficient to prove that

$$(1.3.16) \qquad \forall\, T \in M \quad \text{and} \quad \forall\, x_0 \in G, \qquad T_\gamma(\gamma, x_0) \in M,$$

where $T_\gamma(\gamma, x_0) \in A'(\Gamma)$ is obviously well-defined by

$$\langle T_\gamma(\gamma, x_0), \varphi(\gamma)\rangle = \langle T_\gamma, \varphi(\gamma)(\gamma, x_0)\rangle;$$

in fact, if $\hat{T} = \Phi \in \mathscr{T}$ then $\langle (T_\gamma(\gamma, x_0))\hat{}, g\rangle = \langle \Phi(x+x_0), g(x)\rangle$, and so $\tau_{-x_0}\Phi \in \mathscr{T}$. Let $\{\psi_\alpha\} \subseteq A(\Gamma)$ be an approximate identity for $A(\Gamma)$ as in Proposition 1.2.1. Thus, since M is a module and $\psi_\alpha(\cdot)(\cdot, x_0) \in A(\Gamma)$, $T_\gamma \psi_\alpha(\gamma)(\gamma, x_0) \in M$; and we have (1.3.16) from (1.2.1).

q.e.d.

1.3.12 A basic duality technique. For $I \subseteq A(\Gamma)$ and $M \subseteq A'(\Gamma)$, we define

$$(1.3.17) \qquad M(I) = \{T \in A'(\Gamma) : \forall\, \varphi \in I,\ T\varphi = 0\},$$

and

$$(1.3.18) \qquad I(M) = \{\varphi \in A(\Gamma) : \forall\, T \in M,\ T\varphi = 0\},$$

respectively. It is trivial to check that:

if $I \subseteq A(\Gamma)$ is a closed ideal then $M(I) \subseteq A'(\Gamma)$ is a weak $*$ closed submodule,

and

if $M \subseteq A'(\Gamma)$ is a weak $*$ closed submodule then $I(M) \subseteq A(\Gamma)$ is a closed ideal.

The following is a fundamental fact in functional analysis:

If X is a normed space and X', taken with the weak $*$ topology, is denoted by X'_*, then

$$(X'_*)' = X. \tag{1.3.19}$$

Because of (1.3.19) we can formulate the dual form of the Hahn–Banach theorem (for the locally convex space X'_* and its dual X) which we now use.

Proposition 1.3.9. *Let I, $I_j \subseteq A(\Gamma)$ denote closed ideals and let M, $M_j \subseteq A'(\Gamma)$ denote weak $*$ closed submodules. Then*

a) $I_1 = I_2 \quad \Leftrightarrow \quad M(I_1) = M(I_2)$.

b) $M_1 = M_2 \quad \Leftrightarrow \quad I(M_1) = I(M_2)$.

c) $M = M(I(M))$.

d) $I = I(M(I))$.

Proof. a) If $\varphi \in I_2 \setminus I_1$ then apply the Hahn–Banach theorem to obtain a contradiction.

b) If $T \in M_2 \setminus M_1$ then apply the dual Hahn–Banach theorem, which we have from the setting of (1.3.19), to obtain a contradiction.

c) Clearly, $M \subseteq M(I(M))$ and $J \subseteq I(M(J))$, for $J = I(M)$.

On the other hand, if we apply "I" to $M \subseteq M(I(M))$ we see that

$$J \supseteq I(M(J)),$$

from which we can conclude that

$$I(M) = I(M(I(M))).$$

c) then follows from a).

A similar argument works for d).

q.e.d.

1.3.13 Helson sets and S-sets. After proving some theoretical results about $A'(\Gamma)$, examples of elements $T \in A'(\Gamma) \setminus M(\Gamma)$ are in order. Bochner's theorem and the fact that $\widehat{A'(\Gamma)} = L^\infty(G)$ obviously provide a means—although a bit too hygienic—to compute examples. We promise hair-raising calculations, subsequently, for the "mathochist".

For the time being we introduce a certain subclass of $A'(\Gamma) \setminus M(\Gamma)$ in a non-constructive way. In order to do this let $E \subseteq \Gamma$ be closed, and write

$$A'(E) = \{T \in A'(\Gamma) : \operatorname{supp} T \subseteq E\} = \overline{j(E)}^{\perp},$$

$$A'_S(E) = \{T \in A'(E) : \forall\, \varphi \in k(E), \langle T, \varphi \rangle = 0\} = k(E)^{\perp} = A(E)',$$

and

$$M(E) = \{\mu \in M(\Gamma) : \operatorname{supp} \mu \subseteq E\} = C_0(E)',$$

where $C_0(E)$ is the space of continuous functions defined on E which vanish at infinity. From Proposition 1.3.2 and Theorem 1.3.2a, $M(E) \subseteq A'_S(E)$ and, of course, $A(E) \subseteq C_0(E)$. By a standard Banach space argument we have

Proposition 1.3.10. *Let $E \subseteq \Gamma$ be a closed set. $A(E) = C_0(E)$ if and only if $A'_S(E) = M(E)$.*

A closed set $E \subseteq \Gamma$ is an *Helson set* if $A(E) = C_0(E)$ (cf. the remark at the end of Exercise 1.1.3). Consequently if E is not Helson we always have non-measures in $A'(E)$. To satisfy your curiosity for the moment we note that a closed interval in $\mathbb{R}$ is not an Helson set. [Körner, 1], and afterwards [Kaufman, 2], have provided examples of a different kind and on a much deeper level by proving that Helson sets exist for which $A'(E) \setminus A'_S(E) \neq \emptyset$. As we'll see, there are uncountable Helson sets E such that

$$A'(E) = M(E); \tag{1.3.20}$$

and diverse conditions on an Helson set to establish (1.3.20) have been given in [Benedetto, 8].

The above approach to find elements in $A'(\Gamma) \setminus M(\Gamma)$ is justified because of its relationship to spectral synthesis. In fact, $A'_S(E)$ is the space of *synthesizable pseudo-measures supported by E*; and a closed set $E \subseteq \Gamma$ is a *set of spectral synthesis*, or, briefly, an *S*-set, if $A'(E) = A'_S(E)$.

Exercises 1.3

1.3.1 Wiener's characterization of continuous measures

a) Recall that $\mu \in D^0(\Gamma)$ is *continuous* if $\mu(\{\gamma\}) = 0$ for each $\gamma \in \Gamma$. Prove Wiener's theorem [Wiener, 1; Lozinski, 1]:

$$\forall\, \mu \in M(\mathbf{T}), \quad \sum_{\gamma \in \mathbf{T}} |\mu(\{\gamma\})|^2 = \lim_N \frac{1}{2N+1} \sum_{-N}^{N} |\hat{\mu}(n)|^2, \tag{E1.3.1}$$

and so $\mu \in M(\mathbf{T})$ is in $M_c(\mathbf{T})$ (e.g. Section 1.3.4) if and only if

$$\lim_N \frac{1}{2N+1} \sum_{-N}^{N} |\hat{\mu}(n)|^2 = \lim_N \frac{1}{2N+1} \sum_{-N}^{N} |\hat{\mu}(n)| = 0.$$

Thus, $\mu \notin M_c(\mathbf{T})$ if $\mu \in M(\mathbf{T})$ and $|\hat{\mu}| = 1$. It is not difficult to generalize (E1.3.1) to any Γ and for a general class of *means* $\left(\text{not only } \text{“}\frac{1}{2N+1}\sum_{-N}^{N}\text{”}\right)$:

$$\text{(E.1.3.2)} \quad \forall\, \mu \in M(\Gamma), \qquad \sum_{\gamma\in\Gamma} |\mu(\{\gamma\})|^2 = \lim_U \int_G f_U(x)|\hat{\mu}(x)|^2\,dx,$$

where $\hat{f}_U = \varphi_U \in A(\Gamma)$, U goes through a compact neighborhood basis at $0 \in \Gamma$, $\operatorname{supp}\varphi_U \subseteq U$, and $0 \leqslant \varphi_U(\gamma) \leqslant \varphi_U(0) = 1$ (cf. Proposition 1.2.1 and (1.2.3) where the roles of G and Γ are reversed). (*Hint*: Set $\nu = \mu * \tilde{\mu}$ so that $\hat{\nu} = |\hat{\mu}|^2$; hence, from (1.3.1) and (1.3.4),

$$\nu(\{0\}) = \lim_U \int_G f_U(x)|\hat{\mu}(x)|^2\,dx.$$

Then note that $\nu(\{0\}) = \sum_\gamma |\mu(\{\gamma\})|^2$).

Compare this proof with the proof in [Wiener, 7, pp. 146–149] and the methods introduced in Paragraph 2.1.

b) Using a) prove that $M_0(\Gamma) \subseteq M_c(\Gamma)$. (*Hint*: Write the integral in (E1.3.2) as $\int_K + \int_{K^\sim}$ where K is compact and $|\hat{\mu}| < \varepsilon$ on $K^\sim$).

In light of a) $M_c(\Gamma)$ can be viewed as the largest class of complex measures μ such that the arithmetic means of $\hat{\mu}$ tend to 0 at infinity. This is the approach used in [Benedetto, 10] to generalize $A_0'(\Gamma)$, e.g. Exercise 2.1.2.

1.3.2 Unbounded measures in $A'(\mathbf{Я})$ and Fourier series of $L^\infty(\mathbf{T})$

a) Prove that

$$T = \sum{}' \frac{1}{n}\,\delta_n \in (A'(\mathbf{Я}) \cap D^o(\mathbf{Я})) \setminus M(\mathbf{Я}).$$

(*Hint*: $\hat{T}(x) \sim \sum_1^\infty \frac{1}{n} \sin nx$ and, as is well-known, e.g. [Edwards, 5, I, pp. 112–113], this sereis converges pointwise and has uniformly bounded partial sums (i.e., the series is *boundedly convergent*)). Note that $\hat{T}$ is not continuous.

b) Prove that there is no element $\Phi \in L^\infty(\mathbf{T})$ for which

$$\text{(E1.3.3)} \qquad \hat{\Phi}(n) \geqslant 0 \quad \text{and} \quad \sum \hat{\Phi}(n) = \infty.$$

(*Hint*: Let Φ be defined on $\mathbf{T} = \mathbf{R}/2\pi\mathbf{Z}$ and note that $T = \sum \hat{\Phi}(n)\,\delta_n \in D^o(\mathbf{Я})$. Use Proposition 1.3.3 and (E1.3.3) to prove that $T \notin A'(\mathbf{Я})$. Compute $\hat{T}$). (Compare this statement and proof with [Edwards, 5, I, Section 9.2], which is Bochner's theorem for $\mathbf{T}$ and which also yields b)).

On the other hand there are functions $\Phi \in (\bigcap_p L^p(\mathbf{T})) \setminus L^\infty(\mathbf{T})$ which satisfy (E1.3.3), e.g.

$$\text{(E1.3.4)} \qquad \Phi \sim \sum_2^\infty \frac{1}{n[\log n]} \exp ixn[\log n].$$

Incidentally, the first distributional derivative of Φ in (E1.3.4) can be considered as an element of $A'(\mathbf{T})$.

1.3.3 A property of $A_0'(\Gamma)$

Prove that if $T \in A_0'(\Gamma)$ and $\varphi \in A(\Gamma)$ then $T\varphi \in A_0'(\Gamma)$. (*Hint*: Use (1.3.11) and the dominated convergence theorem).

1.3.4 $T\varphi = 0$

a) Prove Proposition 1.3.6. (*Hint*: For $\psi \in A(U)$ choose a sequence $\{\varphi_n : n = 1, \ldots\} \subseteq A(\Gamma)$ such that $\lim_n \|\psi - \psi\varphi_n\|_A = 0$, and write $\langle T, \psi \rangle = \langle T, \psi\varphi_n \rangle + \langle T, \psi - \psi\varphi_n \rangle$).

b) Prove Proposition 1.3.7.

1.3.5 Cesàro summability on G

a) Take $\{\varphi_\alpha\} \subseteq A_c(\Gamma), \hat{f}_\alpha = \varphi_\alpha$, as in Theorem 1.2.1 and define

$$\text{(E1.3.5)} \qquad \forall \hat{f} = \varphi \in A(\Gamma), \qquad C_\alpha \varphi(x) = \int_\Gamma \varphi_\alpha(\gamma)\, \varphi(\gamma)\, \overline{(\gamma, x)}\, d\gamma.$$

Prove that if f is continuous at x then $\lim_\alpha C_\alpha \varphi(x) = f(x)$; further, prove that the convergence is uniform on compact sets $K \subseteq G$ where f is continuous, and that if f is uniformly continuous (e.g. if $f \in C_0(G)$) on G then $\lim_\alpha C_\alpha \varphi = f$, uniformly on G. (*Hint*: If V is a compact symmetric neighborhood of $0 \in G$ then

$$|C_\alpha \varphi(x) - f(x)| \leqslant \sup_{y \in V} |f(x) - f(x-y)|$$

$$\text{(E1.3.6)} \qquad + |f(x)| \int_{G \setminus V} f_\alpha(-y)\, dy + \|f\|_1 \sup_{y \notin V} f_\alpha(-y).$$

The definition of f_α from Proposition 1.2.2 is then used to show that the right-hand side of (E1.3.6) tends to 0).

$C_\alpha \varphi$ is the *Cesàro mean* and φ_α can be compared with Δ_λ in the Exercise 1.2.6. If $f \in C_0(G) \cap L^1(G)$ and $\hat{f} = \varphi \in L^1(\Gamma)$, then (E1.3.5) and the dominated convergence theorem imply that

$$\forall x \in G, \qquad f(x) = \int_\Gamma \varphi(\gamma) \overline{(\gamma, x)}\, d\gamma$$

(cf. Theorem 1.1.1).

b) Given $\varphi \in L^\infty(\Gamma)$. Prove that $\varphi = \hat{\mu} \in B(\Gamma)$ if and only if $\{C_\alpha \varphi\} \subseteq M(G)$ converges to μ in the weak $*$ topology $\sigma(M(G), C_0(G))$. (*Hint*: First show that $\varphi = \hat{\mu} \in B(\Gamma)$ if and only if $\{\|C_\alpha \varphi\|_1\}$ is bounded).

The intrinsic characterization of $A(\Gamma)$ or $B(\Gamma)$ is a very difficult problem even for $A(\mathbf{T})$. The prognosis in [Lévy, 2] for a satisfactory characterization is negative and Lévy's view of such mathematical problems is quite interesting. For the sake of clarification, an "intrinsic characterization" should read something like: φ is the Fourier series of an element in $L^1(\mathbf{Z})$ if and only if $\varphi \in \mathrm{Lip}_\alpha(\mathbf{T})$, $\alpha > 1/2$ (unfortunately, only the sufficient condition is true in this case, cf. Exercise 2.5.2). Because of the difficulty in verifying the conditions, Bochner's theorem, that $\varphi \in A(\mathbf{T})$ if and only if φ is a linear combination of positive definite functions, can not be considered a viable "intrinsic characterization". A survey of known results for the $A(\mathbf{T})$ case is given in [Kahane, 7; 13]; we mention in particular the work of S. Bernstein, Salem, Stečkin, Szasz, and Zygmund (cf. Section 2.4.10 and [Wik, 3]). Related results for $A(\mathbb{R})$, $B(\mathbb{R})$ and $A(\Gamma)$, $B(\Gamma)$ are given in [Berry, 1; Cramér, 1; Dyson, 1; Eberlein, 2; Ryan, 1; Schoenberg, 1; Simon, 1].

1.3.6 Distribution theory

Since $A'(\Gamma)$ is a special class of distributions we quickly recall some of the distribution spaces and refer to the classics, [Schwartz, 5; Gelfand, 1], for anything else.

$C^\infty(\mathbb{R}^n)$ is the space of infinitely differentiable functions on $\mathbb{R}^n$ and $C_c^\infty(\mathbb{R}^n)$ is the subset of $C^\infty(\mathbb{R}^n)$ whose elements have compact support. We write $D^s = \frac{\partial}{\partial \gamma_{s_1}} \cdots \frac{\partial}{\partial \gamma_{s_k}}$, $\gamma_r = \gamma_{r_1} \cdots \gamma_{r_j}$, $|s| = k$, and $|r| = j$. $\varphi \in C_{rd}^\infty(\mathbb{R}^n) \subseteq C^\infty(\mathbb{R}^n)$ if

$$\forall r \quad \text{and} \quad \forall s, \qquad \|\gamma_r D^s \varphi(\gamma)\|_\infty < \infty$$

(rd = rapidly decreasing).

The following convergence criteria are used: $\{\varphi_\alpha\} \subseteq C_c^\infty(\mathbb{R}^n)$ converges to $0 \in C_c^\infty(\mathbb{R}^n)$ if there is a compact set $K \subseteq \mathbb{R}^n$ such that $\operatorname{supp} \varphi_\alpha \subseteq K$ and

$$\forall s, \qquad \lim_\alpha D^s \varphi_\alpha = 0, \text{ uniformly on } K$$

(cf. (1.3.8)); a sequence $\{\varphi_m : m = 1, \ldots\} \subseteq C^\infty(\mathbb{R}^n)$ converges to $0 \in C^\infty(\mathbb{R}^n)$ if

$$\forall\, K, \text{ compact, and } \forall s, \qquad \lim_m D^s \varphi_m = 0, \text{ uniformly on } K;$$

and a sequence $\{\varphi_m : m = 1, \ldots\} \subseteq C_{rd}^\infty(\mathbb{R}^n)$ converges to $0 \in C_{rd}^\infty(\mathbb{R}^n)$ if

$$\forall r \text{ and } \forall s, \qquad \lim_m \gamma^r D^s \varphi_m(\gamma) = 0, \text{ uniformly.}$$

$C^\infty(\mathbb{R}^n)$ and $C_{rd}^\infty(\mathbb{R}^n)$ are Fréchet spaces and $C_c^\infty(\mathbb{R}^n)$ is complete but not metric.

Prove that the Fourier transform (of $L^1(\mathbf{R}^n)$) defines a homeomorphic bijection, $C^\infty_{rd}(\mathbf{R}^n) \to C^\infty_{rd}(\widehat{\mathbf{R}}^n)$ (cf. Exercise 2.5.2).

The duals of $C^\infty_c(\widehat{\mathbf{R}}^n)$, $C^\infty_{rd}(\widehat{\mathbf{R}}^n)$, and $C^\infty(\widehat{\mathbf{R}}^n)$ are, respectively, $D(\widehat{\mathbf{R}}^n)$ the space of *distributions*, $D_t(\widehat{\mathbf{R}}^n)$ the space of *tempered distributions*, and $D_c(\widehat{\mathbf{R}}^n)$ the space of *distributions with compact support*. Also, $A'(\widehat{\mathbf{R}}^n) \subseteq D_t(\widehat{\mathbf{R}}^n) \subseteq D(\widehat{\mathbf{R}}^n)$ and $D_c(\widehat{\mathbf{R}}^n) \subseteq D_t(\widehat{\mathbf{R}}^n)$. If $C(\widehat{\mathbf{R}}^n)$ is the space of complex-valued continuous functions on $\widehat{\mathbf{R}}^n$ taken with the topology of uniform convergence on compact sets (e.g. Paragraph 2.2), then its dual, $D^0_c(\widehat{\mathbf{R}}^n)$, is the space of *Radon measures with compact support*. Clearly, $D^0_c(\widehat{\mathbf{R}}^n) \subseteq M(\widehat{\mathbf{R}}^n)$.

With the above notation, the *distributional derivative* D^sT *of* $T \in D(\widehat{\mathbf{R}}^n)$ is defined as

$$\forall\ \varphi \in C^\infty_c(\widehat{\mathbf{R}}^n), \qquad \langle D^s T, \varphi\rangle = (-1)^{|s|}\langle T, D^s\varphi\rangle$$

(cf. Exercise 1.1.4, Section 2.1.8 and Section 3.2).

The *Fourier transform* $F\colon D_t(\widehat{\mathbf{R}}^n) \to D_t(\mathbf{R}^n)$ is defined as the transpose of the ordinary Fourier transform $C^\infty_{rd}(\mathbf{R}^n) \to C^\infty_{rd}(\widehat{\mathbf{R}}^n)$. Clearly, F is an extension to $D_t(\widehat{\mathbf{R}}^n)$ of the map defined in (1.3.1).

We can formulate the Fourier transform in terms of the Laplace transform in the following way. Let E be the space of entire functions of exponential type which are tempered on vertical lines and assume that it is taken with the compact-open topology (cf. Exercise 2.2.3b). The *Laplace transform*

$$L(T)(s) = \langle T_x, e^{sx}\rangle,$$

where $T \in D_c(\mathbf{R})$ and $s = \sigma + i\gamma \in \mathbf{C}$, defines a bijection

$$L\colon D_c(\mathbf{R}) \;\to\; E$$

whose restriction

$$\text{(E1.3.7)}\qquad \begin{aligned} L_r\colon D_c(\mathbf{R}) &\;\to\; C^\infty(\widehat{\mathbf{R}}) \\ T &\;\mapsto\; L(T)(i\gamma) \end{aligned}$$

is continuous. The restriction L_c of L_r to $C^\infty_c(\mathbf{R})$,

$$L_c\colon C^\infty_c(\mathbf{R}) \;\to\; C^\infty_{rd}(\widehat{\mathbf{R}}),$$

is continuous and extends to a bijective homeomorphism

$$L_e\colon C^\infty_{rd}(\mathbf{R}) \;\to\; C^\infty_{rd}(\widehat{\mathbf{R}})$$

whose transpose is the bijective homeomorphism

$$F\colon D_t(\widehat{\mathbf{R}}) \to D_t(\mathbf{R}).$$

1.3.7 $A'(\{\gamma\}) = M(\{\gamma\})$

In Paragraph 1.2 we saw that a one-point set is an S-set (Theorem 1.2.2b and Exercise 1.2.3). In light of Exercise 1.3.6 and our formulation of spectral synthesis in terms of pseudo-measures, prove the result distributionally in $\mathbb{R}^n$; and, in fact, show that $A'(\{\gamma\}) = M(\{\gamma\})$. (*Hint*: If $0 \in \mathbb{R}^n$ is the support of $T \in D(\mathbb{R}^n)$ then it is an elementary fact from distribution theory (e.g. [Horváth, 1, pp. 343–344; Schwartz, 5, p. 100]) that $T = \sum a_s D^s \delta$, a finite sum).

In 1949, [Riss, 1] proved the theorem for any LCAG using a theory of distributions for groups (cf. [Katznelson, 5, pp. 152–153]).

1.4 The spectral synthesis problem

1.4.1 The existence of the spectrum and Wiener's theorem. $\mathscr{T} \subseteq L^\infty(G)$ (with or without subscripts) will always denote a weak $*$ closed translation invariant subspace. For $\mathscr{T} \subseteq L^\infty(G)$ we define

(1.4.1) $\quad \operatorname{sp}\mathscr{T} = \{\gamma \in \Gamma : (\gamma, \cdot) \in \mathscr{T}\},$

and let $\mathscr{T}_{\text{sp}}$ be generated by $\operatorname{sp}\mathscr{T}$. Clearly

(1.4.2) $\quad \mathscr{T}_{sp} \subseteq \mathscr{T}$

and

(1.4.3) $\quad \operatorname{sp}\mathscr{T}_{\text{sp}} = \operatorname{sp}\mathscr{T}.$

It is conceivable that $\operatorname{sp}\mathscr{T} = \emptyset$ if $\mathscr{T}$ is non-zero; we shall use Wiener's theorem to prove that such can not be the case.

If $M \subseteq A'(\Gamma)$ we write

$$\operatorname{supp} M = \cup\{\operatorname{supp} T : T \in M\}.$$

For the following recall the notation of (1.3.17) and (1.3.18), and note that

(1.4.4) $\quad (\tau_\gamma \delta)^\wedge(x) = (\gamma, x).$

Proposition 1.4.1. *Let $M \subseteq A'(\Gamma)$ be a weak $*$ closed submodule and set $\hat{M} = \mathscr{T}$ (cf. Proposition 1.3.8).*

a) $\gamma \in \operatorname{sp}\mathscr{T} \quad \Leftrightarrow \quad \forall\, \varphi \in I(M), (\tau_\gamma \delta)\, \varphi = 0.$

b) $\operatorname{sp}\mathscr{T} = Z(I(M)).$

c) $\operatorname{sp}\mathscr{T} = \operatorname{supp} M.$

Proof. a) If $\gamma \in \operatorname{sp}\mathscr{T}$ then $\tau_\gamma \delta \in M$ by (1.4.3) and so, by the definition of $I(M)$,

(1.4.5) $\quad \forall\, \varphi \in I(M), \qquad (\tau_\gamma \delta)\varphi = 0.$

Conversely, given (1.4.5) we have

$$\tau_\gamma \delta \in M(I(M));$$

but

$$M(I(M)) = M$$

from Proposition 1.3.9c.

Consequently, $\gamma \in \operatorname{sp}\mathscr{T}$ by (1.4.4).

b) If $\gamma \in Z(I(M))$, (1.4.5) is immediate; and so $\gamma \in \operatorname{sp}\mathscr{T}$ from part a) (the direction that uses Proposition 1.3.9).

Conversely, if $\gamma \in \operatorname{sp}\mathscr{T}$ and $\varphi \in I(M)$ we see that

$$\forall\, \psi \in A(\Gamma), \qquad (\varphi\psi)\,(\gamma) = 0,$$

by the trivial direction in part a). Thus $\varphi(\gamma) = 0$ and hence $\gamma \in Z(I(M))$.

c) We have

(1.4.6) $\qquad Z(I(M)) \subseteq \operatorname{sp}\mathscr{T} \subseteq \operatorname{supp} M$

from b) and the fact that if $\gamma \in \mathscr{T}$ then $\tau_\gamma\delta \in M$. Let $\gamma \in \operatorname{supp} M$ so that $\gamma \in \operatorname{supp} S$ for some $S \in M$. If $\gamma \notin Z(I(M))$ let $\varphi(\gamma) \neq 0$ for some $\varphi \in I(M)$.

Note that $S\varphi = 0$ since $S \in M$ and $\varphi \in I(M)$. Thus $\varphi = 0$ on $\operatorname{supp} S$ by Theorem 1.3.1c, a contradiction.

Therefore $\operatorname{supp} M \subseteq Z(I(M))$ which, when combined with (1.4.6), yields c).

q.e.d.

Consequently,

Proposition 1.4.2. a) *If* $\mathscr{T} \subseteq L^\infty(G)$ *is non-zero then*

(1.4.7) $\qquad \operatorname{sp}\mathscr{T} \neq \emptyset.$

b) *If* $E \subseteq \Gamma$ *is closed there is a weak* $*$ *closed submodule* $M \subseteq A'(\Gamma)$ *such that, for* $\mathscr{T} = \hat{M}$,

$$E = \operatorname{sp}\mathscr{T} = \operatorname{supp} M = Z(I(M)).$$

Proof. a) is clear from Proposition 1.3.8, Proposition 1.4.1 and the hypothesis that $\mathscr{T} \neq \{0\}$.

b) Let $I = k(E)$ so that $ZI = E$ and $I = I(M(I))$ by Proposition 1.3.9d.

Setting $\mathscr{T} = (M(I))^\wedge$ we have $\operatorname{sp}\mathscr{T} = Zk(E)\,(=E)$ because of Proposition 1.4.1b.

q.e.d.

Clearly, $M = A_S'(E)$ in the proof of Proposition 1.4.2b.

Note that if a weak $*$ closed submodule $M_T \subseteq A'(\Gamma)$ is generated by $T \in A'(\Gamma)$ and if $\hat{M}_T = \mathscr{T}_\Phi$, where $\hat{T} = \Phi$, then Proposition 1.4.1c tells us that

$$\operatorname{sp}\mathscr{T}_\Phi = \operatorname{supp} T.$$

In this case we write

$$\text{sp}\,\Phi = \text{sp}\,\mathscr{T}_{\Phi}.$$

Thus the spectrum of a given element $\Phi \in L^\infty(G)$ is $\text{supp}\,T$, where $\hat{T} = \Phi$ (cf. Section 1.3.10). A crucial feature of this definition of spectrum is (1.4.7) which implies that

(1.4.8) $\quad \forall\ \Phi \in L^\infty(G) \setminus \{0\}, \qquad \text{sp}\,\Phi \neq \emptyset.$

It is important to note that Theorem 1.3.1c is the key step to prove (1.4.7).

1.4.2 Heuristics for the problems of spectral analysis and synthesis. The spectral analysis and synthesis problems have a diverse array of ancestors. One that significantly influenced the way we have defined $\text{sp}\,\Phi$ is the theory of convolution integral equations (cf. Section 1.5.4); these equations have the form

$$J(\Phi)(x) = \int_G \Phi(y) f(x-y)\,dy = \Psi(x)$$

where we are given $f \in L^1(G)$ and Ψ and wish to determine a solution Φ satisfying certain conditions. If $\hat{f}$ never vanishes, $\Psi = 0$, and $\hat{T} = \Phi \in L^\infty(G)$, we have $Z(I(M_T)) = \text{sp}\,\Phi = \emptyset$ by Proposition 1.4.1b; hence, by Proposition 1.4.2a, $\Phi = 0$ *is the only* $L^\infty(G)$ *solution to* $J(\Phi) = 0$ *when* $\hat{f}$ *never vanishes* (cf. [Rudin, 6, p. 218]). The chief developers of this approach to the notion of spectrum were [Carleman, 1, pp. 74–78 and pp. 111–116] and [Beurling, 1, p. 346; 2; 5] (cf. the motivation in [Herz, 5, pp. 185–186], noting that the above integral equation with $L^1(G)$ kernel f could just as well have a measure kernel); although [Wiener, 5; 6; 7] was certainly a key figure in the area.

The fact that (1.4.8) is valid for the notion of spectrum defined by (1.4.1) tells us that the "spectral analysis" of the (non-zero) phenomenon Φ yields a non-empty set of elementary "waves", viz.

$$\{(\gamma, \cdot) : \gamma \in \text{sp}\,\Phi\}.$$

The converse problem is the problem of spectral synthesis: when and how can Φ be reconstructed in terms of $\{(\gamma, \cdot) : \gamma \in \text{sp}\,\Phi\}$?

1.4.3 Spectra of representations and the spectral analysis problem for Beurling weighted spaces. Except when specifically mentioned to the contrary our definition of spectrum shall be (1.4.1). Consequently, because of (1.4.8), we shall not generally worry about spectral analysis problems.

Remark. [Domar, 1] (cf. [Vretblad, 1]) solved the spectral analysis problem for Beurling weighted spaces [Beurling, 1]

$$A(p) = \{\hat{f} = \varphi \in A(\Gamma) : \int_G |f|\,p < \infty\},$$

where p satisfies $p(x) \geqslant p(0) = 1$, $p(x+y) \leqslant p(x)p(y)$, and the *non-quasi-analyticity* condition

$$\forall\, x \in G, \qquad \sum_{1}^{\infty} \frac{\log p(nx)}{n^2} < \infty.$$

In the process he introduced the following general notion of spectrum which includes (1.4.1). Let X be a complex commutative Banach algebra with maximal ideal space X^m, let Y be a complex topological vector space, and let

$$R: X \rightarrow L(Y, Y)$$

be a continuous representation; then the *R-spectrum of* $T \in Y$ is

$$\operatorname{supp}_R T = Z\{\hat{x} : x \in X \quad \text{and} \quad (R(x))(T) = 0\} \subseteq X^m.$$

("$L(Y, Y)$" designates the space of continuous linear functions, $Y \rightarrow Y$, cf. "$\mathscr{L}_Y$" in Section 2.4). We have $\operatorname{supp}_R = \operatorname{supp}$ if $X = A(\Gamma)$, $Y = A'(\Gamma)$, and $(R(\varphi))(T) = T\varphi$. The study of spectra of representations has been pursued by Feldman, Lyubich, and Matsaev (cf. the penultimate remark in [Feldman, 1] with [DeVito, 1]).

1.4.4 A technical remark about S-sets. The elements in $\mathscr{T}$ are *synthesizable* if

$$\mathscr{T} = \mathscr{T}_{\text{sp}} \tag{1.4.9}$$

(cf. (1.4.2)). We now relate this notion with those introduced at the very end of Paragraph 1.3. This comes down to being more precise about Proposition 1.4.2b in the following way.

Proposition 1.4.3. *Let $E \subseteq \Gamma$ be closed.*

a) *$\overline{j(E)}$ is the smallest closed ideal and $k(E)$ the largest for which*

$$E = Z(\overline{j(E)}) = Z(k(E)).$$

b) *$A'_S(E)$ is the smallest weak $*$ closed submodule and $A'(E)$ the largest for which*

$$E = \operatorname{supp} A'_S(E) = \operatorname{supp} A'(E).$$

Proof a) is Proposition 1.2.5.

b) Since $\operatorname{supp} T\varphi \subseteq \operatorname{supp} T$ it is trivial to check that $A'(E)$ and $A'_S(E)$ are weak $*$ closed submodules. It remains to prove that $A'_S(E)$ is the smallest weak $*$ closed submodule M for which $E = \operatorname{supp} M$. We have $Z(I(M)) = E$ from Proposition 1.4.2, so that $I(M) \subseteq k(E)$ from a).

Hence, using Proposition 1.3.9,

$$M = M(I(M)) \supseteq M(k(E)) = A'_S(E).$$

q.e.d.

1.4.5 Standard characterizations of S-sets. Define

$$M_f(E) = \{\mu \in M(E): \text{card supp } \mu < \infty\}.$$

Note that the closure of $M_f(E)$ in the $\sigma(A'(\Gamma), A(\Gamma))$ topology is $A'_S(E)$ (as we proved in the proof of Proposition 1.4.3b). Summing up our above remarks, we obtain

Theorem 1.4.1. *The following are equivalent for a closed set $E \subseteq \Gamma$:*

a) *E is an S-set (i.e., $A'(E) = A'_S(E)$).*

b) $\overline{j(E)} = k(E)$.

c) *$E = ZI$ for a unique closed ideal $I \subseteq A(\Gamma)$.*

d) *$E = \text{sp}\,\mathscr{T}$ for a unique $\mathscr{T} \subseteq L^\infty(G)$.*

e) *$M_f(E)$ is weak $*$ dense in $A'(E)$ (with the induced weak $*$ topology from $A'(\Gamma)$).*

f) *For each $T \in A'(E)$ and for all $\varphi \in k(E)$, $\langle T, \varphi \rangle = 0$.*

g) *For each $T \in A'(E)$, where $\hat{T}$ is uniformly continuous, and for all $\varphi \in k(E)$, $\langle T, \varphi\rangle = 0$.*

h) *For each closed ideal $I \subseteq A(\Gamma)$ for which $ZI = E$, we have*

$$I = \bigcap_{\gamma \in E} M_\gamma,$$

where M_γ was defined in Exercise 1.1.6.

Proof. The only parts that require any comment are g) and h).

g) Take $S \in A'(E)$, $\varphi \in k(E)$, and $\psi \in A(\Gamma)$.

Then $\widehat{S\psi}$ is uniformly continuous and so $\langle S\psi, \varphi\rangle = 0$. Thus $\langle S\varphi, \psi\rangle = 0$ for all $\psi \in A(\Gamma)$ and so $S\varphi = 0$. Consequently, $\langle S, \varphi \rangle = 0$ by (1.3.12).

h) For any closed set $E \subseteq \Gamma$, observe that $k(E) = \bigcap_{\gamma \in E} M_\gamma$.

q.e.d.

We shall continue our characterization of S-sets in Theorem 1.4.2 and Theorem 1.4.3. These results require a different type of proof.

1.4.6 An approximation in G to determine S-sets in Γ. Consider the following approximation condition for a closed set E in a compact group Γ:

$\forall\, T \in A'(E)$, $\forall\, \{r_n : n = 1, \ldots\} \subseteq \mathbf{R}^+$ *increasing to infinity, and* $\forall\, \{x_n : n = 1, \ldots\} \subseteq G$, $\exists\, \mu \in M_f(E)$ *such that* $\forall\, n = 1, \ldots,$

$$|\hat{\mu}(x_n) - \hat{T}(x_n)| < r_n. \tag{1.4.10}$$

Theorem 1.4.2. *Let Γ be compact and let $E \subseteq \Gamma$ be closed. E is an S-set* $\Leftrightarrow$ (1.4.10) *is valid.*

Proof. ($\Leftarrow$) Given $T \in A'(E)$, $\hat{f} = \varphi \in k(E)$, and $\varepsilon > 0$. We shall prove that $|\langle T, \varphi\rangle| < \varepsilon$.

By Dini's theorem in infinite series take $\{r_n : n = 1, \ldots\} \subseteq \mathbf{R}^+$ increasing to infinity such that

$$\sum r_n |f(x_n)| < \infty,$$

where $\{x_n : n = 1, \ldots\} \subseteq G$ is precisely the set where f is non-zero. Choose $\varepsilon_1 > 0$ for which $\varepsilon_1 \sum r_n |f(x_n)| < \varepsilon$, and let $s_n = \varepsilon_1 r_n$. Then taking $\mu \in M_f(E)$ as in (1.4.10) (with r_n replaced by s_n there) we compute

$$|\langle T, \varphi\rangle| = |\langle T, \varphi\rangle - \langle \mu, \varphi\rangle| = \Big|\sum_n (\hat{T}(x_n) - \hat{\mu}(x_n)) f(x_n)\Big| \leqslant \sum_n s_n |f(x_n)| < \varepsilon.$$

($\Rightarrow$) Let E be an S-set. Assume there is $T \in A'(E)$, $\{x_n : n = 1, \ldots\} \subseteq G$, and $\{r_n : n = 1, \ldots\} \subseteq \mathbf{R}^+$ increasing to infinity such that for each $\mu \in M_f(E)$,

$$\left|\frac{\hat{\mu}(x_n)}{r_n} - \frac{\hat{T}(x_n)}{r_n}\right| \geqslant 1 \tag{1.4.11}$$

for some n. We shall obtain a contradiction.

Define

$$\forall\, n = 1, \ldots, \qquad \mu_n^* = \hat{\mu}(x_n)/r_n \quad \text{and} \quad T_n^* = \hat{T}(x_n)/r_n,$$

and set $\mu^* = \{\mu_n^* : n = 1, \ldots\}$ and $T^* = \{T_n^* : n = 1, \ldots\}$. Clearly, $\mu^*, T^* \in C_0(\mathbf{Z})$ by the hypothesis on $\{r_n : n = 1, \ldots\}$. From (1.4.11),

$$\|\mu^* - T^*\|_\infty \geqslant 1,$$

and so

$$T^* \notin \overline{\{\mu^* : \mu \in M_f(E)\}},$$

where the closure is taken in the sup norm. Thus, from the Hahn–Banach theorem, choose $\{a_n : n = 1, \ldots\} \in L^1(\mathbf{Z})$ which annihilates $\{\mu^* : \mu \in M_f(E)\}$ and such that $\langle \{a_n\}, T^*\rangle \neq 0$. Set $\varphi(\gamma) = \sum a_n(\gamma, x_n)$, so that $\varphi \in A(\Gamma)$. By the above observation with the Hahn–Banach theorem we have that

$$\forall\, \mu \in M_f(E), \quad 0 = \sum a_n \frac{\hat{\mu}(x_n)}{r_n} = \int_E \left(\sum \frac{a_n}{r_n}(\gamma, x_n)\right) \mathrm{d}\mu(\gamma)$$

(1.4.12) and

$$\sum a_n \frac{\hat{T}(x_n)}{r_n} \neq 0.$$

Since $\varphi \in A(\Gamma)$,

$$\psi(\gamma) = \sum \frac{a_n}{r_n}(\gamma, x_n) \in A(\Gamma);$$

and $\psi \in k(E)$ from (1.4.12).

From the hypothesis that E is an S-set we take $\{\psi_n : n=1, \ldots\} \subseteq j(E)$ for which $\lim_n \|\psi - \psi_n\|_A = 0$.

Thus $0 = \langle T, \psi_n \rangle \to \langle T, \psi \rangle$, and this contradicts (1.4.12).

q.e.d.

Theorem 1.4.2 can be proved more generally with a proper modification of (1.4.10); in fact, with such a modification it is trivial to prove the sufficient conditions that E be an S-set for any LCAG Γ.

1.4.7 A characterization of S-sets in terms of principal ideals. If $\varphi \in A(\Gamma)$ the closure of the principal ideal generated by φ is denoted by I_φ and is referred to as the *closed principal ideal generated by* φ.

Theorem 1.4.3. *Let Γ be σ-compact and metric (so that G is metric and σ-compact) and let $E \subseteq \Gamma$ be closed.*

a) *$\overline{j(E)}$ is a closed principal ideal.*

b) *E is an S-set $\Leftrightarrow$ for each closed principal ideal I_φ for which $ZI_\varphi = E$, we have*

$$I_\varphi = \bigcap_{\gamma \in E} M_\gamma.$$

Proof. a) will be proved in the sufficient conditions of b).

b) ($\Rightarrow$) This is clear from Theorem 1.4.1.

($\Leftarrow$) We'll prove that $\overline{j(E)} = I_\varphi$ for some $\varphi \in A(\Gamma)$. This is sufficient since then $ZI_\varphi = E$ and so $k(E) = I_\varphi$.

Because Γ is metric, E is a closed G_δ set. Thus $E^\sim$ is an open F_σ set, and hence we can write

$$E^\sim = \bigcup_1^\infty K_n,$$

where each K_n is compact. We take $K_n \subseteq \operatorname{int} K_{n+1}$ for each n; this can be done by the "Baire category" property of locally compact spaces. Consequently we can choose $\varphi_n \in j(E)$ such that $\varphi_n = 1$ on K_n and $0 \leqslant \varphi_n \leqslant 1$.

Set

$$\varphi = \sum_1^\infty \frac{\varphi_n}{2^n \|\varphi_n\|_A}$$

so that $\varphi \in \overline{j(E)}$, and hence $I_\varphi \subseteq \overline{j(E)}$. Clearly, $ZI_\varphi = E$; and therefore $I_\varphi = \overline{j(E)}$ by Proposition 1.2.5b.

q.e.d.

1.4.8 Non-synthesis and the principal ideal problem. If G is not σ-compact then $k(E)$ need not be a closed principal ideal. Generally, $E = ZI_\varphi$ if and only if E is a G_δ set and $E^\sim$ is an F_σ set.

If G is not compact then Γ contains non-S-sets as we'll prove in Paragraph 3.1. With this in mind we see that if E is an S-set in a metric σ-compact Γ then $k(E)$ is a closed principal ideal (from Theorem 1.4.3a). On the other hand (and this is not the converse situation), using Malliavin's technique to determine non-S-sets, Rudin proved the existence of $\varphi \in A(\Gamma)$ such that the elements of $\{I_{\varphi^n}: n = 1, \ldots\}$ are all distinct, noting that $ZI_{\varphi^n} = ZI_\varphi$ (e.g. Section 3.1.5). A special case of Rudin's approach was first given for $\Gamma = \mathbb{R}^n$, $n \geqslant 3$, in [Reiter, 5] using the original counter-example to synthesis discovered by [Schwartz, 2]. Reiter's work coupled with [Varopoulos, 7, Theorem 3] yield the fact that $k(E) \subseteq A(\mathbb{R}^3)$ can be a closed principal ideal whereas E is a non-S-set (e.g. Exercise 1.4.4).

The above observations do not preclude the possibility that every closed ideal in $A(\Gamma)$ is a closed principal ideal. For perspective, note, from Theorem 1.1.3 (Wiener's Tauberian theorem) and Proposition 1.2.6, that $A(\Gamma)$ is a closed principal ideal if Γ is σ-compact (recall Exercise 1.1.1). [Atzmon, 2] has settled this principal ideal problem by proving: *if Γ is not discrete then $A(\Gamma)$ contains a closed ideal which is not finitely generated.* The above remarks indicate that such non-principal ideals have non-S-zero sets in σ-compact metric Γ.

1.4.9 A characterization of C-sets. We mentioned C-sets earlier ("C" is for [Calderón, 1]). Now let's be more explicit. A closed set $E \subseteq \Gamma$ is a C-set if

$$\forall\ \varphi \in k(E), \qquad \varphi \in \overline{\varphi j(E)}.$$

Obviously each C-set is an S-set. We mention C-sets at this point, when we are giving characterizations of S-sets, since it is not known if every S-set is a C-set. We shall see in Section 2.5.2 that a very extensive generalization of Wiener's Tauberian theorem is closely related to C-sets. For perspective recall that one version of the Tauberian theorem states that closed proper ideals are contained in maximal ideals (Theorem 1.2.5) whereas every S-set is characterized by Theorem 1.4.1h. For now we characterize C-sets as follows:

Theorem 1.4.4. *Let $E \subseteq \Gamma$ be closed. E is a C-set $\Leftrightarrow$ whenever $T \in A'(\Gamma)$ and $\varphi \in A(\Gamma)$* satisfy

$$\varphi \in k(E \cup \operatorname{supp} T) \qquad \text{and} \qquad \operatorname{supp} T\varphi \subseteq E$$

we can conclude that $T\varphi = 0$.

Proof. ($\Leftarrow$) Let $\varphi \in k(E)$. We shall show that $I_\varphi \subseteq \overline{\varphi j(E)}$, i.e., $M(I_\varphi) \supseteq M(\overline{\varphi j(E)})$. If $T \in M(\overline{\varphi j(E)})$ then for each $\psi \in j(E)$, $\psi = 0$ on $\operatorname{supp} T\varphi$ (by Theorem 1.3.1), and so $\operatorname{supp} T\varphi \subseteq E$.

We now use Theorem 1.3.1 to obtain that $\varphi\psi = 0$ on $\operatorname{supp} T$ for each $\psi \in j(E)$. Consequently, if $\gamma \in (\operatorname{supp} T) \setminus E$ and $\psi \in j(E)$ doesn't vanish at γ we have $\varphi(\gamma) = 0$. Therefore $\varphi = 0$ on $(\operatorname{supp} T) \setminus E$. Thus we can use our hypothesis since $\varphi = 0$ on E. Hence $T\varphi = 0$, i.e., $T \in M(I_\varphi)$.

($\Rightarrow$) Take $\varphi \in k(E)$ and $\operatorname{supp} T\varphi \subseteq E$. We'll prove that $T\varphi = 0$.
Since E is a C-set we can choose $\{\psi_n : n = 1, \ldots\} \subseteq j(E)$ for which $\lim_n \|\varphi - \varphi\psi_n\|_A = 0$.
Thus if $\psi \in A(\Gamma)$

$$\langle T\varphi, \psi \rangle = \langle T, \varphi\psi \rangle = \lim_n \langle T\varphi, \psi\psi_n \rangle = 0.$$

q.e.d.

Note that in the necessary conditions we really prove that if $\varphi \in k(E)$ and $\operatorname{supp} T\varphi \subseteq E$ then $T\varphi = 0$. In fact, the proof of Theorem 1.4.4 shows that the conditions $\varphi \in k(E \cup \operatorname{supp} T)$ and $\operatorname{supp} T\varphi \subseteq E$, are equivalent to the conditions, $\varphi \in k(E)$ and $\operatorname{supp} T\varphi \subseteq E$; that is, if $\varphi \in k(E)$ and $\operatorname{supp} T\varphi \subseteq E$, then $\varphi \in k(E \cup \operatorname{supp} T)$.

1.4.10 Local synthesis. We now discuss "local synthesis". $T \in A'(\Gamma)$ (resp. $\varphi \in A(\Gamma)$) is *synthesizable* if for all $\varphi \in k\,(\operatorname{supp} T)$ (resp. for all $T \in A'(Z\varphi)$)

$$\langle T, \varphi \rangle = 0.$$

Naturally this definition is consistent with the definition of synthesizable elements in $\mathscr{T}$. By duality, the characterization of synthesizable pseudo-measures is intimately related to a characterization of $A(\Gamma)$ (cf. the remarks accompanying Exercise 1.3.5).
Clearly,

Proposition 1.4.4. *Given a closed set $E \subseteq \Gamma$.*

a) *If each $T \in A'(E)$ is synthesizable then E is an S-set.*

b) *If each $\varphi \in k(E)$ is synthesizable then E is an S-set.*

Obviously, in general, each possible converse in Proposition 1.4.4 is false since S-sets may contain non-S-subsets. Note that if $Z\varphi$, $\varphi \in A(\Gamma)$, (resp. $\operatorname{supp} T$, $T \in A'(\Gamma)$) is an S-set then φ (resp. T) is synthesizable.

Proposition 1.4.5 a) *Given $T \in A'(\Gamma)$, where $\hat{T} = \Phi$. Then*

$$A_S'(\operatorname{supp} T)^\wedge = (\mathscr{T}_\Phi)_{\mathrm{sp}}. \tag{1.4.13}$$

b) *Given $T \in A'(\Gamma)$, where $\hat{T} = \Phi$. T is synthesizable $\Leftrightarrow \mathscr{T}_\Phi = (\mathscr{T}_\Phi)_{\mathrm{sp}}$.*

Proof. a) Clearly $(\mathscr{T}_\Phi)_{\mathrm{sp}} \subseteq A_S'(\operatorname{supp} T)$ since $\operatorname{sp} \Phi = \operatorname{supp} T$ (e.g. Proposition 1.4.1). If $S \in A_S'(\operatorname{supp} T)$ let $\lim_\alpha \mu_\alpha = S$ in the weak $*$ topology $\sigma(A'(\Gamma), A(\Gamma))$, where $\mu^\alpha \in M_f(\operatorname{supp} T)$. When $\gamma \in \operatorname{sp} \Phi$ then $(\tau_\alpha \delta)^\wedge \in (\mathscr{T}_\Phi)_{\mathrm{sp}}$, and so $\mu_\alpha \in (\mathscr{T}_\Phi)_{\mathrm{sp}}$ since $\operatorname{sp} \Phi = \operatorname{supp} T$.
b) ($\Rightarrow$) Let $S \in M_T$; from a), we must prove that $S \in A_S'(\operatorname{supp} T)$.
We have $\lim_\alpha S_\alpha = S$ in $\sigma(A'(\Gamma), A(\Gamma))$, where $S_\alpha = T\psi_\alpha$ and $\psi_\alpha \in A(\Gamma)$.
If $\varphi \in k(\operatorname{supp} T)$ then $\langle T, \varphi \rangle = 0$ by hypothesis, and so $\langle S_\alpha, \varphi \rangle = 0$.
Thus $S \in A_S'(\operatorname{supp} T)$.

($\Leftarrow$) $T \in A'_S(\operatorname{supp} T)$ by hypothesis and a); hence $\langle T, \varphi \rangle = 0$ if $\varphi \in k(\operatorname{supp} T)$.

q.e.d.

Proposition 1.4.5b ties in with the remark at the end of Section 1.4.2: the search for synthesizable pseudo-measures $T \in A'(\Gamma)$ is precisely the problem of finding elements $\Phi \in L^\infty(G)$ which can be reconstructed as "weak $*$ convergent Fourier series" $\sum_\gamma a_\gamma(\gamma, x)$ with frequencies $\gamma \in \operatorname{sp} \Phi$.

1.4.11 The role of uniformly continuous functions in $L^\infty(G)$. We'll see in Paragraph 3.1 that there are non-synthesizable pseudo-measures T, $\hat{T} = \Phi$ (as we mentioned in Section 1.4.8). Consequently, for such T, $\mathscr{T}_\Phi$ is "not determined" by $\operatorname{sp} \Phi$. Noting that $\operatorname{sp} \Phi$ is a family of uniformly continuous functions in $\mathscr{T}_\Phi$, we prove

Proposition 1.4.6. *Given $\mathscr{T} \subseteq L^\infty(G)$ and let $\mathscr{T}_{UC}$ be the weak $*$ closed translation invariant subspace generated by the uniformly continuous elements in $\mathscr{T}$. Then $\mathscr{T} = \mathscr{T}_{UC}$.*

Proof. Let $\hat{M} = \mathscr{T}$ (recall Proposition 1.3.8). Since $M(A(\Gamma)) \subseteq M$ it is sufficient to prove that X, the weak $*$ closure of $M(A(\Gamma))$, is M.

If $T \in M \setminus X$ we employ the dual Hahn–Banach theorem (e.g. (1.3.19)) and have that

$$\exists\, \varphi \in A(\Gamma) \qquad \text{such that } \langle T, \varphi \rangle \neq 0$$

and

$$\forall\, \psi \in A(\Gamma), \qquad \langle T, \varphi\psi \rangle = 0.$$

We obtain a contradiction by letting $\lim_n \|\varphi - \varphi\psi_n\|_A = 0$.

q.e.d.

Another important feature of uniformly continuous functions is

Proposition 1.4.7. *Given $T \in A'(\Gamma)$. $\hat{T}$ is uniformly continuous $\Leftrightarrow$ there is a directed system $\{T_\alpha\} \subseteq A'(\Gamma)$, for which each* $\operatorname{supp} T_\alpha$ *is compact, such that*

$$\lim_\alpha \|T - T_\alpha\|_{A'} = 0.$$

Proof. ($\Leftarrow$) It is only necessary to check that $\hat{T}$ is uniformly continuous if $\operatorname{supp} T$ is compact.

If $\varphi \in A(\Gamma)$ equals 1 on a neighborhood of $\operatorname{supp} T$ then $T = T\varphi$ and $\widehat{T\varphi}$ is uniformly continuous.

($\Rightarrow$) Let $\Phi = \hat{T}$ be uniformly continuous. Thus

$$\forall\, \varepsilon > 0\ \exists\, U \subseteq G, \text{ a neighborhood of } 0 \in G, \text{ such that}$$

$$\forall\, x \in U,\ \|\Phi - \tau_x \Phi\|_\infty < \frac{\varepsilon}{2}. \tag{1.4.14}$$

Take $\{f_\alpha\}$ as in Proposition 1.2.2 so that $T_\alpha = T\varphi_\alpha$ has compact support. Recall that $f_\alpha \geqslant 0$ and $\int_G f_\alpha(x)\,dx = 1$.

Choose α_0 so that

$$(1.4.15) \qquad \forall\, \alpha \geqslant \alpha_0, \qquad \int_{U^\sim} f_{\alpha}\text{-}(x)\,dx < \varepsilon/(4\|T\|_{A'}).$$

Then for all $\alpha \geqslant \alpha_0$

$$\|T - T_\alpha\|_{A'} \leqslant \sup_{y \in G} \left| \int_U (\Phi(y) - \Phi(y-x)) f_{\alpha}\text{-}(x)\,dx \right| + \sup_{y \in G} \left| \int_{U^\sim} (\Phi(y) - \Phi(y-x)) f_{\alpha}\text{-}(x)\,dx \right|.$$

(1.4.14) and (1.4.15) conclude the proof.

q.e.d.

Using Proposition 1.4.7 and an extension of the Cohen factorization theorem, [Curtis and Figá-Talamanca, 1, pp. 174–176] have observed that *the uniformly continuous functions on G are precisely the elements of* $\{(T\varphi)^\wedge : T \in A'(\Gamma)$ *and* $\varphi \in A(\Gamma)\} = L^\infty(G) * L^1(G)$. Because of this we conclude that *if* $\hat{T} \in L^\infty(G)$ *is uniformly continuous then* supp T *is* σ*-compact.*

Clearly,

Proposition 1.4.8. $\varphi \in A(\Gamma)$ *is synthesizable* $\Leftrightarrow$ *for all* $T \in A'(Z\varphi)$, *for which* $\hat{T}$ *is uniformly continuous, we have* $\langle T, \varphi \rangle = 0$.

1.4.12 Remarks on non-weak $*$ analysis and synthesis. We have defined "spectral synthesis for the weak $*$ topology". We chose to define the spectrum in (1.4.1) since we were able to solve the spectral analysis problem by guaranteeing that sp $\mathscr{T} \neq \emptyset$ if $\mathscr{T} \neq 0$; consequently, our main effort in this book will be to see when we can synthesize this spectrum into the given phenomenon. There is, however, a next step in the study of spectral analysis and synthesis. Let X be a translation invariant subspace of functions on G. The problem of spectral analysis can be viewed as finding the strongest topology T_a in which sp $\bar{X} \neq \emptyset$ (the closure being taken in T_a), where sp $\bar{X}$ is defined as in (1.4.1) with $\mathscr{T}$ replaced by $\bar{X}$. The problem of spectral synthesis is to find the strongest topology T_s for which X can be synthesized from sp $\bar{X}$, i.e., such that X is contained in the T_s-closed translation invariant subspace generated by sp $\bar{X}$. For example, [Beurling, 2] originally envisaged doing analysis and synthesis in terms of the narrow topology, which we'll discuss in Paragraph 2.2; and, also, because of the influence from integral equations to analysis and synthesis, he was led to introduce the β topology (e.g. [Beurling, 2, p. 133]). The β topology has been widely developed, its fundamental properties being given in [Buck, 1; 2]. Physical considerations warrant this "topological approach"; obviously, a given phenomenon might not be synthesized in a real world model by a weak $*$ procedure since its construction in terms of its frequency components (its spectrum) might demand a more restrictive piecing-together process (topology).

Translation invariant phenomena fall naturally into the realm of Fourier analysis. On the other hand, analysis and synthesis form a general framework in which to study many physical and social phenomena. We present some examples in Paragraph 1.5.

Exercises 1.4

1.4.1 Ideals in $A(\Gamma)$ for which $IJ = I \cap J$

Let $I, J \subseteq A(\Gamma)$ be closed ideals and assume that I contains a bounded approximate identity. Prove that

$$IJ = I \cap J$$

(cf. Remark 1 in Exercise 1.2.5). (*Hint*: The proof of (1.1.5) in [Cohen, 1] yields the statement that if X is a commutative Banach algebra with bounded approximate identity and $\varphi \in X$ then $\varphi = \psi\theta$ where $\psi, \theta \in X$ and $\theta \in I_\varphi$. Thus if $\varphi \in I \cap J$ let $X = I$ and note that $I_\theta \subseteq I \cap J \subseteq J$).

Observe that (1.1.5) is a special case of $IJ = I \cap J$.

Using this exercise it can be shown that if $J, K \subseteq A(\Gamma)$ are ideals such that $ZJ = ZK$, where $J \subsetneq K$ and either J or K is closed, then there is an ideal I for which $J \subsetneq I \subsetneq K$ (e.g. [Dietrich, 1]). When both J and K are closed we can also choose I to be closed; we shall discuss this phenomenon in Section 3.1.7.

1.4.2 Estimates with the Fejér kernel on G

Let $U \subseteq \Gamma$ have positive Haar measure and define the continuous map $\rho_U : \Gamma \to \mathbf{R}^+$ as

$$\forall\, x \in G, \qquad \rho_U(x) = \sup_{\gamma \in U} |1 - (\gamma, x)|.$$

$\rho_U(x, y) = \rho_U(x - y)$ defines a pseudo-metric on $\Gamma \times \Gamma$ which is invariant in the sense that $\rho_U(x + z, y + z) = \rho_U(x, y)$. Take the Fejér kernel $\{f_\alpha\}$ from Proposition 1.2.2 and Theorem 1.2.1; and recall that

$$\varphi_\alpha = \frac{1}{mU}\chi_U * \tilde{\chi}_U, \qquad \hat{f}_\alpha = \varphi_\alpha,$$

where $U \subseteq \Gamma$ is a compact neighborhood of $0 \in \Gamma$. Writing $f_\alpha = f_U$ and $\varphi_\alpha = \varphi_U$ for this exercise, prove that

a) $\dfrac{1}{mU}|f_U(x - z) - f_U(y - z)| \leqslant 2\,\rho_U(x - y)$, uniformly in z;

b) $\displaystyle\int_G |f_U(x - z) - f_U(y - z)|\,dz \leqslant 2\,\rho_U(x - y)$.

(*Hint*: Since $f_U(x) = \frac{1}{mU} |\hat{\chi}_U(-x)|^2$,

$$\text{(E1.4.1)} \qquad \frac{1}{mU} |f_U(x-z) - f_U(y-z)| \leqslant \left(\frac{1}{mU} |\hat{\chi}_U(z-x) + \hat{\chi}_U(z-y)| \right)_1 \times \left(\frac{1}{mU} |\hat{\chi}_U(z-x) - \hat{\chi}_U(z-y)| \right)_2 .$$

For a), $(\ldots)_1$ is bounded by 2 and

$$(\ldots)_2 \leqslant \frac{1}{mU} \int |\chi_U(\gamma)(\gamma, z)((\gamma, -x) - (\gamma, -y))| \,d\gamma \leqslant \sup_{\gamma \in U} \overline{|(\gamma, x) - (\gamma, y)|}.$$

For b), use Hölder's inequality on (E1.4.1) and estimate the L^2-norms of $(\ldots)_1$ and $(\ldots)_2$).

1.4.3 All subsets of discrete Γ are S-sets

Let G be compact. Prove that every closed set $E \subseteq \Gamma$ is an S-set. (*Hint*: Let $I \subseteq A(\Gamma)$ be a closed ideal with $ZI = E$; if $\varphi \in k(E)$ we have $ZI \subseteq Z\varphi$. Apply Exercise 1.2.2).

1.4.4 The Fourier transform of a radial function

In Section 1.4.8 we mentioned that there are non-S-sets $E \subseteq \mathbb{R}^3$ for which $k(E)$ is a closed principal ideal. Setting $S^{n-1} = \{\gamma \in \mathbb{R}^n : |\gamma| = 1\}$ for $n > 1$, it turns out that we can take $E = S^2 \subseteq \mathbb{R}^3$ and that there is $\varphi \in A(\mathbb{R}^3)$ for which $\overline{j(S^2)} = I_{\varphi^2} \subsetneq I_\varphi = k(S^2)$ [Varopoulos, 7, Theorem 3]. In Exercise 2.5.5 we shall indicate the proof of the earlier (and weaker) result that $I_{\varphi^2} \subsetneq I_\varphi$ for some $\varphi \in k(S^{n-1})$, $n \geqslant 3$ [Reiter, 5, pp. 469–470]. Both results depend on properties of radial functions.

A function $\varphi : \mathbb{R}^n \to \mathbf{C}$ is *radial* if there is a function $\psi : [0, \infty) \to \mathbf{C}$ such that

$$\forall\, \gamma \in \mathbb{R}^n, \qquad \varphi(\gamma) = \psi(|\gamma|).$$

In this exercise we calculate the Fourier transform of a radial function. Our computation is found in [Bochner and Chandrasekharan, 1, pp. 69–74].

a) Let $f \in L^1(\mathbf{R}^n)$, $n \geqslant 2$, be a radial function and write $F(r) = f(x) = f(x_1, \ldots, x_n)$ where $r = (x_1^2 + \ldots + x_n^2)^{1/2}$. Prove that $\hat{f}$ is radial. This is equivalent to proving that if a linear map $R : \mathbb{R}^n \to \mathbb{R}^n$ is an *orthogonal transformation*, i.e., R preserves the Euclidean inner product and the determinant of R is 1, then

$$\hat{f}(\gamma) = \hat{f}(R\gamma).$$

Thus we write $\hat{f}(\gamma) = \varphi(\gamma_1, \ldots, \gamma_n) = \Phi(\rho)$ where $\rho = (\gamma_1^2 + \ldots + \gamma_n^2)^{1/2}$. Prove that

$$\text{(E1.4.2)} \qquad \Phi(\rho) = \int_0^\infty F(r)\, r^{n-1} K(\rho r)\, dr,$$

where

(E1.4.3) $$K(r) = c\int_0^{\pi} e^{ir\cos\theta}(\sin\theta)^{n-2}\,d\theta$$

and c is a constant we'll be able shortly to forget. The calculation is straightforward using the definition of the parametric surface describing S^{n-1} and the definition of surface integral.

It is also true that if $f \in L^1(\mathbf{R}^n)$ and $\hat{f}$ is radial then f is radial.

b) Letting $r \in \mathbf{C}$ in (E1.4.3), prove that $K(r)$ has a non-trivial power series expansion, $\sum_{k=0}^{\infty} a_k r^k$; in fact, note that $|K(r)| \leqslant e^{|r|}$, i.e., K is an entire function of exponential type (cf. Exercise 2.2.3b).

c) Prove that

(E1.4.4) $$\pi^{n/2}e^{-\rho^2/4} = \sum_{k=0}^{\infty} \tfrac{1}{2}\Gamma\left(\frac{k+n}{2}\right)a_k\rho^k.$$

(*Hint*: Let $F(r) = e^{-r^2}$ so that $\Phi(\rho) = \pi^{n/2}e^{-\rho^2/4}$ by direct calculation (cf. Exercise 1.1.1a). For the right-hand side of (E1.4.4), use (E1.4.2), the expansion $K(\rho r) = \sum_0 a_k\rho^k r^k$, and the definition of the gamma function (as a Laplace transform)

$$\Gamma(z) = s^z\int_0^{\infty} e^{-st}t^{z-1}\,dt, \qquad \text{Re } s > 0, \qquad \text{Re } z > 0$$

(cf. Exercise 1.1.1a)).

d) Prove that

$$K(\rho r) = 2\pi^{n/2}\sum_{k=0}^{\infty}\frac{(-1)^k(\rho r)^{2k}}{4^k\Gamma(k)\Gamma(\beta+k-1)}, \qquad \beta > -\frac{1}{2}.$$

(*Hint*: Expand $e^{-\rho^2/4}$ as a power series and compare coefficients with (E1.4.4)).

e) Using the fact that $J_\beta(r)$, the *Bessel function* of order $\beta > -1/2$, can be written as

$$J_\beta(r) = \left(\frac{r}{2}\right)^{\beta}\sum_{k=0}^{\infty}\frac{(-1)^k r^{2k}}{4^k\Gamma(k)\Gamma(\beta+k+1)}, \qquad \beta > -\frac{1}{2},$$

prove that

(E1.4.5) $$\Phi(\rho) = \frac{2\pi^{n/2}}{\rho^{(n-2)/2}}\int_0^{\infty} F(r)\,r^{n/2}J_{(n-2/2)}(\rho r)\,dr.$$

Remark 1. It would be interesting to verify if there are non-S sets $E \subseteq \mathbb{R}$ for which $k(E)$ is principal. Even with the tensor algebra techniques of [Varopoulos, 8] and the validity of the result in $\mathbb{R}^3$, the problem presents difficulties.

2. With regard to Atzmon's result, Exercise 1.2.5 and these remarks, it is natural to ask if every closed ideal in $A(\mathbb{R})$ is the intersection of the closed principal ideals containing it.

1.4.5 The classification of closed ideals in algebras of distributions

In the spirit of the "finite dimensional" Exercise 1.2.5 we now consider infinite dimensional Banach algebras X which are synthesizable in the sense that every closed ideal $I \subseteq X$ is an intersection of maximal ideals (e.g. Theorem 1.4.1h); such Banach algebras are sometimes called N-algebras, e.g. [Shilov, 1]. General criteria for a Banach algebra to be an N-algebra are found in [Shilov, 1, p. 108; Rickart, 1, pp. 92–96].

a) Let $X = C_0(\Gamma)$ and prove that every closed proper ideal in the sup norm algebra $C_0(\Gamma)$ is an intersection of maximal ideals. (Obviously, Γ is only required to be a locally compact Hausdorff space). This result was first established by M. Stone in 1937 and there are several well-known proofs, e.g. [Hewitt and Stromberg, 1, pp. 365–366; Loomis, 1, pp. 57–58].

b) For each $\mu \in M_c(\mathbb{R})$ let $F: \mathbb{R} \to \mathbb{C}$ be the unique continuous function of bounded variation vanishing at $-\infty$ with distributional derivative $F' = \mu$ (cf. Exercise 1.1.4b and Exercise 1.3.6). If F corresponds to $\mu \in M_c(\mathbb{R})$ and G corresponds to $\nu \in M_c(\mathbb{R})$ then

$$(FG)' \in M_c(\mathbb{R})$$

defines a multiplication under which $M_c(\mathbb{R})$ is a Banach algebra normed by

$$\forall\, \mu \in M_c(\mathbb{R}), \qquad \|\mu\| = \|F\|_\infty + \|\mu\|_1.$$

Prove that with this norm and multiplication $M_c(\mathbb{R})$ is an N-algebra. Spectral synthesis properties of Helson sets have been studied in this context in [Benedetto, 4], cf. Exercise 3.2.3e.

c) Let $C^n(\mathbf{T})$, $n < \infty$, be the space of n-times continuously differentiable functions φ on $[0, 2\pi]$ such that $\varphi^{(k)}(0) = \varphi^{(k)}(2\pi)$ for $0 \leqslant k \leqslant n$. $C^n(\mathbf{T})$ is a Banach algebra under pointwise multiplication and with norm

$$\|\varphi\| = \sum_{k=0}^{n} \frac{\|\varphi^{(k)}\|_\infty}{k!}.$$

$C^\infty(\mathbf{T})$, considered in Exercise 1.1.4b, can never be a Banach algebra, e.g. [Gelfand, Raikov, and Shilov, 1, p. 67]. Prove that $C^n(\mathbf{T})$ is not an N-algebra for $n \geqslant 1$; in fact, for all $\gamma \in \mathbf{T}$, $\overline{j(\gamma)} \subsetneq k(\gamma)$ and $\overline{j(\gamma)}$ is a closed primary ideal (obviously, j and k are defined in the context of $C^n(\mathbf{T})$). On the other hand Shilov (1940) proved that each closed ideal in $C^n(\mathbf{T})$ is an intersection of closed primary ideals. The extension of this theorem to to regions in $\mathfrak{R}^k$ was made by [Whitney, 1], cf. [Glaeser, 1]. Given this Shilov–Whitney result, what can be said about the C^∞ case?

Remark 1. The ideal structure of the space of functions analytic in the open unit disc and continuous on the boundary has been characterized by Beurling (unpublished) and [Rudin, 2], e.g. [Gelfand, Raikov, and Shilov, 1, pp. 235–239; Hoffman, 1, pp. 82–89]. The study of closed ideals in locally convex spaces of analytic functions is presently a popular sport.

2. The problem to classify the closed ideals in algebras of distributions goes back to [Schwartz, 1]. This problem is closely related to questions about ideals in spaces of analytic functions (e.g. Exercise 1.3.6 and the Laplace transform operator L) and Delsarte's theory of mean-periodic functions. Let $F(G)$ be a topological algebra of functions defined on G. $V \subseteq F(G)$ is a *variety* if it is a closed translation invariant subspace of $F(G)$; V_Φ will designate the variety generated by $\Phi \in F(G)$. For example, if $F(G) = L^\infty(G)$ is taken with the weak $*$ topology then $V_\Phi = \mathscr{T}_\Phi$. $\Phi \in F(G)$ is *mean-periodic* if the variety V_Φ is properly contained in $F(G)$. Expositions of the theory of mean-periodic functions are found in [Kahane, 5; Meyer, 5]; the latter reference follows the outline of the former and contains the more recent work of [Beurling and Malliavin, 1].

For the duality $(C(\mathbf{R}), D_c^0(\mathbf{R}))$, [Schwartz, 1] proved that every closed ideal in $D_c^0(\mathbf{R})$ is an intersection of primary ideals. If $V \subseteq C(\mathbf{R})$ is a variety then V contains *exponential monomials* Φ, i.e., $\Phi(x) = x^m e^{sx}$, where $m \in \mathbf{Z}^+$ and $s \in \mathbf{C}$; this is the spectral analysis of V. Schwartz's theorem can be rephrased as the corresponding spectral synthesis result: if $V \subseteq C(\mathbf{R})$ is a variety then the set of exponential monomials (in V) is total in V. An *exponential polynomial* has the form $\Phi(x) = P(x)e^{sx}$ where $P(x)$ is a polynomial. It is interesting to note that $D_c^0(\mathbf{R})$ has non-maximal primary ideals.

$\Phi \in C(\mathbf{R}^n)$ (resp. $\Phi \in C^\infty(\mathbf{R}^n)$) is *mean-periodic* if and only if

$$\exists\, T \in D_c^0(\mathbf{R}^n) \setminus \{0\} \text{ (resp. } T \in D_c(\mathbf{R}^n) \setminus \{0\})$$

such that

$$T * \Phi = 0.$$

In this context Schwartz's investigation is closely tied in with the theory of partial differential equations since many partial differential operators are elements of $D_c(\mathbf{R}^n)$. [Malgrange, 1, p. 310] proved that any solution, $\Phi \in C^\infty(\mathbf{R}^n)$, of

$$T * \Phi = 0, \tag{E1.4.6}$$

for a fixed $T \in D_c(\mathbf{R}^n)$, belongs to the closure of the set of solutions (of (E1.4.6)) each of whose elements is a linear combination of exponential polynomials. Related work is found in [Ehrenpreis, 1; 2].

In these questions it is important to note that the maximal ideal space of $D_c(\mathbf{R})$, say, is identified with $\mathbf{C}$. In fact, if $\Phi: D_c(\mathbf{R}) \to \mathbf{C}$ is a continuous homomorphism then for each $T \in D_c(\mathbf{R})$, $\langle \delta' * T, \Phi \rangle = \langle T, \Phi \rangle \langle \delta', \Phi \rangle$; thus $\Phi' = \Phi(0)\Phi$ and so $\Phi(x) = e^{sx}$ for some $s \in \mathbf{C}$. Functions of the form $\Phi(x) = e^{sx}$ are mean-periodic. Also, every primary ideal $I = I_{s,n} \subseteq D_c(\mathbf{R})$ has the form

$$I_{s,n} = \{T \in D_c(\mathbf{R}) \colon \forall_j = 0, \ldots, n, (L(T))^{(j)}(s) = 0\}.$$

There are many open problems in this area associated generally with extensions of 1-dimensional results to higher dimensions, cf. [Gilbert, 1]. For example, if $S, T \in D_c(\mathbf{R})$ and $L(S), L(T)$ have disjoint zero sets then the closed ideal generated by S and T is $D_c(\mathbf{R})$ (the converse is obvious for $\mathbf{R}^n$, $n \geqslant 1$); the result is not known for $\mathbf{R}^n$, $n > 1$. On the other hand, Schwartz's result remains true in the setting of an LCAG G if the added hypothesis is made that the *annihilator ideal*, $V^\perp = \{T \in D_c^0(G) \colon \forall\ \Phi \in C(G), \langle T, \Phi \rangle = 0\}$, of the variety V is a closed principal ideal [Elliott, 1]. The techniques related to annihilator ideals were developed by [Ehrenpreis, 1; 2]. In the case that G is discrete Schwartz's result remains true without the hypothesis concerning annihilator ideals [Elliott, 2].

1.5 Classical motivation for the spectral synthesis problem

1.5.1 Analysis and synthesis. The process of analysis and synthesis has proved to be a significant intellectual pattern in the arts and sciences. Generally, the analysis of a phenomenon is the determination of its fundamental components, i.e., its spectrum; and the synthesis is the reconstruction of the phenomenon in terms of these components. The significance of this reconstruction becomes apparent if one views it as a study of the manner in which the given phenomenon is a combination of elements from its spectrum. The basic problem of specifying what this manner should be or how one should define the notion of spectrum is, of course, intimately related to the process of analysis and synthesis itself.

1.5.2 Examples

1.5.2.1 The theory of tides

The ancient mariner was aware that tidal phenomena were related to astronomical factors.

The study of tides received its first scientific basis with Newton's law of gravitation for the sun, moon, and earth. Laplace was able to separate various cyclic influences of the sun and moon (on tides) by defining a model of the sun, moon, and earth having a number of tide-affecting satellites; the Newtonian solution to this model associated an elementary tidal constituent with each satellite in such a way that the tide was viewed as a combination of these constituents. Lord Kelvin (1824-1907) systematized this method and initiated the design of mechanical analyzers and synthesizers which determined the constituents and reconstructed the tides, respectively. [Godin, 1; Macmillan, 1] and exhibits of mechanical harmonic analyzers and synthesizers at various science museums present interesting introductions to tidal analysis.

1.5.2.2 Cubism

The development of cubism dating from the influence of Cézanne and African art on Braque and Picasso to the latters' introduction of *collages* and *papiers collés* (in 1912) forms the background for a pattern of synthesis in the "synthetic" cubism of Gris. In his early cubism of 1911–1912, Gris dissected his subject, examined each part from a different angle, and then reconstructed the total image again from the various parts [Golding, 1, p. 101]. Later, after his own work with *papiers collés*, his explicit "synthetic" cubism evolved to the point where he began a painting by first planning its structure and then superimposing the subject in this framework [Read, 1, p. 86]. The actual composition was made up of "flat colored forms" [Golding, 1, p. 136]. In our language the above structure or framework is the "topology" in which we wish to synthesize, and the colored forms represent the elements of the spectrum; the actual synthesis of the given subject, then, is the painting.

1.5.2.3 The theory of filters (in communication theory)

The theory of filters in communication networks [Cuccia, 1] and general information theory [Goldman, 1] provides our third example. A filter is a circuit which is chosen to produce specific frequency responses $F(\omega)$ given a certain class of inputs (to the circuit). Thus, a filter might be designed to attenuate certain elementary waves (corresponding to given frequencies ω) which go into the composition of an input signal. The design of such filters can be viewed as a synthesis problem in which F is to be synthesized by the circuit in terms of some interval of frequencies. [Wiener, 9, pp. 88–96] referred to such problems in the context of synthesis and posed the corresponding problem of analysis, e.g. [Wiener, 9, pp. 97–100]. Wiener's first and fundamental work, [Wiener, 11], on prediction and filtering problems dates from 1940–1943; and in subsequent years he developed his basic notion of filter to the context of cybernetic problems. In [Wiener, 10, pp. 69–125], Masani has written an important essay, remarkable both in its clarity and scope, which deals with Wiener's filter theory.

1.5.2.4 Music

An origin of spectral analysis and synthesis is the study of the vibrating string problem: the sounds produced by a bowed violin string can be considered as combinations of pure sounds or harmonics. (Actually there are non-trivial non-linear effects associated with bowing as opposed to plucking; we shall not discuss these.) The spectral analysis is the search for these harmonics; the synthesis is the reconstruction of the original sound in terms of its harmonics. With regard to the discussion of filters in Section 1.5.2.3 we note that an important spectral analysis and synthesis result in non-linear filters was given by [Wiener, 9, pp. 88–100] with white noise as the input. In music theory, white noise is sometimes considered as a mathematical model for snare drums or cymbals; actually, given the purely stochastic nature of white noise, such a model is really quite inaccurate.

The propagation of sound in one space dimension was characterized in terms of the wave equation by Euler in 1759; of course, the mathematical model for the motion of the vibrating string with certain restrictions on its behavior is given by this equation. The subsequent analysis led not only to the representation problem in Fourier series (cf. Section 1.5.3) but to a precise definition of function. There was a good deal of intellectual bloodshed before the final resolution of these two intimately related issues.

It was Daniel Bernoulli (1700–1782) who suggested a Fourier series solution to the wave equation. He also did major experimental spectral analysis and synthesis; for example, he showed that in closed organ pipes the frequencies of the overtones are odd multiples of the frequency of the fundamental. Helmholtz devised effective mechanical devices to synthesize complex tones, analogous to the mathematical synthesis associated with solutions to the wave equation in terms of Fourier series. These are described in [Josephs, 1, pp. 70–72]. There are, of course, startling recent developments to Helmholtz's research in the form of electronic music. Now we can listen to the *Toccata and Fugue in D Minor* not on the organ as Bach (1685–1750) originally planned but on the Putney V.C.S.3 synthesizer [Hankinson, 1].

1.5.2.5 The interferometer

The origin of Wiener's work in generalized harmonic analysis was in the study of beams of white light. The well-developed area of Fourier spectroscopy, of which Wiener's work can be considered a part, has its origins in Michelson's invention of the interferometer [Loewenstein, 1; Michelson, 1]. Basically, an interferogram [Loewenstein, 1, p. 846] is the record of a detected signal and corresponds (for us) to some $\Phi \in L^{\infty}(G)$. The interferometer synthesizes the spectrum experimentally to produce the interferogram; and the problem of Fourier spectroscopy is the problem of spectral analysis, viz. to determine the spectrum E of Φ. In optics, the spectroscopy problem is sometimes posed so that the spectral distribution $\mu \in M(\Gamma)$ is sought (cf. Chapter 2.1). In our terminology this is equivalent to finding E since $\operatorname{supp} \mu = E$.

1.5.2.6 The Heisenberg uncertainty principle

The wave theory in quantum mechanics arose since electron beams diffracted through crystals produced an effect analogous to Newton's spectral theory of white light diffracted through a prism (cf. Section 1.3.10). This experiment with electron beams actually followed de Broglie's suggestion (1924) that matter has both a wave and particle representation related by

(1.5.1) $$p = \hbar x,$$

where $\hbar$ is Planck's constant, p is the momentum of the given particle, and x is the frequency of the corresponding wave. The wave function $\Psi(\gamma, t)$, normalized so that

$\int |\Psi(\gamma,t)|^2\,d\gamma = 1$, was introduced by **Schrödinger** (1926) to describe the wave interpretation of matter; and an important aspect of its physical significance is that for a fixed time t,

$$(1.5.2)\qquad \int_A |\Psi(\gamma,t)|^2\,d\gamma$$

is the probability that the given particle is in $A \subseteq \mathbf{Я}$ [**Schiff**, 1, Chapters 1 and 2]. Also, as in the spectral synthesis associated with classical problems characterized by the wave equation, we have formally that

$$(1.5.3)\qquad \Psi(\gamma,t) = \int_{\mathbf{R}} f(x,t)e^{i\gamma x}\,dx.$$

Because of (1.5.2), if at time t the particle is in $\left[-\frac{\lambda}{2},\frac{\lambda}{2}\right]$ then

$$\int_{-\lambda/2}^{\lambda/2} |\Psi(\gamma,t)|^2\,d\gamma = 1.$$

Consequently, in this case, the support of the wave function is "concentrated" in $\left[-\frac{\lambda}{2},\frac{\lambda}{2}\right]$ at time t. Suppose that the support of f is "concentrated" in $\left[-\frac{a}{2},\frac{a}{2}\right]$ in the sense that

$$\int_{-a/2}^{a/2} |f(x,t)|^2\,dx \sim 1$$

(i.e., Plancherel's theorem) and

$$\Psi(\gamma,t) \sim \int_{-a/2}^{a/2} f(x,t)\,e^{i\gamma x}\,dx.$$

Then, by Hölder's inequality,

$$(1.5.4)\qquad 1 \sim \int_{-\lambda/2}^{\lambda/2} \left| \int_{-a/2}^{a/2} f(x,t)\,e^{i\gamma x}\,dx \right|^2 d\gamma \leqslant \lambda \left(\int_{-a/2}^{a/2} |f(x,t)|\,dx \right)^2$$

$$\leqslant \lambda a \int_{-a/2}^{a/2} |f(x,t)|^2\,dx \sim \lambda a;$$

and so a must satisfy the inequality

$$(1.5.5)\qquad a \geqslant 1/\lambda.$$

From (1.5.1), (1.5.4), and the fact that $\left[-\frac{a}{2},\frac{a}{2}\right]$ is an interval of frequencies, we have the

Heisenberg uncertainty principle,

$$\Delta\gamma\Delta p \gtrsim \hbar,$$

where we've written $\Delta\gamma = \lambda$ and $\Delta x = a$. It is important to compare this phenomenon and (1.5.5) with calculations such as Exercise 1.1.1a and Proposition 1.2.2. Thus, "if a function $f \in L^2(\mathbf{R})$ is close to 1 on a long interval of $\mathbf{R}$, then $\hat{f}$ will not only have its support "concentrated" in a small set but also will attain large values there". Because of the previous remarks, this "fact" concerning the relation between f and $\hat{f}$ could be alluded to as the Heisenberg uncertainty principle.

1.5.3 Synthesis and the representation problem in harmonic analysis. We have defined a precise spectral synthesis problem in Paragraph 1.4, but let us now observe, for a moment, that the two major problems in harmonic analysis can be viewed in terms of spectral analysis and synthesis. These two problems are, respectively, to compute Fourier coefficients, for distributions T on $\mathbf{T}$, say, and then to construct T in some way in terms of these coefficients. This latter problem is usually referred to as the representation problem. A typical situation is that we are given a function φ (on $\mathbf{T}$) and are able to solve the spectral analysis problem by computing the Fourier coefficients $\hat{\varphi}(n)$; the problem of representation or synthesis is to investigate if there are any ways (i.e., convergence criteria) of representing φ in terms of $\sum \hat{\varphi}(n)e^{in\gamma}$. For example, [Schwartz, 5, pp. 225–227] gave a spectacularly simple solution to the representation problem for any distribution on $\mathbf{T}^n$ in terms of the topology on $D(\mathbf{T}^n)$; whereas [Carleson, 2] (cf. [Fefferman, 1]) gave a spectacularly difficult solution to the representation problem for any $\varphi \in C(\mathbf{T})$ in pointwise convergence a.e.

1.5.4 Integral equations. Many of the phenomena of classical mathematical physics are approximately and realistically described by linear operators K in the form of ordinary and partial differential equations and integral equations, e.g. [Schwartz, 4; Tricomi, 1]. Note that on $\mathbf{R}$,

$$\frac{df}{dx} = \delta' * f,$$

where "d/dx" is the ordinary operation of differentiation for a continuously differentiable function f. Generally, large classes of the above-mentioned equations can be written in the form

$$K * f = g; \tag{1.5.6}$$

where $K, g \in D(\mathbf{R}^n)$ are given and f is to be found. Naturally, a major problem in the solution of (1.5.6) is to determine the proper range for K.

A special case of (1.5.6) and one for which a very fruitful method exists, e.g. [F. Riesz and Sz.-Nagy, 1; Dunford and Schwartz, 1], is the eigenvalue problem where K

has the form $\lambda\delta - H$ for a compact operator H. This problem can be viewed in terms of spectral analysis and synthesis. The spectral analysis problem has a positive solution since there is a non-empty set of eigenvalues; and for the synthesis problem, solutions are given by eigenfunction expansions where eigenvalues play the role of frequencies in the eigenfunction. Because of our discussion of Riemann's ζ-function in Paragraph 2.3, we note that Carleman (1934) expressed the Green's function of the heat equation, when dealing with a two-dimensional membrane, in terms of a generalized ζ-function; and using standard ζ-function arguments he was able to calculate asymptotic formulas for the eigenfunctions [Kac,1; Weyl, 1].

The study of (1.5.6) for more general kernels K than those of the form $K = \lambda\delta - H$ presents formidable problems. One approach which takes advantage of the convolution in (1.5.6) is to use the Fourier transform and therefore deal with $\hat{K}\hat{f} = \hat{g}$. Thus if $K, g \in L^1(\mathbf{Z})$ and we wish to find a solution $f \in L^1(\mathbf{Z})$ we need only assume that $\hat{K}$ doesn't vanish and then apply Proposition 1.1.5 to assure the existence of such an f. Obviously $Z\hat{K}$ plays a role in any refinement of this technique. Such refinements and the corresponding study of zero sets have been one influence for the development of the spectral synthesis problem as stated in Paragraph 1.4.

If we take $g = 0$ in (1.5.6) we see quite precisely the connection with spectral synthesis; in fact, in order to determine if a given $K \in L^\infty(\mathbf{R})$ is synthesizable we ask if $K * f_- = 0$ for each $\hat{f} \in k(\operatorname{sp} K)$ (cf. Section 1.4.2).

2 Tauberian theorems

2.1 The Wiener spectrum and Wiener's Tauberian theorem

2.1.1 Introductory remark. In this chapter we shall see how the notion of "spectrum" has evolved in the context of Tauberian theorems (Paragraphs 2.1 through 2.3), and then study the intimate relation between Tauberian theorems and spectral synthesis (Paragraphs 2.4 and 2.5). As such, our setting will be $G = \mathbf{R}$ and $\Gamma = \hat{\mathbf{R}}$ for part of the chapter, although much of what we say can be developed in a more abstract framework; for example, the notion of the "Wiener spectrum" is studied generally in [Benedetto, 11; Herz, 5, p. 195].

2.1.2 Heuristics for generalized harmonic analysis. In the late 1920's Wiener took up the problem of characterizing rigorously the spectral analysis of a white light signal $\Phi \in L^\infty(\mathbf{R})$ (cf. Section 1.5.2.5 and Section 1.3.10). Two mathematical problems arose immediately; and these led him to his theory of generalized harmonic analysis [Wiener, 5]. First, Newton's experiment with sunlight demanded a notion of a *continuous* spectrum; as such, Wiener was forced to abandon the theory of Fourier series as a tool since series can only yield a discrete spectrum of frequencies. Second, sunlight is conveniently considered as a signal which is supposed to last for an indefinite time and which does not decrease to zero with time; as such, Wiener could not use the classical L^1 or L^2 Fourier transform theory on $\mathbf{R}$ as a tool since

$$\forall\, a > 0, \qquad \lim_{t \to \infty} \int_t^{t+a} |\Phi(x)|^2 \, dx = 0,$$

when $\Phi \in L^2(\mathbf{R})$ (even though the spectrum of Φ could be defined reasonably as "supp $\hat{\Phi}$", a non-discrete set).

Heuristically it is desirable to associate an energy to each frequency of a given signal $\Phi \in L^\infty(\mathbf{R})$ so that the total energy is distributed among all of the frequencies making up the signal. Thus, if

$$\Phi(x) = \sum_1^n a_k e^{i\lambda_k x},$$

then the spectral distribution of energy for the signal Φ is

$$\mu = \mu_\Phi = \sum_1^n |a_k|^2 \, \delta_{\lambda_k},$$

where (as would be expected physically) each coefficient $|a_k|^2$ represents an energy. Note that in this case,

$$\hat{\mu}_\Phi(x) = \lim_{R\to\infty} \frac{1}{2R} \int_{-R}^{R} \overline{\Phi(y)}\, \Phi(x+y)\, dy.$$

Consequently, for $\Phi \in L^\infty(\mathbf{R})$, Wiener defined the *autocorrelation* or *covariance function*

$$\overset{W}{\Phi}(x) = \lim_{R\to\infty} \frac{1}{2R} \int_{-R}^{R} \overline{\Phi(y)}\, \Phi(x+y)\, dy,$$

when the limit exists pointwise everywhere. From the point of view of optics, Wiener's generalized harmonic analysis of an arbitrary signal $\Phi \in L^\infty(\mathbf{R})$ was designed to define reasonably the spectral distribution of energy for Φ and to relate analytically the *brightness* (or *power*) of the signal Φ, viz.

$$\lim_{R\to\infty} \frac{1}{2R} \int_{-R}^{R} |\Phi(x)|^2\, dx$$

(assuming the limit exists), to this spectral distribution; this program was accomplished in terms of the autocorrelation function $\overset{W}{\Phi}$. The statistical terminology in defining $\overset{W}{\Phi}$ is justified in [Wiener, 11, pp. 4–5], and some physical examples of the Wiener theory are given in [Arsac, 1, Chapter 11].

2.1.3 The Wiener spectrum and generalized harmonic analysis. We now present a formal, brief, and non-historical introduction to generalized harmonic analysis.

Given $T \in A'(\hat{\mathbf{R}})$, where $\hat{T} = \Phi$. Define

$$\overset{W}{\Phi}_R(x) = \frac{1}{2R} \int_{-R}^{R} \overline{\Phi(y)}\, \Phi(x+y)\, dy.$$

The set, $\{\overset{W}{\Phi}_R : R > 0\}$, is obviously $\| \ \|_\infty$-bounded. Let $\Psi \in L^\infty(\mathbf{R})$ be a weak $*$ limit point of $\{\overset{W}{\Phi}_R : R > 0\}$. We shall show that

(2.1.1) $\quad \Psi = \hat{\mu},$

where $\mu \in M(\hat{\mathbf{R}})$ and $\mu \geqslant 0$.

We first make some remarks on positive definite distributions. It is a routine calculation, using Theorem 1.3.2b, to prove that *a continuous function $\Phi : \mathbf{R} \to \mathbf{C}$ is positive definite if and only if*

$$\forall f \in C_c^\infty(\mathbf{R}), \qquad \int_{\mathbf{R}} \Phi(x) f * \tilde{f}(x)\, dx \geqslant 0$$

[Schwartz, 5, pp. 274–275]. As such, a distribution $P \in D(\mathbf{R})$ is defined to be *positive definite* if

$$\forall f \in C_c^\infty(\mathbf{R}), \qquad \langle P, f * \tilde{f} \rangle \geqslant 0.$$

[Schwartz, 5, pp. 276–277] has proved Bochner's theorem for positive definite distributions which, when combined with Proposition 1.3.3, yields the following statement of *Bochner's theorem* (cf. [Cramér, 1]): if $\Phi \in L^\infty(\mathbf{R})$ is a positive definite distribution then there is $\nu \in M(\mathbf{\Gamma})$, for which $\nu \geqslant 0$, such that $\hat{\nu} = \Phi$ a.e. (in particular, there is a continuous positive definite function which equals Φ a.e.).

In order to verify (2.1.1), we take $f \in C_c^\infty(\mathbf{R})$, consider it as an element of $M(\mathbf{R})$, and approximate it in the weak $*$ topology by elements of $M_f(\mathbf{R})$. Then, since $\overset{W}{\Phi}_R$ is a continuous positive definite function, we have that

$$\forall\, R > 0, \qquad \int_{\mathbf{R}} \overset{W}{\Phi}_R(x) f * \tilde{f}(x)\, dx \geqslant 0.$$

Hence, Ψ is a positive definite distribution and (2.1.1) is true.

By standard measure theoretic results, e.g. [Bourbaki, 1, Chapitre II, Section 1, no. 3, Prop. 1 and Chapitre II, Section 2, no. 2, Théorème 1], we can conclude the following from (2.1.1) for a fixed pseudo-measure $T \in A'(\mathbf{\Gamma})$, where $\hat{T} = \Phi$: *there is a unique element* $\mu_\Phi \in M(\mathbf{\Gamma})$, $\mu_\Phi \geqslant 0$, *such that for each* σ $(L^\infty(\mathbf{R}), L^1(\mathbf{R}))$ *limit point* $\Psi = \hat{\mu}$ *of* $\{\overset{W}{\Phi}_R : R > 0\}$, *we have*

$$\mu - \mu_\Phi \geqslant 0,$$

and if $\mu - \nu \geqslant 0$ *for each such* μ *then*

$$\mu_\Phi - \nu \geqslant 0$$

(where $\nu \in M(\mathbf{\Gamma})$, $\nu \geqslant 0$). μ_Φ is the *spectral distribution of energy* for Φ. The *Wiener spectrum of* $\hat{T} = \Phi \in L^\infty(\mathbf{R})$, $\mathrm{sp}_W\, \Phi$, and the *Wiener support of* $T \in A'(\mathbf{\Gamma})$, $\mathrm{supp}_W T$, are defined as

(2.1.2) $$\mathrm{sp}_W\, \Phi = \mathrm{supp}_W T = \mathrm{supp}\, \mu_\Phi.$$

For technical convenience in our presentation we say that $T \in A'_*(\mathbf{\Gamma}) \subseteq A'(\mathbf{\Gamma})$, where $\hat{T} = \Phi$, if $\{\overset{W}{\Phi}_R : R > 0\}$ converges (as opposed to convergence of a *sub*-directed system) in the $\sigma(L^\infty(\mathbf{R}), L^1(\mathbf{R}))$ topology.

2.1.4 Examples

2.1.1 a) The space, $A'_c(\mathbf{\Gamma})$, of *continuous pseudo-measures* consists of pseudo-measures $T \in A'(\mathbf{\Gamma})$ for which

(2.1.3) $$\lim_{R \to \infty} \frac{1}{2R} \int_{-R}^{R} |\hat{T}(x)|^2\, dx = 0.$$

Clearly, $A_c'(\mathfrak{R}) \subseteq A_*'(\mathfrak{R})$. Because of Hölder's inequality, (2.1.3) is equivalent to the condition that

$$\lim_{R\to\infty} \frac{1}{2R} \int_{-R}^{R} |\hat{T}(x)|\,\mathrm{d}x = 0.$$

Thus if $T \in A_c'(\mathfrak{R})$ then

$$\left| \frac{1}{2R} \int_{-R}^{R} \overline{\hat{T}(y)}\,\hat{T}(x+y)\,\mathrm{d}y \right| \leqslant \|T\|_{A'} \frac{1}{2R} \int_{-R}^{R} |\hat{T}(z)|\,\mathrm{d}z;$$

and so $\mathrm{supp}_W T = \emptyset$.

b) Define

$$\hat{T}(x) = \Phi(x) = \mathrm{e}^{\mathrm{i}x^2}$$

(which can be considered as a solution to a Schrödinger equation). Clearly, $T \notin A_c'(\mathfrak{R})$.

On the other hand,

$$\overset{W}{\Phi}_R(x) = \frac{\mathrm{e}^{\mathrm{i}x^2}}{2R} \int_{-R}^{R} \mathrm{e}^{2\mathrm{i}xy}\,\mathrm{d}y = \begin{cases} 1 & \text{for } x = 0, \\ \dfrac{\mathrm{e}^{\mathrm{i}x^2}}{2Rx} \sin 2Rx & \text{for } x \neq 0; \end{cases}$$

and so $\mathrm{supp}_W T = \emptyset$. The same result is valid for $\hat{T}(x) = \mathrm{e}^{\mathrm{i}x^r}$, $r > 1$.

c) Define the discontinuous measure

$$T = \sum_{\gamma\in X} a_\gamma \delta_\gamma, \qquad \text{where} \qquad \sum_{\gamma\in X} |a_\gamma| < \infty.$$

If $\hat{T} = \Phi$ then

$$\overset{W}{\Phi}_R(x) = \sum_{\gamma\in X} |a_\gamma|^2 \mathrm{e}^{\mathrm{i}\gamma x} + \frac{1}{R} \sum_{\substack{\lambda,\gamma\in X\\ \gamma\neq\lambda}} a_\gamma \bar{a}_\lambda \frac{\mathrm{e}^{\mathrm{i}\gamma x}}{\gamma-\lambda} \sin R(\gamma-\lambda).$$

Thus, $\mathrm{supp}_W T = \mathrm{supp}\, T = \bar{X}$ if

$$\sum_{\substack{\lambda,\gamma\in X\\ \gamma\neq\lambda}} \frac{|a_\gamma \bar{a}_\lambda|}{|\gamma-\lambda|} < \infty.$$

For example, this series converges if $X = \left\{\frac{1}{n} : n = 1, \ldots\right\}$ and $a_n = 1/2^n$. Actually, a much stronger result is true, e.g. Remark 3 after Theorem 2.1.1.

Remark. Assume that the sequence $\{\rho(N) > 0 : N = 1, \ldots\}$ satisfies the condition that

$$\lim_{N\to\infty} \rho(N+1)/\rho(N) = 1;$$

and define

$$\overset{W}{\Phi}_{\rho(N)}(x) = \frac{1}{\rho(N)} \int_{-N}^{N} \overline{\Phi(y)}\, \Phi(x+y)\, dy$$

for $\hat{T} = \Phi \in L^\infty(\mathbf{R})$. Starting with the sequence $\{\overset{W}{\Phi}_{\rho(N)} : N = 1, \ldots\}$ (instead of $\{\overset{W}{\Phi}_N : N = 1, \ldots\}$) and using an ingenious trick due to M. Riesz, [Bochner, 1, pp. 328 ff.] was able to prove the existence of $\mu_\Phi^\rho \in M(\mathbf{Я})$, $\mu_\Phi^\rho \geqslant 0$, in the same way that μ_Φ was determined.

2.1.5 Outline of the following sections. Given $M \subseteq A'(\mathbf{Я})$. Analogous to (1.3.18) we define

$$I_W(M) = \left\{ \varphi \in A(\mathbf{Я}) : \forall\, T \in M,\ \lim_{R \to \infty} \frac{1}{2R} \int_{\mathbf{R}} |\widehat{T\varphi}(x)| \Delta_R(x)\, dx = 0 \right\} \tag{2.1.4}$$

("Δ_R" was defined in Exercise 1.2.6a). It is easy to check that $I_W(M) \subseteq A(\mathbf{Я})$ is a closed ideal and that $ZI_W(M) \subseteq ZI(M)$. Setting

$$\operatorname{supp}_W M = \cup \{\operatorname{supp}_W T : T \in M\},$$

we shall see in Theorem 2.1.2 that

$$\forall\, M \subseteq A'_*(\mathbf{Я}), \qquad \operatorname{supp}_W M = ZI_W(M), \tag{2.1.5}$$

and so $\operatorname{supp}_W M \subseteq \operatorname{supp} M$ (cf. [Benedetto, 11, Section 3]).

It is in the verification of results centering around (2.1.5) that Wiener was led to introduce his Tauberian techniques (although presumably he did not think of (2.1.5)); as such we shall give the background for the use of Tauberian results in Theorem 2.1.3 as well as outlining the proof of (2.1.5). Our outline is taken from [Meyer, 1]; modifications and extensions are found in [Benedetto, 11].

2.1.6 A characterization of the Wiener spectrum in terms of summability kernels. If $T \in A'(\mathbf{Я})$ we set

$$t_R(\gamma) = \int_{\mathbf{R}} e^{-i\gamma x}\, \hat{T}(x) \Delta_R(x)\, dx.$$

It is easy to check that $\lim_{R \to \infty} t_R = T$ in the weak $*$ topology $\sigma(A'(\mathbf{Я}), A(\mathbf{Я}))$. Next, by some routine calculations, we see that if $T \in A'(\mathbf{Я})$ and $\varphi \in A_c(\mathbf{Я})$ then

$$\lim_{R \to \infty} \frac{1}{2R} \int_{|x| > R} |\widehat{t_R \varphi}(x)|\, dx = 0 \tag{2.1.6}$$

and

$$\text{(2.1.7)} \qquad \lim_{R\to\infty} \frac{1}{2R} \int_{-R}^{R} |\Delta_R(x)\, \widehat{T\varphi}(x) - \widehat{t_R\varphi}(x)|\,dx = 0.$$

Consequently, $\varphi \in A(\mathfrak{R})$ is an element of $I_W(M_T)$ if and only if

$$\text{(2.1.8)} \qquad \lim_{R\to\infty} \left\| \frac{1}{2R} \widehat{t_R\varphi} \right\|_1 = 0.$$

Armed with (2.1.8) we have the following characterization (Exercise 2.1.4) of the Wiener support.

Theorem 2.1.1. *Given $M \subseteq A'(\mathfrak{R})$. $ZI_W(M)$ is the smallest closed set such that for any compact neighborhood F which satisfies*

$$\text{(2.1.9)} \qquad F \cap ZI_W(M) = \emptyset,$$

we can conclude that

$$\text{(2.1.10)} \qquad \forall\, T \in M, \qquad \lim_{R\to\infty} \left\| \frac{1}{\sqrt{2R}}\, t_R \right\|_{L^2(F)} = 0,$$

or equivalently, that

$$\text{(2.1.11)} \qquad \forall\, T \in M, \qquad \lim_{R\to\infty} \left\| \frac{1}{\sqrt{2R}}\, t_R \right\|_{L^1(F)} = 0.$$

Remark 1. In order to prove Theorem 2.1.1, proceed as follows:

a) Observe (from Proposition 1.2.5) that

$$\text{(2.1.12)} \qquad j(ZI_W(M)) \subseteq I_W(M);$$

b) Use Plancherel's theorem to prove that

$$\text{(2.1.13)} \qquad \left\| \frac{1}{\sqrt{2R}}\, t_R \right\|^2_{L^2(F)} \leqslant \frac{1}{2R} \|t_R\varphi\|_{A'} \int_{\mathbb{R}} |\widehat{t_R\varphi}(x)|\,dx,$$

where $\varphi \in A_c(\mathfrak{R})$ is 1 on F;

c) Deduce (2.1.10) from (2.1.8), (2.1.12), and (2.1.13).

d) Prove that $\varphi \in I_W(M_T)$ if and only if

$$\text{(2.1.14)} \qquad \lim_{R\to\infty} \left\| \frac{1}{\sqrt{2R}}\, t_R\varphi \right\|_2 = 0,$$

by using (2.1.8) again.

e) Suppose that $ZI_W(M)$ is not the smallest such set and obtain a contradiction by employing (2.1.14).

2. Results such as Theorem 2.1.1 were developed by [Beurling, 5] for $ZI(M_T)$ (instead of $ZI_W(M_T)$) and used by him in [Beurling, 6, p. 229] to prove that all pseudo-measures are synthesizable for a certain weaker than weak $*$ topology. [Pollard, 2] then used similar results to characterize a class of synthesizable elements in $A(\mathbf{Я})$, e.g. Section 3.2.5. The technique has also been refined in [Herz, 2; Kinukawa, 3; Pollard, 1]. We refer to Exercise 2.1.4 for further remarks on this type of characterization of the support.

3. Using a result similar to Theorem 2.1.1 as well as the basic properties of almost periodic functions (e.g. Theorem 2.2.3) we can show [Benedetto, 11] that $ZI(M_T) = ZI_W(M_T)$ if $\hat{T}$ is almost periodic. In light of Theorem 2.2.3, this strengthens Example 2.1.1c.

2.1.7 A characterization of the Wiener spectrum in terms of ideals

Theorem 2.1.2. *Given* $M \subseteq A'_*(\mathbf{Я})$. (2.1.5) *is valid.*

Proof. i) Given $T \in M$, where $\hat{T} = \Phi$, we know that $\lim_{R\to\infty} \overset{W}{\Phi}_R = \hat{\mu}_\Phi$ in the weak $*$ topology $\sigma(L^\infty(\mathbf{R}), L^1(\mathbf{R}))$, and we must prove that $\operatorname{supp} \mu_\Phi = ZI_W(M_T)$.

ii) Let

$$d_R(x) = \overset{W}{\Phi}_R(x) - \frac{1}{2R}[(\hat{T}\Delta_R) * (\hat{T}\Delta_R)^\sim(x)].$$

It is easy to check that $\lim_{R\to\infty} d_R = 0$, uniformly on compact sets, and so

$$\lim_{R\to\infty} \frac{1}{2R}[(\hat{T}\Delta_R) * (\hat{T}\Delta_R)^\sim] = \hat{\mu}_\Phi$$

in the weak $*$ topology, $\sigma(L^\infty(\mathbf{R}), L^1(\mathbf{R}))$.

Thus,

$$\lim_{R\to\infty} \frac{1}{2R}|t_R|^2 = \mu_\Phi \qquad \text{in } \sigma(A'(\mathbf{Я}), A(\mathbf{Я})). \tag{2.1.15}$$

iii) If $F \subseteq (\operatorname{supp} \mu_\Phi)^\sim$ is a compact neighborhood then we take $\varphi = 1$ on F and $\varphi \in k(\operatorname{supp} \mu_\Phi)$; from (2.1.15) we compute that

$$\lim_{R\to\infty} \left\| \frac{1}{\sqrt{2R}} t_R \right\|_{L^2(F)} = 0,$$

and so $F \subseteq (ZI_W(M_T))^\sim$ by Theorem 2.1.1. Hence, $ZI_W(M_T) \subseteq \operatorname{supp} \mu_\Phi$.

iv) Conversely, if $F \subseteq (ZI_W(M_T))^\sim$ then (2.1.10) holds and we wish to show that μ_Φ restricted to F is 0.

This follows by (2.1.10), (2.1.15), and by restricting the measures $\left\{\frac{1}{2R}|t_R|^2 : R > 0\right\}$ to F.

q.e.d.

Remark. An interesting fact about $I_W(M)$, where $M \subseteq A'(\mathbf{R})$, is that if $E = ZI(M) = ZI_W(M)$ then

$$k(E) = I_W(M);$$

and so the elements of M are synthesizable if and only if $I(M) = I_W(M)$ [Benedetto, 11].

2.1.8 The primitive of a pseudo-measure on $\mathbf{R}$. Given a distribution $S \in D(\mathbf{R})$. As in Exercise 1.3.6, we define the *distributional derivative*,

$$S' : C_c^\infty(\mathbf{R}) \to \mathbf{C},$$

of S by

$$\forall\, \varphi \in C_c^\infty(\mathbf{R}), \qquad \langle S', \varphi\rangle = -\langle S, \varphi'\rangle.$$

Auspiciously enough, $S' \in D(\mathbf{R})$. Note that if $F \in L^1_{\text{loc}}(\mathbf{R})$ (i.e., $F \in L^1(K)$ for each compact set $K \subseteq \mathbf{R}$) then $F \in D(\mathbf{R})$ and

$$\forall\, \varphi \in C_c^\infty(\mathbf{R}), \qquad \langle F, \varphi\rangle = \int F(\gamma)\,\varphi(\gamma)\,\mathrm{d}\gamma.$$

Proposition 2.1.1. *Given $T \in A'(\mathbf{R})$, where $\hat{T} = \Phi$.*

a) *Define*

$$F_1(\gamma) = -\int_{\mathbf{R}} \chi_{\mathbf{R}\setminus[-1,1]}(x)\,\frac{\Phi(x)}{\mathrm{i}x}\,\mathrm{e}^{-\mathrm{i}\gamma x}\,\mathrm{d}x, \tag{2.1.16}$$

$$F_2(\gamma) = -\int_{-1}^{1} \frac{\Phi(x)}{\mathrm{i}x}\,[\mathrm{e}^{-\mathrm{i}\gamma x} - 1]\,\mathrm{d}x, \tag{2.1.17}$$

and

$$F(\gamma) = F_1(\gamma) + F_2(\gamma). \tag{2.1.18}$$

Then $F \in L^1_{\text{loc}}(\mathbf{R})$ and T is the distributional derivative of F.

b) *Define*

$$F_\lambda(\gamma) = \frac{1}{2\sqrt{\pi}}\,(F(\gamma+\lambda) - F(\gamma-\lambda)). \tag{2.1.19}$$

Then $F_\lambda \in L^2(\mathbf{R})$ and

$$\hat{F}_\gamma(x) = \frac{1}{\sqrt{\pi}}\,\frac{\sin \lambda x}{x}\,\Phi(x). \tag{2.1.20}$$

Proof. a) $F_1 \in L^2(\mathbb{R})$ by the Plancherel theorem and $F_2 \in L^\infty(\mathbb{R})$ by the differentiability of the exponential function at the origin; consequently, $F \in L^2_{loc}(\mathbb{R}) \subseteq L^1_{loc}(\mathbb{R})$.
If $\hat{f} = \varphi \in C_c^\infty(\mathbb{R})$ we compute, using Theorem 1.1.2, that

$$\langle F_1', \varphi\rangle = -\int_{\mathbb{R}} F_1(\gamma)\,\varphi'(\gamma)\,d\gamma = \int_{|x|>1} \Phi(x) f(x)\,dx.$$

Because of Fubini's theorem and the fact that $\int_{\mathbb{R}} \varphi'(\gamma)\,d\gamma = 0$ (cf. [Benedetto, 12, Theorem 4.16]) we compute

$$\langle F_2', \varphi\rangle = \int_{-1}^{1} \Phi(x)\left(\int_{\mathbb{R}} \varphi'(\gamma)\frac{e^{-i\gamma x}}{ix}\,d\gamma\right)dx = \int_{-1}^{1} \Phi(x) f(x)\,dx.$$

b) (2.1.20) is immediate by formally comparing the terms of the difference,

$$\tfrac{1}{2}(F(\gamma+\lambda) - F(\gamma-\lambda)) - \int \Phi(x)\frac{\sin \lambda x}{x} e^{-ix\gamma}\,dx,$$

using the identity $\sin \lambda x = (e^{i\lambda x} - e^{-i\lambda x})/2i$, noting the cancellation of the "constant" terms when dealing with F_2, and applying the definition of the L^2-Fourier transform.

q.e.d.

We shall have more to say about the primitives of pseudo-measures in Paragraph 3.2.

2.1.9 Generalized harmonic analysis and Wiener's motivation for his Tauberian theorem. (2.1.22) establishes the relation between the power of a signal and its spectral distribution of energy; in order to see this, let $\nu = \delta$ in (2.1.22) (although technically we shouldn't) and recall (2.1.15) and the discussion after (2.1.1).

Wiener had to introduce Tauberian arguments to prove (2.1.25), which in turn yields (2.1.22) in a relatively easy way. He was led to these techniques by A. E. Ingham [Wiener, 6, p. 6; 8, p. 115] in 1926. Instead of following the results of Hardy and Littlewood to which Ingham directed him, Wiener devised a general Tauberian theorem which combined some of his own trigonometric methods with a new general Tauberian method due to R. Schmidt. Schmidt further suggested the problem to prove Hardy and Littlewood's Tauberian theorem for Lambert series (1921) independently of the prime number theorem; and Wiener's solution was considered one of the main successes of his Tauberian theory. We shall discuss this in detail in Paragraph 2.3.

In any case, the *raison d'être* for Wiener's Tauberian theory is (2.1.22) [Wiener, 5, pp. 141–153, esp. p. 152]. In Theorem 2.1.3 we'll provide the details only for the implication which requires the Tauberian theorem; this is the more difficult direction.

If $T \in A'(\mathfrak{R})$, $\hat{T} = \Phi$, and F_λ is defined as in (2.1.16)–(2.1.19) then

$$\sup_{0<\lambda\leqslant 1} \left\| \frac{1}{\lambda} |F_\lambda|^2 \right\|_1$$

is bounded since

$$\text{(2.1.21)} \qquad \frac{1}{\lambda} \int_{\mathfrak{R}} |F_\lambda(\gamma)|^2 \, \mathrm{d}\gamma = \frac{\lambda}{\pi} \int_{\mathbf{R}} \left| \frac{\sin x}{x} \Phi\left(\frac{x}{\lambda}\right) \right|^2 \mathrm{d}x$$

because of Plancherel's theorem, (2.1.19), and a change of variable. Consequently, by the Alaoglu theorem, $\left\{\frac{1}{\lambda} |F_\lambda|^2 : 0 < \lambda \leqslant 1\right\}$ has weak $*$ limit points in $M(\mathfrak{R})$.

Theorem 2.1.3. *Given $T \in A'(\mathfrak{R})$, where $\hat{T} = \Phi$. $T \in A'_*(\mathfrak{R})$ if and only if*

$$\lim_{\lambda\to 0} \frac{1}{\lambda} |F_\lambda|^2 = \mu_\Phi \qquad \text{in } \sigma(A'(\mathfrak{R}), A(\mathfrak{R})).$$

In this case we have

$$\text{(2.1.22)} \qquad \forall\, \nu \in M_0(\mathbf{R}), \qquad \int_{\mathfrak{R}} |\hat{\nu}|^2 \, \mathrm{d}\mu_\Phi = \lim_{\lambda\to 0} \frac{1}{\lambda} \int_{\mathfrak{R}} |F_\lambda(\gamma)\, \hat{\nu}(\gamma)|^2 \, \mathrm{d}\gamma$$

$$= \lim_{R\to\infty} \frac{1}{2R} \int_{\mathfrak{R}} |t_R(\gamma)\, \hat{\nu}(\gamma)|^2 \, \mathrm{d}\gamma$$

$$= \lim_{R\to\infty} \frac{1}{2R} \int_{-R}^{R} |\Phi * \nu(x)|^2 \, \mathrm{d}x.$$

Proof. i) *We shall assume that* $\lim_{\lambda\to 0} \frac{1}{\lambda} |F_\lambda|^2 = \mu_\Phi$ *in* $\sigma(A'(\mathfrak{R}), A(\mathfrak{R}))$ *and verify* (2.1.22). By Plancherel's theorem,

$$\text{(2.1.23)} \qquad \frac{1}{\lambda} \int_{\mathfrak{R}} |F_\lambda(\gamma)\, \hat{\nu}(\gamma)|^2 \, \mathrm{d}\gamma = \frac{1}{\lambda\pi} \int_{\mathbf{R}} \left| \int_{\mathbf{R}} \frac{\sin\lambda(x-y)}{x-y} \Phi(x-y) \, \mathrm{d}\nu(y) \right|^2 \mathrm{d}x$$

and

$$\text{(2.1.24)} \qquad \frac{1}{2R} \int_{\mathfrak{R}} |t_R(\gamma)\, \hat{\nu}(\gamma)|^2 \, \mathrm{d}\gamma = \frac{1}{2R} \int_{\mathbf{R}} |(\Delta_R \Phi) * \nu(x)|^2 \, \mathrm{d}x.$$

Also, it is straightforward to check (e.g. [Meyer, 1, p. 194]) that the difference between the right-hand side of (2.1.23) (resp. (2.1.24)) and

$$\frac{1}{\lambda\pi}\int_{\mathbf{R}} \frac{\sin^2 \lambda x}{x^2}\,|\Phi * v(x)|^2\,\mathrm{d}x$$

$$\left(\text{resp. } \frac{1}{R}\int_{-R}^{R} |\Phi * v(x)|^2\,\mathrm{d}x\right)$$

tends to zero.

Thus, setting $\Psi(x) = |\Phi * v(x)|^2 + |\Phi * v(-x)|^2$, we shall have verified (2.1.22) once we show that

$$\text{(2.1.25)} \qquad \lim_{\lambda\to 0}\frac{1}{\pi\lambda}\int_0^{\infty} \Psi(x)\,\frac{\sin^2\lambda x}{x^2}\,\mathrm{d}x = \lim_{R\to\infty}\frac{1}{2R}\int_0^{R} \Psi(x)\,\mathrm{d}x.$$

i) By the above remarks, our hypothesis is that

$$\text{(2.1.26)} \qquad \lim_{\lambda\to\infty}\frac{1}{\pi\lambda}\int_0^{R} \Psi(x)\,\frac{\sin^2\lambda x}{x^2}\,\mathrm{d}x = r,$$

where $\Psi \in L^{\infty}(\mathbf{R})$.

ii) Let $R = 1/\lambda = \mathrm{e}^y$ and let $x = \mathrm{e}^t$ so that (2.1.25) becomes

$$\text{(2.1.27)} \qquad \lim_{y\to\infty}\frac{2}{\pi}\int_{-\infty}^{\infty} \mathrm{e}^{y-t}\sin^2(\mathrm{e}^{-(y-t)})\,\Psi(\mathrm{e}^t)\,\mathrm{d}t = \lim_{y\to\infty}\int_{-\infty}^{y} \mathrm{e}^{-(y-t)}\,\Psi(\mathrm{e}^t)\,\mathrm{d}t.$$

Note that in order to find the upper bound of integration in the second integral it is necessary to consider first the substitution $x = \mathrm{e}^t$.

v) Setting

$$f(x) = \frac{2}{\pi}\,\mathrm{e}^x \sin^2(\mathrm{e}^{-x}),$$

we now assert that

$$\text{(2.1.28)} \qquad \forall\, \gamma \in \widehat{\mathbb{R}}, \qquad \hat{f}(\gamma) \neq 0.$$

The verification of (2.1.28) is left as Exercise 2.1.5.

Also, by an easy residue argument,

$$\text{(2.1.29)} \qquad \hat{f}(0) = \frac{2}{\pi}\int_0^{\infty}\frac{\sin^2 x}{x^2}\,\mathrm{d}x = 1.$$

Note that $\int_0^\infty \frac{\sin^2 x}{x^2} dx = \int_0^\infty \frac{\sin x}{x} dx$.

v) Consequently, if we write $\Theta(t) = \Psi(e^t) - r$ then (2.1.26) becomes

$$\lim_{y \to \infty} \Theta * f(y) = 0. \tag{2.1.30}$$

Let I consist of all elements $g \in L^1(\mathbf{R})$ such that $\lim_{y \to \infty} \Theta * g(y) = 0$.

Then I is a closed translation invariant subspace of $L^1(\mathbf{R})$. Thus, from Proposition 1.2.6 and Theorem 1.2.3 (Wiener's Tauberian theorem), $I = L^1(\mathbf{R})$.
Hence, if we let $g(x) = e^{-x} \chi_{[0,\infty)}(x)$, the right-hand side of (2.1.27) becomes

$$\lim_{y \to \infty} \Theta * g(y) + r \int_0^\infty e^{-x} dx = r,$$

and we are done.

q.e.d.

2.1.10 Riemann's summability methods, Tauberian theorems, and generalized harmonic analysis

Remark. The basis of Riemann's theory of trigonometric series is the following result, e.g. [Zgymund, 2, I, pp. 319–320]: assume that $\sum_1^\infty a_n = s$; then

$$\lim_{\lambda \to 0} \sum_1^\infty a_n \left(\frac{\sin n\lambda}{n\lambda} \right)^2 = s \tag{2.1.31}$$

and

$$\lim_{\lambda \to 0} \frac{2}{\pi} \sum_1^\infty s_n \frac{\sin^2 n\lambda}{n^2 \lambda} = s, \tag{2.1.32}$$

where $s_n = \sum_1^n a_j$. The summability technique embodied in (2.1.31) (resp. (2.1.32)) is denoted by $(R, 2)$ (resp. R_2); it is $(R, 2)$ to which we referred in our reference to Riemann after (1.3.15). In 1935, [Marcinkiewicz, 1] showed that a trigonometric series may be summable $(R, 2)$ without its series of partial sums being summable R_2 and that the series of partial sums of a trigonometric series may be summable R_2 without the original series being summable $(R, 2)$. In 1947, Hardy and Rogosinski proved the same result for Fourier series.

In the verification of (2.1.25) (in the direction we proved) Wiener really showed that if Ψ is bounded and $\frac{2}{\pi} \int_0^\infty \Psi(x) dx$ is R_2 summable to $2r$ then the arithmetic means

$\left\{\frac{1}{R}\int_0^R \Psi(x)\,dx : R > 0\right\}$, converge to $2r$. This is the classical format of any Tauberian theorem: a boundedness (or some such) condition and summability for a certain method yield summability by other means (e.g. Paragraph 2.3). Note that the R_2 kernel in (2.1.25) is the Fejér kernel f_λ (e.g. Exercise 1.2.6a) but that in Theorem 2.1.3 we let "$\lambda \to 0$" (instead of "$\lambda \to \infty$") and that we have an integral of a product (instead of a convolution).

2.1.11 Wiener's original proof of his Tauberian theorem. Our proof of (2.1.25) followed directly (after some indirect substitutions) from Theorem 1.2.3. The first proofs of (2.1.25), including the non-Tauberian implication, are found in [Bochner and Hardy, 1; Wiener, 2; 3]. Wiener's original proof [Wiener, 4, pp. 163–172] (1928) of (2.1.25) *in the general form of Theorem* 1.1.3 did not have the algebraic flavor one finds in the statement of Proposition 1.1.5 (cf. Section 1.1.8 and Paragraph 2.4). On the other hand this first proof clearly shows the necessity of the condition that "φ never vanishes", whereas the elegant intrinsic algebraic property of Proposition 1.1.5 was replaced by added integrability conditions on f (e.g. (2.1.33)), for $\hat{f} = \varphi$. The following *outline* of Wiener's original proof is due to [Levinson, 2; 3]. We say that $f \in L^1(\mathbf{R})$ is *normalized* if $\int f(x)\,dx = 1$.

Theorem 2.1.4. (*i.e., Theorem* 1.1.3) *Let* $f \in L^1(\mathbf{R})$ *have a non-vanishing Fourier transform* $\hat{f} = \varphi$ *and let* $\Phi \in L^\infty(\mathbf{R})$. *Assume that*

(2.1.33) $$\int_{\mathbf{R}} |xf(x)|\,dx < \infty.$$

If

(2.1.34) $$\lim_{x\to\infty} f * \Phi(x) = r\int_{\mathbf{R}} f(y)\,dy,$$

then for each $g \in L^1(\mathbf{R})$,

(2.1.35) $$\lim_{x\to\infty} g * \Phi(x) = r\int_{\mathbf{R}} g(y)\,dy.$$

Proof. i) Let f be normalized and choose a normalized element $h \in L^1(\mathbf{R})$ which is "close" to δ; to be specific, suppose h is f_λ for some "large" λ (f_λ is the Fejér kernel and $\hat{f}_\lambda = \Delta_\lambda$).
We'll show that

(2.1.36) $$\exists\, w \in L^1(\mathbf{R}), \text{ normalized, such that } w * f = h.$$

Let $\psi = \Delta_\lambda/\varphi$, which, by our hypothesis on φ, is an element of $C_c(\text{Я})$.
Because of (2.1.33), φ' exists and is a continuous function.

Thus $\psi' \in C_c(\hat{\mathbf{R}})$.
Consequently, $\|\psi\|_2 < \infty$ and $\|\psi'\|_2 < \infty$; and so, by Plancherel's theorem,

(2.1.37) $$\int_{\mathbf{R}} |w(x)|^2 \, dx < \infty \quad \text{and} \quad \int_{\mathbf{R}} |xw(x)|^2 \, dx < \infty,$$

where

$$w(x) = \int_{\hat{\mathbf{R}}} \psi(\gamma) \, e^{-ix\gamma} \, d\gamma.$$

Therefore, by Hölder's inequality and (2.1.37),

$$\left(\int_{\mathbf{R}} \left| \frac{w(x)\sqrt{1+x^2}}{\sqrt{1+x^2}} \right| dx \right)^2 \leqslant \int_{\mathbf{R}} |w(x)|^2 (1+x^2) \, dx \int_{\mathbf{R}} \frac{dx}{1+x^2} < \infty;$$

and we have (2.1.36).

ii) With (2.1.36) it is easy to verify (2.1.35) heuristically.

From the definition of h,

(2.1.38) $$g * (h * \Phi) \approx g * \Phi,$$

and because of (2.1.34) and (2.1.36),

(2.1.39) $$\lim_{x \to \infty} h * \Phi(x) = r.$$

By the normalization of g,

$$\lim_{x \to \infty} g * (h * \Phi)(x) = r,$$

and thus, by (2.1.38), we can conclude with (2.1.35).

"q.e.d."

Remark 1. We'll discuss the classical distinction between Abelian and Tauberian theorems in Paragraph 2.3. Let us note now that the implication "(2.1.34) $\Rightarrow$ (2.1.39)" is Abelian since the asymptotic behavior (2.1.34), implies the same asymptotic behavior for the mean $w * (f * \Phi)$ of $f * \Phi$, e.g. Exercise 2.1.3c.

2. Motivation for Theorem 1.1.3 in terms of Karamata's extension process is given in [Bochner and Chandrasekharan, 1, pp. 171–174]. Karamata's work was published in 1930–1931. An enlightening observation about Theorem 1.1.3 is that $\varphi(x)$ somehow approximates $f * \varphi(x)$ (in (2.1.34)) since f is normalized, integrable, and the weight of f is at x in the convolution. Thus when the limit in (2.1.34) exists it is reasonable that (2.1.35) would also be true for all normalized elements of $L^1(\mathbf{R})$.

2.1.12 An algebraic proof of Wiener's Tauberian theorem. Using results from Paragraph 1.2 we'll now prove Theorem 2.1.4 without the hypothesis (2.1.33); we give a slightly jazzed-up statement for use in Paragraph 2.3.

Theorem 2.1.5. *Let $f \in L^1(G)$ have a non-vanishing Fourier transform $\hat{f} = \varphi$ (and so Γ is σ-compact) and let $\Phi \in L^\infty(G)$. Assume that*

$$\lim_{x\to\infty} f * \Phi(x) = r \int_G f(y)\,\mathrm{d}y.$$

Then for each sequence $\{g_n : n = 1, \ldots\} \subseteq L^1(G)$ which satisfies the properties that

$$\lim_{n\to\infty} g_n * \Phi(x) = L(x) \text{ a.e.}$$

and

$$\lim_{n\to\infty} \int_G g_n(x)\,\mathrm{d}x = L,$$

we have that

$$\lim_{x\to\infty} L(x) = rL.$$

Proof. We prove the case that $g_n = g$ for each n. It is then trivial to prove the result as we've stated it.

Without loss of generality (cf. (2.1.30)) assume that $r = 0$.

Let $I = \{g \in L^1(G) : \Phi * g \in C_0(G)\}$.

Clearly, I is a closed ideal by Proposition 1.2.6 (cf. the last part of the proof of Theorem 2.1.3) and $f \in I$.

Since φ never vanishes we obtain $I = L^1(G)$ by Theorem 1.2.3 and this completes the proof.

q.e.d.

Remark 1. If $g_n = g$ for each n we have Wiener's original result [Wiener, 4; 6, Theorem VIII, p. 25; 7, Theorem 4, pp. 73–74]. Wiener derived this asymptotic form of the Tauberian theorem from Theorem 1.1.3.

2. Recall that "$\lim_{x\to\infty} \Psi(x) = s$" (in G) means that $\Psi - s$ vanishes at infinity.

2.1.13 Concluding remark. From the Hahn–Banach theorem and Proposition 1.2.6 we see that the condition

$$I_\Phi = A(\Gamma)$$

is equivalent to the statement,

(2.1.40) whenever $T\varphi = 0$ we can conclude that $T = 0$.

Clearly, a spectral synthesis result would have the form

(2.1.41) if $\varphi = 0$ on supp$T$ then $T\varphi = 0$.

We shall see in Section 3.2.5 that if $\varphi' \in A(\mathbf{Я})$ then (2.1.41) is valid. We mention this now because the condition (2.1.33), in the statement of Theorem 2.1.4, tells us that $\varphi' \in A(\mathbf{Я})$.

Exercises 2.1

2.1.1 Sets of strict multiplicity

In Exercise 1.3.1 we saw that if $\mu \in M_c(\Gamma)$ then the arithmetic means of $\hat{\mu}(x)$ converge to 0 at infinity. In light of this fact and the Riemann–Lebesgue theorem, a natural problem is to determine if $M_0(\Gamma) \setminus L^1(\Gamma)$, or even $A_0'(\Gamma) \setminus L^1(\Gamma)$, can be empty (cf. Exercise 1.1.4c). A strong response in the negative is given below in part b).

A closed set $E \subseteq \Gamma$ is a set of *multiplicity* (resp. *strict multiplicity*) if

$$A_0'(E) = A_0'(\Gamma) \cap A'(E) \neq \{0\}$$

(resp. $M_0(E) \neq \{0\}$); E is a set of *uniqueness* or *U-set* (resp. *uniqueness in the wide sense*) if it is not a set of multiplicity (resp. strict multiplicity). The study of such sets began with Riemann; and historical–bibliographical notes and detailed proofs of some of the major results are found in [Bary, 2, II, Chapter 14; Benedetto, 6, Chapter 3; 12, Appendix 3.1].

a) Prove that if μ is a *discontinuous measure* (i.e., $\mu = \sum a_\gamma \delta_\gamma$ and $\sum |a_\gamma| < \infty$) then $\mu \notin M_0(\Gamma) \setminus \{0\}$. Also, show that if $\mu \in M(\mathbf{T})$ is discontinuous and $\lim_{n \to \infty} \hat{\mu}(n) = \alpha$, then $\mu = \alpha\delta$, e.g. [Benedetto, 12, Exercise 5.16b].

b) Prove Men'shov's result (1916): *there is a closed set $E \subseteq \mathbf{T}$ and $\mu \in M_0(E) \setminus \{0\}$ such that $mE = 0$ and $\hat{\mu}(n) = 0((\log|n|)^{-1/2})$*, $|n| \to \infty$. (*Hint*: Let $E \subseteq \mathbf{T}$ be the perfect symmetric set determined by $\xi_k = \dfrac{1}{2}\dfrac{k}{k+1}$; from Exercise 1.1.4b we see that $mE = 0$. After k steps in the construction, $1 + 2^1 + \ldots + 2^{k-1} = 2^k - 1$ contiguous intervals have been removed and we list them from left to right in $[0, 2\pi]$ as $I_1, \ldots, I_{2^k-1}$. Define the continuous function $F_k : [0, 2\pi] \to \mathbf{R}$ by the conditions that $F_k(0) = F_k(2\pi) = 0$,

$$F_k(\gamma) = \begin{cases} \dfrac{2\pi j}{2^{k-1}} & \text{for } \gamma \in I_j \text{ and } j \leqslant 2^{k-1} \\[2ex] \dfrac{2\pi(2^k - j)}{2^{k-1}} & \text{for } \gamma \in I_j \text{ and } j \geqslant 2^{k-1}, \end{cases}$$

and F_k is monotonic and linear between each I_j and I_{j+1}. The sequence $\{F_k : k = 1, \ldots\}$ converges uniformly to a continuous function F and we let $\mu = F'$ distributionally (cf. Exercise 1.1.4b)).

In 1927, Bary posed the problem to see if there was any lower limit to the rate at which the sequence

$$\{\hat{\mu}(n): n \in \mathbf{Z}, \mu \in M(\mathbf{T}), \quad \text{and} \quad m(\operatorname{supp} \mu) = 0\}$$

can tend to 0. Obviously we can not have $\hat{\mu}(n) = 0(1/|n|^{(1/2)+\varepsilon})$, $|n| \to \infty$, when $\varepsilon > 0$, for such μ because of the Plancherel theorem. In 1936, Littlewood showed the existence of $\mu \in M(\mathbf{T})$, for which $m(\operatorname{supp} \mu) = 0$, such that $\hat{\mu}(n) = 0(1/|n|^{\varepsilon})$, $|n| \to \infty$, for some $\varepsilon > 0$; and, in 1942, [Salem, 1] proved that for each $\varepsilon > 0$ there is $\mu \in M(\mathbf{T})$, for which $m(\operatorname{supp} \mu) = 0$, such that

(E2.1.1) $\quad \hat{\mu}(n) = 0(1/|n|^{(1/2)-\varepsilon}), \quad |n| \to \infty.$

Later, [Salem, 2] extended this result in the following way: given $d \in (0, 1)$ and $\varepsilon > 0$, there is $\mu \in M(\mathbf{T})$, for which $m(\operatorname{supp} \mu) = 0$, such that $\hat{\mu} \in L^p(\mathbf{Z})$ if $p \geqslant \dfrac{2}{d} + \varepsilon$ and d is the Hausdorff dimension of $\operatorname{supp} \mu$. This theorem shows that Beurling's upper bound, mentioned in Remark 1 after Exercise 2.1.6, can not generally be replaced by a smaller number, cf. [Donoghue, 1, Section 52]. (For properties of Hausdorff dimension see [Kahane and Salem, 4, Chapitre 2]).

It is interesting to note that in 1938 [Wiener and Wintner, 1], cf. [Zygmund, 2, II, pp. 146–147], found measures $\mu = F'$ (distributional derivative) which satisfied (E2.1.1) and the condition $F' = 0$ a.e. (ordinary derivative) but for which F was strictly increasing (thus $m(\operatorname{supp} \mu) = m\mathbf{T}$); in a sequel, which contains an error in the first part related to Salem's 1942 result, [Wiener and Wintner, 2] show that every positive bounded measure on $\mathfrak{R}$ is the spectral distribution of energy for some signal (not necessarily essentially bounded) on $\mathbf{R}$.

The most complete results on Bary's problem are due to Ivašev-Mucatov (1957) [Bary, 2, II, pp. 404 ff.; Kahane and Salem, 4, pp. 110–111] who proved, in particular, that $\varepsilon = 0$ in (E2.1.1) still produces a valid result.

2.1.2 Continuous pseudo-measures

As in (2.1.3) we say that $T \in A'_c(\mathbf{T})$ if

$$\lim_{N \to \infty} \frac{1}{2N+1} \sum_{-N}^{N} |\hat{T}(n)| = 0;$$

and we also define

$$\forall\, T \in A'(\mathbf{T}), \qquad \|T\|^{\infty} = \sup_{N} \frac{1}{2N+1} \sum_{-N}^{N} |\hat{T}(n)|.$$

Assume for this exercise that $\hat{T}(0) = 0$. Thus, if

$$F(\gamma) \sim \sum{}' \frac{\hat{T}(n)}{in} e^{in\gamma},$$

we have $F \in L^2(\mathbf{T})$ (even more is true by the Hausdorff–Young theorem as we'll discuss in Section 3.2.10) and $F' = T$ distributionally.

a) Prove that $T \in A_c'(\mathbf{T})$ if and only if

$$\int_{\mathbf{T}} |F(\gamma+\lambda) - F(\gamma)|^2 \, d\gamma = o(\lambda), \qquad \lambda \to 0$$

(cf. Theorem 2.1.3, Exercise 1.3.1, and Wiener's original proof [Wiener, 1; 7, p. 140 and pp. 146–149; Zygmund, 1, p. 221]).

b) Prove that $T \in A_c'(\mathbf{T})$ if and only if

$$\|T - \tau_\lambda T\|^\infty = o(1), \qquad \lambda \to 0$$

(cf. [Goldberg and Simon, 1]).

c) Prove that $T \in A_0'(\mathbf{T})$ if and only if

$$\|T - \tau_\lambda T\|_{A'} = o(1), \qquad \lambda \to 0.$$

d) Prove that $A_0'(\mathbf{T}) \subseteq A_c'(\mathbf{T})$.

e) Prove that $T \in A_c'(\mathbf{T})$ if and only if

$$\text{(E2.1.2)} \qquad \lim_{N\to\infty} (2N+1) \sum_{|n|\geq N} \left|\frac{\hat{T}(n)}{n}\right|^2 = 0$$

or, equivalently,

$$\text{(E2.1.3)} \qquad \lim_{N\to\infty} \frac{1}{2N+1} \sum_{N\leq |n| \leq 2N} |\hat{T}(n)|^2 = 0.$$

(*Hint*: (E2.1.2) follows from a technique with dyadic sums which goes back to S. Bernstein and O. Szasz (cf. [Kahane, 13, pp. 13–14]). (E2.1.3) follows since

$$\frac{1}{2N+1} \sum_{N\leq|n|\leq 2N} |\hat{T}(n)|^2 = \frac{4N+1}{2N+1}\frac{1}{4N+1}\sum_{-2N}^{2N} |\hat{T}(n)|^2 - \frac{2N-1}{2N+1}\frac{1}{2N-1}\sum_{-(N-1)}^{N-1} |\hat{T}(n)|^2).$$

2.1.3 An Abelian theorem

a) Given $\Phi \in L^\infty(\mathbf{R})$ and $f \in L^1(\mathbf{R})$. Prove that for each $b > 0$,

$$\lim_{x\to\infty} \int_{-\infty}^{b} f(x-t)\,\Phi(t)\,dt = 0.$$

b) Prove that if $\lim_{x\to\infty} \Phi(x) = A$ and $\Phi \in L^\infty(\mathbf{R})$ then

$$\forall f \in L^1(\mathbf{R}), \qquad \lim_{x\to\infty} f * \Phi(x) = A \int_{\mathbf{R}} f(y)\,dy.$$

Consequently, if $g = f * \Phi$ then the behavior of g at infinity is determined by the behavior of Φ at infinity (g is a mean of Φ). This result is the converse of Theorem 1.1.3 and is an Abelian theorem (cf. Paragraph 2.3). (*Hint*: Let $A = 0$ and consider the two integrals

$$I_1(x) = \int_{-\infty}^{x/2} f(x-t)\,\Phi(t)\,dt$$

and

$$I_2(x) = \int_{x/2}^{\infty} f(x-t)\,\Phi(t)\,dt).$$

2.1.4 A comparison of supp T *and* supp$_W T$

a) Provide the details for the proof of Theorem 2.1.1 using Remark 1 which follows its statement (cf. [Benedetto, 11, Section 3; Meyer, 1]).

b) Prove the following characterization of the support: given $M \subseteq A'(\mathfrak{R})$; supp M is the smallest closed set such that for any compact neighborhood F which satisfies

$$F \cap \operatorname{supp} M = \emptyset,$$

we can conclude that

$$\forall\, T \in M, \qquad \lim_{R \to \infty} \|t_R\|_{L^2(F)} = 0,$$

or, equivalently, that

$$\forall\, T \in M, \qquad \lim_{R \to \infty} \|t_R\|_{L^1(F)} = 0$$

(e.g. [Benedetto, 11, Theorem 2.1]).

c) Prove that part b) is not true if $t_R(\gamma)$ is replaced by

$$\frac{1}{R} \int_{-R}^{R} \hat{T}(x)\, e^{-i\gamma x}\, dx$$

(e.g. [Benedetto, 11, Section 1]).

2.1.5 $\int \exp x(1+i\gamma) \sin^2 e^{-x} dx$

Prove that

$$\int_{\mathbb{R}} e^{x(1+i\gamma)} \sin^2 e^{-x}\, dx = \frac{\Gamma(1-i\gamma)}{2^{1-i\gamma}\gamma(1+i\gamma)}\,(e^{\frac{\pi\gamma}{2}} - e^{-\frac{\pi\gamma}{2}}),$$

and so (2.1.28) is valid (e.g. [Wiener, 7, pp. 142–143]).

2.1.6 The Wiener closure problem for $L^p(G)$

a) Let $f \in L^2(G)$. Prove that the variety generated by f (in the L^2-norm) is $L^2(G)$ if and only if

(E2.1.4) $\quad m\{\gamma \in \Gamma : \hat{f}(\gamma) = 0\} = 0.$

From Remark 2 after Exercise 1.4.5 we see that this is equivalent to the statement that $f \in L^2(G)$ is mean periodic if and only if $m(Z\hat{f}) > 0$. (*Hint*: Analogous to the verification of (2.1.40) we see that the variety generated by f is $L^2(G)$ if and only if

$$\text{whenever } \hat{f}\hat{g} = 0, \text{ for } g \in L^2(G), \text{ we can conclude that } \hat{g} = 0 \text{ a.e.}$$

This is obviously equivalent to (E2.1.4)).

b) Given $r > 1$ and define

$$f(x) = \begin{cases} 2 & \text{for } x \in (0,1), \\ 1 & \text{for } x \in (1,r), \\ 0 & \text{otherwise.} \end{cases}$$

Prove that $V_f = L^1(\mathbf{R})$ if and only if r is irrational; and that $V_f = L^2(\mathbf{R})$ for every $r > 1$. (The closures in V_f are taken in the L^1 and L^2-norms, respectively.)

c) Let $f \in L^1(\mathbf{R}) \setminus \{0\}$. Prove that $V_f = L^1(0, \infty)$, e.g. [Newman, 2].

Remark 1. Because of Exercise 2.1.6a and Theorem 1.1.3, Wiener posed the problem to find conditions that $V_f = L^p(G)$ in terms of $Z\hat{f}$ when $1 < p < 2$ [Wiener, 6, p. 93]. [Segal, 2] proved that for each $p \in (1,2)$ there is $f \in L^p(\mathbf{R})$ with $m(Z\hat{f}) = 0$ such that

$$V_f \subsetneq L^p(\mathbf{R}),$$

where the closure in V_f is taken with the L^p-norm (cf. the Remark in Section 3.1.5). Note that if $f \in \cap L^p(\mathbf{R})$ and $V_f \subsetneq L^p(\mathbf{R})$ then

$$\forall\, r \in [1,p), \qquad V_f \subsetneq L^r(\mathbf{R}).$$

Using this observation [Beurling, 8] defined the *closure exponent* $c = \sup\{p : V_f \subsetneq L^p(\mathbf{R})\}$ of $f \in \cap L^p(\mathbf{R})$ and was able to determine an upper bound for c in terms of the Hausdorff dimension of $Z\hat{f}$. [Newman, 1] gave an extension of Beurling's work.

Another aspect of Wiener's translation problem was taken up by Pollard [Beurling, 3; Pollard, 1; Kahane and Salem, 4, pp. 111–112] (also, see the discussion in [Segal, 3]). Along with the closure property (2.1.40) and the synthesis property (2.1.41), consider the uniqueness property

$$\varphi = 0 \text{ on } \operatorname{supp} T \quad \Rightarrow \quad T = 0$$

and the Tauberian property (Theorem 1.3.1c)

$$T\varphi = 0 \quad \Rightarrow \quad \varphi = 0 \text{ on } \operatorname{supp} T.$$

Thus the uniqueness (respectively, the closure) property implies the closure (respectively, the uniqueness) property because of the Tauberian (respectively, the synthesis) property. The extension of these results to the L^p case is the subject matter of [Herz, 3; Kinukawa, 3; Pollard, 1]. Instead of the condition "$\varphi = 0$ on supp$T$", Pollard considered criteria for sets of uniqueness in the form of pointwise convergence to 0 on $(Z\varphi)^{\sim}$ of an Abelian mean of T. The equivalence of these points of view obviously plays an important role in his proofs, as do certain smoothness conditions on φ to ensure that the synthesis property (2.1.41) is valid for the implication, "closure property implies the uniqueness property". [Beurling, 4; 5; 6] form a background for this work.

2. Besides the sort of question we discussed in Remark 1, we can also ask if every variety of $L^p(\mathbf{R})$, $1 \leqslant p < \infty$, is generated by a single element. In light of the discussion in Section 1.4.8 we know that the answer to this question is negative for $p = 1$. On the other hand the answer is affirmative for $p = 2$; and [Atzmon, 3] has shown that the answer is negative for $p \in [1, 4/3)$!

3. We mention [Carleson, 1, pp. 341 ff.] and [Beurling, 13] for two closure problems which are related to but not directly concerned with the above discussion. Further, [Beurling, 7] is the source of a class of closure problems related to function theory; [Helson, 4] provides both background and bibliography.

2.2 Beurling's spectrum

2.2.1 Fundamental properties of the narrow (i.e., α) and strict (i.e., β) topologies. Let $C_b(G)$ be the space of $\mathbf{C}$-valued bounded continuous functions on G. For each $\Theta \in C_0(G)$ we define the semi-norm

(2.2.1) $$\forall\ \Phi \in C_b(G), \qquad \|\Phi\|_\Theta = \|\Phi\Theta\|_\infty$$

and the pseudo-metric

(2.2.2) $$\forall\ \Phi, \Psi \in C_b(G), \qquad \rho_\Theta(\Phi, \Psi) = \|\Phi - \Psi\|_\Theta + |\|\Phi\|_\infty - \|\Psi\|_\infty|.$$

The locally convex topology on $C_b(G)$ generated by the family $\{\|\ \ \|_\Theta : \Theta \in C_0(G)\}$ of semi-norms is called the *strict* or β (for Beurling) topology on $C_b(G)$. If $\Theta \in C_0(G)$ never vanishes then the metric topology on $C_b(G)$ generated by ρ_Θ is called the Θ-*narrow* or α_Θ topology on $C_b(G)$. In the space $C(G)$ we let κ denote the topology of uniform convergence on compact sets (of G).

Proposition 2.2.1 a) *$C_b(G)$ is complete in the β topology and in any α_Θ topology.*

b) *Given a sequence $\{\Phi_n : n = 1, \ldots\} \subseteq C_b(G)$. $\{\Phi_n : n = 1, \ldots\}$ converges to $\Phi \in C_b(G)$ in the β topology if and only if $\{\Phi_n : n = 1, \ldots\}$ converges to $\Phi \in C(G)$ in the κ topology and is uniformly bounded.*

c) *Given a sequence* $\{\Phi_n : n = 1, \ldots\} \subseteq C_b(G)$. $\{\Phi_n : n = 1, \ldots\}$ *converges to* $\Phi \in C_b(G)$ *in the* α_Θ *topology if and only if* $\{\Phi_n : n = 1, \ldots\}$ *converges to* $\Phi \in C(G)$ *in the* κ *topology and* $\lim_{n \to \infty} \|\Phi_n\|_\infty = \|\Phi\|_\infty < \infty$.

Proof. i) We first note that *the* κ *topology is weaker than the* β *topology.*

Let $K \subseteq G$ be compact and let $\Theta \in C_0(G)$ equal 1 on K. Then for each $\Phi \in C_b(G)$,

$$\sup_{x \in K} |\Phi(x)| \leqslant \|\Phi\|_\Theta.$$

ii) We now prove that *the* β *topology and the topology of uniform convergence yield the same class of bounded sets in* $C_b(G)$.

Clearly any uniformly bounded set is β bounded since $\|\Phi\|_\Theta \leqslant \|\Phi\|_\infty \|\Theta\|_\infty$, an estimate which also tells us that the β topology is weaker than the topology of uniform convergence on G.

If $B \subseteq C_b(G)$ is β bounded and not uniformly bounded we choose sequences $\{\Phi_n : n = 1, \ldots\} \subseteq B$ and $\{x_n : n = 1, \ldots\} \subseteq G$ such that $|\Phi_n(x_n)| = \lambda_n \to \infty$ as $n \to \infty$. From the definition of β boundedness, $\lim x_n = \infty$ in the sense that

$$\forall\, K \subseteq G, \text{ compact}, \exists\, N \text{ such that } \forall\, n \geqslant N,\ x_n \in K^\sim.$$

Next choose $\Theta \in C_0(G)$ for which $\Theta(x_n) = \lambda_n^{-1/2}$. Thus

$$\|\Phi_n\|_\Theta \geqslant \lambda_n^{1/2},$$

and this contradicts the hypothesis that the sequence $\{\Phi_n : n = 1, \ldots\}$ is β bounded.

iii) We now show that *the* κ *and* β *topologies are identical on uniformly bounded sets* $B \subseteq C_b(G)$.

There is $R > 0$ such that for all $\Phi \in B$, $\|\Phi\|_\infty \leqslant R$.

Taking Ψ in the κ closure of B, a non-vanishing $\Theta \in C_0(G)$, and $\varepsilon > 0$, it is sufficient to prove that

$$\exists\, \Phi \in B \quad \text{such that } \|\Phi - \Psi\|_\Theta < \varepsilon \tag{2.2.3}$$

(note that $\Psi \in C_b(G)$ and $\|\Psi\|_\infty \leqslant R$ since B is uniformly bounded by R).

Choose $K \subseteq G$, compact, such that

$$\sup_{x \in G \setminus K} |\Theta(x)| < \frac{\varepsilon}{2(R + \|\Psi\|_\infty)}; \tag{2.2.4}$$

and, by hypothesis, take $\Phi \in B$ for which

$$\sup_{x \in K} |\Phi(x) - \Psi(x)| < \frac{\varepsilon}{2\|\Theta\|_\infty}. \tag{2.2.5}$$

From (2.2.4) and (2.2.5) we obtain

$$\|\Theta\|_\infty \|\Phi - \Psi\|_\infty \leqslant \|\Theta\|_\infty \sup_{x \in K} |\Phi(x) - \Psi(x)| + \sup_{x \in G \setminus K} |\Theta(x)(\Phi(x) - \Psi(x))| < \varepsilon,$$

and (2.2.3) follows.

iv) We prove that $C_b(G)$ *is a complete locally convex space in the* β *topology.*

If $\{\Phi_\alpha\} \subseteq C_b(G)$ is a β Cauchy directed system, then for each $\Theta \in C_0(G)$, $\{\Theta\Phi_\alpha\} \subseteq C_0(G)$ is Cauchy in the uniform norm, and so converges to some $\Psi_\Theta \in C_0(G)$. Also, by choosing $\Theta \in C_0(G)$ equal to 1 on compact sets we see that Φ_α converges to $\Phi \in C(G)$ in the κ topology.

Thus, for each $\Theta \in C_0(G)$, $\Theta\Phi = \Psi_\Theta \in C_0(G)$.

If $\Phi \notin C_b(G)$ we choose $\{x_n : n = 1, \ldots\} \subseteq G$ such that for each n, $|\Phi(x_n)| \geqslant n$. Without loss of generality we assume that if $n \neq m$ then $\Phi(x_n) \neq \Phi(x_m)$. Consequently, $\{x_n : n = 1, \ldots\}$ is discrete since Φ is continuous.

Since G is locally compact it is straightforward to construct $\Theta \in C_0(G)$ such that $\Theta(x_n) = 1/n$ for each n. Thus $|\Psi_\Theta(x_n)| \geqslant 1$ for each n, a contradiction. Therefore $C_b(G)$ is β complete.

The α_Θ case is clear since $\|\Phi_n\|_\infty \leqslant M$ for an α_Θ Cauchy sequence $\{\Phi_n : n = 1, \ldots\}$.

v) We now prove part b). If $\{\Phi_n : n = 1, \ldots\} \subseteq C_b(G)$ is β Cauchy then $\lim_n \Phi_n = \Phi \in C_b(G)$ in the β topology by part iv). By part i), $\lim_n \Phi_n = \Phi$ in the κ topology; and since $\{\Phi_n : n = 1, \ldots\}$ is a β convergent *sequence* we apply ii) to obtain its uniform boundedness.

For the converse, first note that $\Phi \in C_b(G)$ since $\{\|\Phi_n\|_\infty : n = 1, \ldots\}$ is bounded (by M).

Let $\Theta \in C_0(G)$. Take $\varepsilon > 0$ and choose a compact set $K \subseteq G$ such that

$$\forall\, x \notin K, \qquad |\Theta(x)| < \varepsilon/(2M).$$

Then, by hypothesis,

$$\exists\, N \text{ such that } \forall\, n \geqslant N \text{ and } \forall\, x \in K,\ |\Phi_n(x) - \Phi(x)| < \varepsilon/\|\Theta\|_\infty;$$

and so $\lim_{n\to\infty} \|\Theta(\Phi_n - \Phi)\|_\infty = 0$.

vi) Part c) follows by an argument analogous to that in part v) and by the definition of ρ_Θ.

q.e.d.

2.2.2 Remarks on the α and β topologies

Remark 1. We make some comments on the α_Θ topology. Any α_Θ (sequential) convergence criterion is stronger than β sequential convergence.

For each subset $X \subseteq C_b(G)$ we let cl X consist of those elements Φ of $C_b(G)$ for which there is $\{\Phi_n : n = 1, \ldots\}$ such that $\lim_{n\to\infty} \Phi_n = \Phi$ in the κ topology and $\lim_{n\to\infty} \|\Phi_n\|_\infty = \|\Phi\|_\infty$.

Clearly, $\mathrm{cl}\,\emptyset = \emptyset$, $X \subseteq \mathrm{cl}X$, *and* $\mathrm{cl}(X \cup Y) = \mathrm{cl}X \cup \mathrm{cl}\,Y$, where $X, Y \subseteq C_b(G)$. Take $X \subseteq C_b(G)$ and $\Phi \in \mathrm{cl}(\mathrm{cl}\,X)$. Choose $\{\Phi_n : n = 1, \ldots\} \subseteq \mathrm{cl}\,X$ such that $\lim_{n\to\infty} \Phi_n = \Phi$ in the κ topology and $\lim_{n\to\infty} \|\Phi_n\|_\infty = \|\Phi\|_\infty$; and for each n choose $\{\Phi_{m,n} : m = 1, \ldots\} \subseteq X$ such that $\lim_{m\to\infty} \Phi_{m,n} = \Phi_n$ in the κ topology and $\lim_{m\to\infty} \|\Phi_{m,n}\|_\infty = \|\Phi_n\|_\infty$. Fix any Θ to define ρ_Θ and for each n choose m_n, by Proposition 2.2.1c, such that $\rho_\Theta\,(\Phi_{m_n, n}, \Phi_n) < 1/n$. Pick $\varepsilon > 0$. Thus, there is $N_1 > 0$ for which $\rho_\Theta(\Phi_{m_n, n}, \Phi_n) < \varepsilon/2$ when $n \geqslant N_1$. There is also N_2 such that $\rho_\Theta(\Phi_n, \Phi) < \varepsilon/2$ if $n \geqslant N_2$. Let $N = \max(N_1, N_2)$. Then if $n \geqslant N$ we have $\rho_\Theta(\Phi_{m_n, n}, \Phi) < \varepsilon$; and so $\Phi \in \mathrm{cl}\,X$. Hence $\mathrm{cl}(\mathrm{cl}\,X) = \mathrm{cl}\,X$. Therefore "cl" is a closure operator, and $\mathrm{T}'_\alpha = \{U \subseteq C_b(G) : \mathrm{cl}(C_b(G) \setminus U) = C_b(G) \setminus U\}$ is a topology on $C_c(G)$ with the property that $\overline{X} = \mathrm{cl}X$, where $X \subseteq C_b(G)$ and "$-$" represents closure in T'_α. Clearly, T'_α is independent of any Θ.

We now observe that each metric space $(C_b(G), \rho_\Theta)$ generates the same topology T_α on $C_b(G)$. In fact, if $X \subseteq C_b(G)$ is α_{Θ_1} closed and Φ is in the α_{Θ_2} closure of X then it is easy to check, using Proposition 2.2.1c, that $\Phi \in X$.

From Proposition 2.2.1 we see that the $\sigma(L^\infty(G), L^1(G))$ topology on $C_b(G) \subseteq L^\infty(G)$ is weaker than T_α. Compare T_α and T'_α (obviously, T_α is weaker than T'_α).

2. The κ (resp. the sup norm) topology on $C_b(G)$ is formed in the same way as the β topology except that the family $C_c(G)$ (resp. $C_b(G)$) is used instead of $C_0(G)$. A systematic treatment of spectral synthesis in terms of the β topology is given by [Herz, 5]. We also mention [Domar, 4].

In light of Remark 1 we refer to any α_Θ topology as the *narrow* or α topology. Recall the remarks made in Section 1.4.12 on the narrow and strict topologies.

2.2.3 Examples

Example 2.2.1 All of the functions in this example are defined on **R**.

a) From the above discussion we know that the topology of uniform convergence is stronger than the α topology. Note that if $\Phi_n(x) = \frac{1}{n} \sin \frac{x}{n}$ then $\lim_{n\to\infty} \Phi_n = 0$ in the sup norm topology. Now if $\Phi_n(x) = \sin \frac{x}{n}$ then $\lim_{n\to\infty} \Phi_n = 0$ in the β topology, but since $\|\Phi_n\|_\infty = 1$, $\{\Phi_n : n = 1, \ldots\}$ does not converge in the α topology. As another example of this type, define

$$\Phi_n(x) = \begin{cases} 2(x-n) & \text{for } x \in [n, n+\frac{1}{2}], \\ -2(x-n-1) & \text{for } x \in [n+\frac{1}{2}, n+1], \\ 0 & \text{elsewhere.} \end{cases}$$

Finally we show that the α topology is strictly weaker than the sup norm topology. Let

$\{\gamma_n : n = 1, \ldots\} \subseteq Я$ be a sequence of distinct points which converges to $\gamma \in Я$, and define

$$\Phi_n(x) = e^{ix\gamma_n}.$$

Clearly, $\lim_{n\to\infty} \Phi_n(x) = e^{ix\gamma} = \Phi(x)$ in the α topology. On the other hand,

$$|e^{ix\gamma_n} - e^{ix\gamma}| = \left| 2 \sin \frac{x}{2}(\gamma_n - \gamma)\right|,$$

and so $\|\Phi_n - \Phi\|_\infty = 2$.

b) If $\Phi(x) = \sin x$ then $\Psi(x) = \sin x + \cos x$ belongs to the translation invariant subspace generated by Φ since $\cos x = \sin\left(x - \frac{3\pi}{2}\right)$. On the other hand, if $\Phi(x) = e^{ix}$ then $\Psi(x) = e^{2ix}$ does not belong to the translation invariant subspace generated by Φ. Now if $\Phi(x) = e^{-|x|}$ then $\Psi(x) = (|x| + \frac{1}{2})e^{-|x|} \in V_\Phi$ where the closure of the variety is taken in the sup norm topology (note that $\Phi * \Phi = \Psi$). In this regard see Proposition 2.2.2. Also, it is clear that if $\Phi \in C_b(G)$ then

$$\Psi(x) = \sum_1^\infty \frac{1}{k^2} \Phi(x + x_k) \in V_\Phi,$$

where we are again considering the sup norm closure of the variety.

c) Because of Beurling's theorem, which we shall prove shortly, let us recall some examples concerning uniformly continuous functions. $\Phi(x) = x$ is unbounded and uniformly continuous on $\mathbf{R}$, whereas $\Phi(x) = x^2$ is neither bounded nor uniformly continuous. $\Phi(x) = \sin x$ is bounded and uniformly continuous, whereas $\Phi(x) = \sin e^x$ is bounded but not uniformly continuous.

d) In the proof of Theorem 2.2.1 we'll also use the Arzelà–Ascoli theorem (e.g. [Benedetto, 12]) and so we make the following observations. If $\Phi_n'(x)$ exists for each $x \in \mathbf{R}$ and each $n = 1, \ldots$, and if $\sup_n \|\Phi_n'\|_\infty < \infty$, then $\{\Phi_n : n = 1, \ldots\}$ is an equicontinuous family. Thus, $\{\Phi_n(x) = n \sin \frac{x}{n} : n = 1, \ldots\}$ is an equicontinuous family. Now, $\{\Phi_n(x) = \sin \frac{x}{n} : n = 1, \ldots\}$ is an uniformly bounded equicontinuous family which converges pointwise to $\Phi = 0$; but $\{\Phi_n : n = 1, \ldots\}$ does not converge uniformly on $\mathbf{R}$.

2.2.4 Beurling's theorem: $\mathrm{supp}_\alpha\, T = \mathrm{supp}\, T$. Take $T \in A'(\Gamma)$ for which $\hat{T} = \Phi \in C_b(G)$, and let V_Φ be the α closed variety generated by Φ. We define

$$\mathrm{supp}_\alpha T = \{\gamma \in \Gamma : (\gamma, \cdot) \in V_\Phi\}.$$

Proposition 2.2.2. *Let $\Phi \in C_b(\mathbf{R}) \setminus \{0\}$ be uniformly continuous and take $f \in L^1(\mathbf{R})$. Then $f * \Phi$ belongs to the* sup *norm closed variety generated by Φ.*

Proof. Fix $\varepsilon > 0$. Since Φ is uniformly continuous,

$\exists\, \delta > 0$ such that $\forall\, y \in [k, (k+1)\delta]$ and $\forall\, x \in \mathbf{R}$, $|\Phi(x-y) - \Phi(x-k\delta)| < \varepsilon/(2\|f\|_1)$.

Define

$$\Psi(x) = \sum_{k=-\infty}^{\infty} \Phi(x-k\delta) \int_{k\delta}^{(k+1)\delta} f(t)\,\mathrm{d}t.$$

This series converges uniformly on $\mathbf{R}$ by the Weierstrass M test.
Note that

$$f * \Phi(x) - \Psi(x) = \sum_{k=-\infty}^{\infty} \int_{k\delta}^{(k+1)\delta} [\Phi(x-y) - \Phi(x-k\delta)] f(y)\,\mathrm{d}y,$$

and so $\|f * \Phi - \Psi\|_\infty < \varepsilon/2$. Set

$$\Psi_n(x) = \sum_{k=-n}^{n} \Phi(x-k\delta) \int_{k\delta}^{(k+1)\delta} f(t)\,\mathrm{d}t,$$

and thus

$$\exists\, N \text{ such that } \forall\, n \geqslant N, \qquad \|\Psi_n - \Psi\|_\infty < \varepsilon/2.$$

Consequently, $\|\Psi_n - f * \Phi\|_\infty < \varepsilon$ for all $n \geqslant N$.

q.e.d.

Theorem 2.2.1 ([Beurling, 2]). *Let $\hat{T} = \Phi \in C_b(\mathbf{R})$ be uniformly continuous. Then*

(2.2.6) $\quad \operatorname{supp}_\alpha T = \operatorname{supp} T.$

Proof. i) If $T \neq 0$ we know $\operatorname{supp} T \neq \emptyset$ by Proposition 1.4.2; also,

(2.2.7) $\quad \operatorname{supp}_\alpha T \subseteq \operatorname{supp} T,$

since the α topology is stronger than the weak $*$ topology. Take $\lambda \in \operatorname{supp} T$. Because of Proposition 2.2.2 it is sufficient to find a sequence $\{f_n : n = 1, \ldots\} \subseteq L^1(\mathbf{R})$ such that $\lim_{n\to\infty} f_n * \Phi(x) = \mathrm{e}^{\mathrm{i}x\lambda}$ in the α topology.

ii) Using the notation of Exercise 1.2.6 let $\varphi_n(\gamma) = \tau_\lambda \Delta_{1/n}(\gamma)$, so that $\hat{g}_n = \varphi_n$ where $g_n(x) = \mathrm{e}^{-\mathrm{i}\lambda x} f_{1/n}(x)$.

Note that $T\varphi_n \neq 0$, for if it were we'd have $\varphi_n = 0$ on $\operatorname{supp} T$ by Theorem 1.3.1c (but $\lambda \in \operatorname{supp} T$ and $\varphi_n(\lambda) = 1$). Thus $\Phi * g_{n-}$ is not identically zero and so we can choose $x_n \in \mathbf{R}$ for which

(2.2.8) $\quad |\Phi * g_{n-}(x_n)| \geqslant \left(1 - \dfrac{1}{n}\right) \|T\varphi_n\|_{A'}.$

Define

$$h_n(x) = \frac{g_n(x - x_n)}{\Phi * g_{n-}(x_n)} \in L^1(\mathbf{R}), \qquad \text{where } \hat{h}_n = \psi_n.$$

Clearly

(2.2.9) $\quad h_{n-} * \Phi(0) = 1$

and

(2.2.10) $\quad \forall\, n \geqslant 2, \qquad \|T\psi_n\|_{A'} \leqslant \dfrac{n}{n-1},$

since

$$|\Phi * h_{n-}(x)| \leqslant \frac{n}{n-1} \frac{|\widehat{T\varphi_n(x + x_n)}|}{\|T\varphi_n\|_{A'}}.$$

iii) We now prove that $\{\Phi * h_{n-} : n = 1, \ldots\}$ is a uniformly bounded equicontinuous family so that we can apply the Arzelà–Ascoli theorem.
The uniform boundedness is clear from (2.2.10). If $\hat{g} = \tau_\lambda \nabla_1$, where ∇_1 is the de la Vallée–Poussin kernel, we see that

$$\forall\, n \geqslant 2, \qquad g_- * (\Phi * h_{n-}) = \Phi * h_{n-}.$$

Thus,

$$|\Phi * h_{n-}(x) - \Phi * h_{n-}(y)| \leqslant \frac{n}{n-1} \int_{\mathbf{R}} |g(t - x) - g(t - y)| \, \mathrm{d}t$$

(2.2.11)
$$\leqslant 2 \int_{\mathbf{R}} |g(u + (y - x)) - g(u)| \, \mathrm{d}u$$

$$= \mathrm{o}(1), \qquad |y - x| \to 0,$$

by a basic property of the Lebesgue integral. (2.2.11) obviously entails the equicontinuity. Therefore, by the Arzelà–Ascoli theorem, there is a subsequence $\{f_n : n = 1, \ldots\}$ of $\{h_{n-} : n = 1, \ldots\}$ such that $\Phi * f_n$ converges in the κ topology to a continuous function Ψ.

$\hat{S} = \Psi \in C_b(\mathbf{R})$ by (2.2.10) and $\|\Psi\|_\infty = \lim \|\Phi * f_n\|_\infty$ because of (2.2.9) and (2.2.10).
Thus $\Phi * f_n$ converges to Ψ in the α topology.

iv) Since $\Psi(0) = 1$ and Ψ is continuous, we know that $\operatorname{sp} \Psi \neq \emptyset$ (Proposition 1.4.2).
We'll prove that $\operatorname{supp} S = \{\lambda\}$. Thus, $\Psi(x) = \mathrm{e}^{\mathrm{i}x\lambda}$ because sets consisting of a single point are S sets (cf. Exercises 1.2.3, 1.2.4, and 1.3.7).

Take $\gamma \in Я$, where $\gamma \neq \lambda$. Hence we can choose $\varepsilon > 0$ and $N > 0$ so that $\psi_n \varphi = 0$, where $\hat{f} = \varphi = \tau_\gamma \Delta_\varepsilon$ and $n \geqslant N$ (note that $\varphi(\gamma) \neq 0$). Therefore $(T\psi_n)\varphi = 0$ if $n \geqslant N$.

Because of the α convergence of $\{\Phi * f_n : n = 1, \ldots\}$ and the fact that $\widehat{(T\psi_n)\varphi} = (T * h_{n-}) * f_-$ we can use the dominated convergence theorem to conclude that $S\varphi = 0$ (obviously we do not use the full strength of α convergence at this point).

Consequently, if $\gamma \in \operatorname{supp} S$ we deduce that $\varphi(\gamma) = 0$ by Theorem 1.3.1c; on the other hand, as we pointed out above, $\varphi(\gamma) \neq 0$.

q.e.d.

2.2.5 Wiener's Tauberian theorem and Beurling's theorem. Beurling's original proof of Theorem 2.2.1 used complex variable techniques and did not use the Tauberian theorem; a detailed presentation of this proof is found in [Benedetto, 2, pp. 58–65 and pp. 70–80] (in this reference, the word "strict" should be replaced by "narrow"). The extension of Theorem 2.2.1 to a LCAG with a proof dependent on Wiener's Tauberian theorem was first stated in [Godement, 1]; unfortunately, an error was involved on this point [Godement, 1, Section 4; 2, p. 131] which was not cleared up until the mid-1960's, e.g. Section 2.2.7. The above proof of Theorem 2.2.1 is due to [Koosis, 2, p. 122] and depends on Wiener's Tauberian theorem and the fact that points are S-sets. The latter fact, although proved at the time of Beurling's paper in 1945, had not yet been properly advertised (cf. Section 2.5.1). Other proofs of Beurling's theorem, including generalizations to the LCAG case, are found in [Domar, 1; 2; Garsia, 1].

In light of the fact that Beurling's original proof did not use Wiener's Tauberian theorem, it is interesting that Wiener's Tauberian theorem (Theorem 2.2.2b = Theorem 1.3.1c) can be deduced from the statement of Theorem 2.2.1. Thus, *in the following result we only assume the statement of Theorem* 2.2.1 and the elementary definitions of certain operations with pseudo-measures; in particular, we do not use Theorem 1.3.1c in the proof.

Theorem 2.2.2. a) *Let $\hat{T} = \Phi \in C_b(\mathbf{R}) \setminus \{0\}$ be uniformly continuous and take $\varphi \in A(\mathbf{Я})$.* If $T\varphi = 0$ then $\varphi = 0$ on $\operatorname{supp}_\alpha T$.

b) *Given $T \in A'(\mathbf{Я})$ and $\varphi \in A(\mathbf{Я})$. If $T\varphi = 0$ then $\varphi = 0$ on* $\operatorname{supp} T$.

Proof. a) Since $T\varphi = 0$ we know that if $\hat{S} = \Psi \in V_\Phi$, the α variety generated by Φ, then $S\varphi = 0$.

From the statement of Theorem 2.2.1, there is an element $(\gamma, \cdot) \in V_\Phi$, and so, for each x, $(\gamma, x)\,\varphi(\gamma) = 0$. Thus, $\varphi = 0$ on $\operatorname{supp}_\alpha T$.

b) Since $T\varphi = 0$ we have that $(T\psi)\varphi = 0$ for each $\psi \in A(\mathbf{Я})$.

Therefore, since $\widehat{T\psi}$ is uniformly continuous, we apply part a) and conclude that $\varphi = 0$ on $\operatorname{supp}_\alpha T\psi = \operatorname{supp} T\psi$ (we've used Theorem 2.2.1 again).

Consequently, $\varphi = 0$ on $\operatorname{supp} T$ because ψ is arbitrary.

q.e.d.

[Pollard, 2, Section 3 and Section 5; Korevaar, 2] give other proofs of Theorem 2.2.2b.

2.2.6 Spectral synthesis in the α topology. Since Theorem 2.2.1 is a positive solution for a strong form (i.e., the α topology) of spectral analysis on the uniformly continuous elements of $C_b(\mathbf{R})$, it is natural to investigate the corresponding spectral synthesis problem in the α topology. Because of (2.2.6) and the results of Chapter 1, we have

Proposition 2.2.3. *Let* $\hat{T} = \Phi \in C_b(\mathbf{Я})$ *be uniformly continuous. If* card $\operatorname{supp}_\alpha T < \infty$ *then*

$$T = \sum_{\gamma \in \operatorname{supp}_\alpha T} a_\gamma \delta_\gamma$$

(cf. Theorem 2.2.4).

We also note Beurling's and Garsia's proofs of Proposition 2.2.3 [Benedetto, 2, pp. 90–91; Beurling, 2, p. 132; Garsia, 1, Section 3].

2.2.7 Unbounded spectral analysis.

Remark. One of the important features of Beurling's spectral analysis for the α topology is that an element of $\operatorname{supp}_\alpha T$ can be approximated by a sup *norm bounded* sequence of linear combinations of translates of $\hat{T}$. (Norm boundedness, in one form or another, is the crucial assumption in problems of "switching limits", e.g. Moore–Smith or Lebesgue dominated convergence theorems.)

The weak $*$ spectral analysis of $L^\infty(\mathbf{R})$ which comes from the Tauberian theorem does not provide any such boundedness property for those elements of $L^\infty(\mathbf{R})$ which are not uniformly continuous (we have the boundedness for uniformly continuous functions because of Theorem 2.2.1). [Benedetto, 12, Appendix I; Kahane and Salem, 4, Appendix IV] describe what the weak $*$ topology will and will not do for you. Generally, it will never provide you with norm boundedness when you need it because *directed systems* of linear combinations of translates of Φ are required to find elements $(\gamma, \cdot) \in \mathcal{T}_\Phi$.

In light of Theorem 2.2.1 it is reasonable to try to prove the following: *if* $\Phi \in C_b(G)$ *and* $(\gamma, \cdot) \in \mathcal{T}_\Phi$ *then* $(\gamma, \cdot)$ *is the weak $*$ limit point of a sup norm bounded family of linear combinations of translates of* Φ. [Koosis, 2; Kahane, 9] have provided counterexamples to this statement; Koosis notes that such counterexamples also provide counterexamples to [Godement, 2, Théorème D, p. 131].

2.2.8 The spectral analysis and synthesis of almost periodic functions. We'll now make a refinement of Proposition 2.2.3 (in Theorem 2.2.4), to do this it is convenient to outline some facts about almost periodic functions.

We consider $C_b(G)$ with the supnorm topology. The *orbit* of $\Phi \in C_b(G)$ is the set $O(\Phi) = \{\tau_x \Phi : x \in G\}$. Generally an *orbit* of $C_b(G)$ is a translation invariant set. $\Phi \in C_b(G)$ is *almost periodic* (resp. *weakly almost periodic*) if $O(\Phi)$ is relatively compact (resp.

relatively weakly compact) in $C_b(G)$; in this case we write $\Phi \in AP(G)$ (resp. $\Phi \in WAP(G)$). We state the main facts that we need about $AP(G)$ in the following theorem.

Theorem 2.2.3. *Given $\hat{T} = \Phi \in AP(G) \setminus \{0\}$ and let V_Φ be the sup norm closed variety generated by Φ.*

a) (*Spectral analysis*) $\operatorname{supp}_{AP} T = \{\gamma \in \Gamma : (\gamma, \cdot) \in V_\Phi\}$ *is countable and non-empty.*

b) (*Spectral synthesis*) *Φ can be written as a uniformly convergent series*

(2.1.12) $$\sum_{\gamma \in \operatorname{supp}_{AP} T} a_\gamma(\gamma, x),$$

where each $a_\gamma \neq 0$.

c) $\overline{\operatorname{supp}_{AP} T} = \operatorname{supp} T.$

Part c) follows immediately by the definition of support and part b).

It is easy to check that finite linear combinations of characters are almost periodic, and that uniform limits of almost periodic functions are almost periodic. Clearly, $\operatorname{supp}_{AP} T = \emptyset$ if $T \in A'_0(\Gamma)$; and, in fact, if $\hat{T} \in AP(\mathbf{R})$ and $T \in A'_c(\mathbf{Я})$ then $T = 0$. On the other hand it is possible to find a uniformly continuous function $\hat{T} = \Phi \in L^\infty(\mathbf{R}) \setminus AP(\mathbf{R})$ for which card $\operatorname{supp}_{AP} T > \aleph_0$ [Eggleston, 1; Wallin, 1].

Expositions of Bohr's theory of almost periodic functions, including a proof of Theorem 2.2.3, are found in [Katznelson, 5; Loomis, 1]. The technique developed after (2.1.1) to define the spectral distribution of energy can be used to verify Theorem 2.2.3a in a very neat way [Katznelson, 5, p. 163]. Also, as we indicated in Remark 3 of Section 2.1.6, $\operatorname{supp}_{AP} T \subseteq ZI_W(M_T)$.

2.2.9 The extension to G of Bohr's differentiation criterion for almost periodicity. If $\hat{T} = \Phi \in AP(G)$ we say that T is *almost periodic*. $T \in A'(\Gamma)$ is *almost periodic at* $\gamma \in \Gamma$ if there is a non-negative function $\varphi \in A(\Gamma)$ which is equal to 1 on a neighborhood of γ and which has the property that $T\varphi$ is almost periodic. Clearly,

$$\{\gamma \in \Gamma : T \text{ is almost periodic at } \gamma\}$$

is an open set.

Proposition 2.2.4. *Given $T \in A'(\Gamma)$ and assume that* supp T *is compact.*

a) *T is almost periodic $\Leftrightarrow$ T is almost periodic at each point of Γ.*

b) *If T is almost periodic at each $\gamma \in \Gamma \setminus \{0\}$ then T is almost periodic.*

Proof. a) ($\Rightarrow$) We'll prove

(2.2.13) *if T is almost periodic and $\hat{f} = \varphi \in A(\Gamma)$ then $T\varphi$ is almost periodic.*

Because of the remarks made in Section 2.2.8 we see from (2.2.12) that

$$\widehat{T\varphi}(x) = \sum_{\gamma \in \operatorname{supp}_{AP} T} a_\gamma \hat{f}(\gamma)(\gamma, x)$$

is a uniformly convergent series; and so $T\varphi$ is almost periodic.

($\Leftarrow$) By our hypothesis of almost periodicity at each point and the fact that $\operatorname{supp} T$ is compact, we can find $\varphi_1, \ldots, \varphi_n \in A(\Gamma)$ such that $T\varphi_i$ is almost periodic for each $i = 1, \ldots, n$ and $\sum_1^n \varphi_i > 0$ on a neighborhood of $\operatorname{supp} T$.

Thus $(Z(\sum \varphi_i)) \cap \operatorname{supp} T = \emptyset$.

By Proposition 1.1.5b there is $\varphi \in A(\Gamma)$ for which $\sum \varphi\varphi_i = 1$ on a neighborhood of $\operatorname{supp} T$.

Using (2.2.13) we see that T is almost periodic since $T\varphi_i$ is almost periodic and

$$T = T(\sum \varphi\varphi_i) = \varphi(\sum T\varphi_i).$$

b. i) We'll first prove that *if T is almost periodic at each $\gamma \in \Gamma \setminus \{0\}$ then*

(2.2.14) $\quad \forall x \in G, \quad T_\lambda - (\lambda, x) T_\lambda$

is almost periodic.

Take $\varphi \in A(\Gamma)$ equal to 1 on a neighborhood of $\operatorname{supp} T$; and so $T = T\varphi$. Also, fix $x \in G$. We have

$$T_\lambda - (\lambda, x) T_\lambda = T_\lambda(\varphi(\lambda) - (\lambda, x)\varphi(\lambda)),$$

noting that $\psi(\lambda) = \varphi(\lambda) - (\lambda, x)\varphi(\lambda)$ vanishes at $\lambda = 0$.

Since points are S-sets, there is a sequence $\{\psi_n : n = 1, \ldots\} \subseteq j(\{0\})$ such that $\lim_{n\to\infty} \|\psi - \psi_n\|_A = 0$.

Thus

(2.2.15) $\quad \lim_{n\to\infty} \|(T - (\cdot, x)T) - T\psi_n\|_{A'} = 0.$

Since T is almost periodic at each $\gamma \in \Gamma \setminus \{0\}$ and since ψ_n vanishes on a *neighborhood* of 0, we use (2.2.13) and the sufficient condition for almost periodicity in part a) to conclude that $T\psi_n$ is almost periodic.

Therefore, from (2.2.15) and the comment after Theorem 2.2.3 we have that $T_\lambda - (\lambda, x) T_\lambda$ is almost periodic.

ii) Assuming that T is almost periodic at each $\gamma \in \Gamma \setminus \{0\}$ we now use part i) to prove the almost periodicity of $\hat{T} = \Phi$. Without loss of generality assume that Φ is real-valued, and set

$$M = \sup\{\Phi(x) : x \in G\} \quad \text{and} \quad m = \inf\{\Phi(x) : x \in G\}.$$

Take $\varepsilon > 0$. Choose $a, b \in G$ such that

$$\text{(2.2.16)} \qquad \Phi(-a) > M - \frac{\varepsilon}{4} \quad \text{and} \quad \Phi(-b) < m + \frac{\varepsilon}{4}.$$

Because of part i) we set $x = a - b$ in (2.2.14) and obtain $\Phi - \tau_{b-a}\Phi \in AP(G)$; thus

$$\text{(2.2.17)} \qquad \tau_a\Phi - \tau_b\Phi = \tau_a(\Phi - \tau_{b-a}\Phi) \in AP(G).$$

From (2.2.17) there is a finite set $F \subseteq G$ such that

$$\text{(2.2.18)} \qquad \forall\, \Psi \in O(\tau_a\Phi - \tau_b\Phi)\ \exists\, y \in F \quad \text{for which} \quad \|\Psi - \tau_y(\tau_a\Phi - \tau_b\Phi)\|_\infty < \varepsilon/2.$$

In particular, for each fixed $x \in G$ we set $\Psi = \tau_{x+a}(\tau_a\Phi - \tau_b\Phi)$; and for this Ψ there is $y \in F$ for which we compute, from (2.2.18), that

$$\Psi(x + a) = \Phi(-a) - \Phi(-b)$$

and

$$\text{(2.2.19)} \qquad |(\Phi(-a) - \Phi(-b)) - (\Phi(x - y) - \Phi(x + a - b - y))| < \varepsilon/2.$$

Thus, given $\varepsilon > 0$ we have found a finite set $F \subseteq G$ such that for each $x \in G$ there are $y, z \in F$ for which

$$\text{(2.2.20)} \qquad \Phi(x - y) > M - \varepsilon \quad \text{and} \quad \Phi(x - z) < m + \varepsilon;$$

in fact, (2.2.20) follows, for the case $\Phi(x - y) > M - \varepsilon$, from (2.2.16), (2.2.19), and the computation

$$\begin{aligned} \Phi(x - y) &> \Phi(-a) - \Phi(-b) + \Phi(x + a - b - y) - \frac{\varepsilon}{2} \\ &> M - m - \left(\frac{\varepsilon}{4} + \frac{\varepsilon}{4}\right) + m - \frac{\varepsilon}{2} = M - \varepsilon. \end{aligned}$$

In order to prove that $\Phi \in AP(G)$ we take $\eta > 0$ and shall find a finite set $K \subseteq G$ such that

$$\text{(2.2.21)} \qquad \forall\, \Psi \in O(\Phi)\ \exists\, u \in K \quad \text{for which} \quad \|\Psi - \tau_u\Phi\|_\infty < \eta.$$

For convenience let $\eta = 2\varepsilon$ (ε as above).

It is straightforward to check that there is a finite set $K \subseteq G$ for which

$$\forall\, t \in G\ \exists\, u \in K \quad \text{such that} \quad \forall\, y \in F\ (F \text{ as above})$$

$$\text{(2.2.22)} \qquad \|\tau_t(\Phi - \tau_y\Phi) - \tau_u(\Phi - \tau_y\Phi)\|_\infty < \varepsilon.$$

(2.2.22) is obviously true for each single $y \in F$ since $\Phi - \tau_y\Phi \in AP(G)$; the point is that it is also true uniformly as y ranges through the finite set F.

We now prove that K is the set required to obtain (2.2.21). In order to do this we take $t, x \in G$ and use the $u \in G$ of (2.2.22) and M to calculate that

(2.2.23) $\quad \tau_t \Phi(x) - \tau_u \Phi(x) < 2\varepsilon;$

an analogous argument using m works to check the inequality $\tau_t \Phi(x) - \tau_u \Phi(y) > -2\varepsilon$. From (2.2.20) we pick $y \in F$ such that $\Phi(x - u - y) > M - \varepsilon$.

Then we employ (2.2.22) to compute

$$\tau_t \Phi(x) - \tau_u \Phi(x) < \varepsilon + \tau_t \tau_y \Phi(x) - \tau_u \tau_y \Phi(x) < 2\varepsilon - M + \tau_t \tau_y \Phi(x) \leqslant 2\varepsilon;$$

this is (2.2.23).

q.e.d.

Part ii) of the proof of Proposition 2.2.4b is the generalization to G of Bohr's theorem: *if $\Phi \in L^\infty(\mathbf{R})$ is everywhere differentiable and $\Phi' \in AP(\mathbf{R})$ then $\Phi \in AP(\mathbf{R})$.* An elegant classical treatment of Proposition 2.2.4, including this result on differentiability is found in [Katznelson, 5, pp. 166–168]. The proof of Proposition 2.2.4 is due to [Loomis, 2].

2.2.10 Scattered sets, almost periodicity, and strongly synthesizable pseudo-measures. $E \subseteq \Gamma$ is *perfect* if it is a closed set without isolated points. $E \subseteq \Gamma$ is *scattered* if it is a closed set which does not contain any non-empty perfect subsets. We'll discuss scattered sets in Paragraph 2.5. For now note that in $\hat{\mathbf{R}}$ E is scattered if and only if it is countable.

Theorem 2.2.4. *Let $\hat{T} = \Phi \in C_b(G)$ be uniformly continuous. If* supp T *is scattered then*

$$T = \sum_{\gamma \in \operatorname{supp} T} a_\gamma \delta_\gamma,$$

where convergence is in the $\| \ \|_{A'}$-topology (i.e., T is almost periodic).

Proof. i) Because of Proposition 1.4.7, there is a directed system $\{T_\alpha\} \subseteq A'(\Gamma)$ such that $\lim_\alpha \|T - T_\alpha\|_{A'} = 0$ and supp T_α is compact.

If we prove that each T_α is almost periodic we can therefore conclude that T is almost periodic.

Thus *we assume that* supp T *is a compact scattered set* and shall prove that T is almost periodic.

ii) Since $\{\gamma \in \Gamma : T$ is almost periodic at $\gamma\}$ is an open set and since T is almost periodic at each $\gamma \notin$ supp T we see that the set $X \subseteq \Gamma$ of points at which T is *not* almost periodic is a closed subset of supp T.

We now observe that X contains no isolated points (as a subset of Γ).

In fact, if $\gamma \in X$ is isolated we can choose $\varphi \in A(\Gamma)$ equal to 1 on a neighborhood of γ such that $T\varphi = 0$ on an open neighborhood of $X \setminus \{\gamma\}$. Thus $T\varphi$ is almost periodic at each point of $\Gamma \setminus \{\gamma\}$, and so we conclude that $T\varphi$ is almost periodic because of Proposition 2.2.4b; this contradicts the fact that $\gamma \in X$.

Since $\operatorname{supp} T$ is scattered and X contains no isolated points we see that $X = \emptyset$.

Therefore T is almost periodic at each point of Γ. Using the compactness of $\operatorname{supp} T$ we apply Proposition 2.2.4a to obtain the almost periodicity of T. q.e.d.

We knew from the trivial part of Proposition 1.4.7 that if $\operatorname{supp} T$ is compact then $\hat{T}$ is uniformly continuous. It is an elementary property of almost periodic functions that if $\Phi \in AP(G)$ then Φ is uniformly continuous. The proof of Theorem 2.2.4 tells us further that

Corollary 2.2.4. *If* $T \in A'(\Gamma)$ *and* $\operatorname{supp} T$ *is a compact scattered set then* T *is almost periodic.*

2.2.11 An historical remark. Theorem 2.2.4 is due to [Loomis, 2], but the idea of the proof, once we have Proposition 2.2.4, goes back to Ditkin and is used to obtain general Tauberian theorems, e.g. Section 2.5.2. [Veech, 1] has given an alternative proof of Theorem 2.2.4 employing a characterization of $AP(G)$ in terms of the more general notion of almost automorphic functions (such functions were introduced by Bochner). For the case $G = \mathbf{R}$ the result was proved by [Levitan, 1] in 1937 and [Beurling, 5] in 1947 in the context of integral equations. For example, Levitan's theorem is: *let* $S \in A'(\text{Я})$ *be almost periodic and assume that* $\varphi \in A(\text{Я})$ *is continuously differentiable; if the set of limit points of* $Z\varphi$ *consists of isolated points then every bounded uniformly continuous solution* $\hat{T}$ *of*

$$T\varphi = S$$

is almost periodic. Beurling removed the hypothesis of differentiability and used characterizations of $\operatorname{supp} T$ in terms of Abelian summability in his proof (e.g. Remark 2 in Section 2.1.6 and Exercise 2.1.4). [Reiter, 1, p. 423; 2] made contributions to this problem for the case of G prior to Loomis' work.

Exercises 2.2

2.2.1 Properties of the β topology

a) For each $\Phi \in C_b(G)$ define the family $\tilde{C}_b(G)$ of operators

$$\begin{aligned} L_\Phi : C_0(G) &\to C_0(G) \\ \Psi &\mapsto \Phi\Psi \end{aligned}$$

(cf. Proposition 1.1.6). A *strong* (resp. *weak*) *operator neighborhood* of $\tilde{O} \in \tilde{C}_b(G)$ (i.e., $\tilde{O} = L_O$) has the form

$$\{L_\Phi : \| \Phi\Psi_j \|_\infty < \varepsilon, \quad j = 1, \ldots, n\}$$

$$(\text{resp. } \{L_\Phi : | < \mu_j, \Phi\Psi_j > | < \varepsilon, \quad j = 1, \ldots, n\})$$

where ε, $\{\Psi_j : j = 1, \ldots, n\} \subseteq C_0(G)$, and $\{\mu_j : j = 1, \ldots, n\} \subseteq M(G)$ vary. We identify

$C_b(G)$ and $\tilde{C}_b(G)$ in the obvious way, and thus canonically define the *strong* and *weak operator* topologies on $C_b(G)$; these topologies were introduced by von Neumann in 1929. Prove that the β topology is the strong operator topology on $C_b(G)$; and that the canonical locally compact topology on Γ is the induced β topology on $\{(\gamma, \cdot):\gamma \in \Gamma\}$. The β topology has been studied for $C_b(X)$ for any Hausdorff space X [Hoffman-Jørgensen, 1].

b) Prove that $M(G)$ is the dual of $C_b(G)$ taken with the β topology [Buck, 2, pp. 99–100] (cf. [Muraz, 1]). Of course, this result is true for any topology between $\sigma(C_b(G), M(G))$ and the Mackey topology on $C_b(G)$, e.g. [Horváth, 1, pp. 203–206]. [Buck, 2, p. 100] asks if the Mackey topology is equivalent to the β topology; contributions to this problem are found in [Collins, 1; Conway, 1; Rubel and Ryff, 1].

A survey of recent work on the β topology is found in the Zentralblatt fur Mathematik **244** (1973) 46027, 46028; and there is interesting new work by Sentilles and Wheeler.

Note that if $\{\mu_n : n = 1,\ldots, \mu_n \geqslant 0, \|\mu_n\|_1\} \subseteq M(\mathbf{R})$ satisfies the condition that

$$\forall\ \Phi \in C_b(\mathbf{R}),\ \lim_{n\to\infty} < \mu_n, \Phi > = < \mu, \Phi >,$$

then $\mu \in M(\mathbf{R})$ is non-negative *and* $\|\mu\|_1 = 1$ (convergence on $C_0(\mathbf{R})$ would allow the possibility that $\mu = 0$, e.g. $\mu_n = \delta_n$).

c) Let $X \subseteq C_b(G)$ be a convex orbit which is β closed. Prove that X is closed in the weak $*$ topology $\sigma(L^\infty(G), L^1(G))$ induced on $C_b(G)$. Thus, if $\hat{T} = \Phi \in C_b(G)$, V_Φ is the β closed variety generated by Φ, and

$$\operatorname{supp}_\beta T = \{\gamma \in \Gamma : (\gamma, \cdot) \in V_\Phi\},$$

then

(E2.2.1) $\quad \operatorname{supp} T = \operatorname{supp}_\beta T$

(cf. the observation in [Godement, 2, top of p. 131]). (*Hint*: The proof proceeds in the same manner as the proof of the Kreiň-Šmul'yan theorem and uses part b). Since part b) is not true for the α topology the same proof does not work for $\operatorname{supp}_\alpha T$, where $\hat{T} = \Phi \in C_b(G)$ is uniformly continuous.

Also, with regard to Section 2.2.7 we can ask about extending the β topology to $L^\infty(G)$ and preserving (E2.2.1).

2.2.2 *Properties of the α topology*

a) Prove Theorem 2.2.1 for the case of G, not $\mathbf{R}$.

b) We write $f \in C_p(\mathbf{R}^n) \subseteq C^\infty(\mathbf{R}^n)$, $1 \leqslant p \leqslant \infty$, if f and all of its derivatives are elements of $L^p(\mathbf{R}^n)$. The following convergence criterion gives rise to a metrizable locally convex topology on $C_p(\mathbf{R}^n)$: $f_k \to 0$ as $k \to \infty$ if for each s

$$\lim_{k\to\infty} \|D^s f_k\|_p = 0.$$

$D_p(\mathbf{R}^n)$, $1 < p \leqslant \infty$, is the dual of $C_q(\mathbf{R}^n)$, $\frac{1}{p} + \frac{1}{q} = 1$; and if we define $f \in C_{\infty,0}(\mathbf{R}^n) \subseteq C_\infty(\mathbf{R}^n)$ as

$$\forall\, s, \qquad D^s f \in C_0(\mathbf{R}^n),$$

then $D_1(\mathbf{R}^n)$ is the dual of $C_{\infty,0}(\mathbf{R}^n)$. Prove that

$$\text{(E2.2.2)} \qquad D_p(\mathbf{R}^n) \subseteq D_r(\mathbf{R}^n) \subseteq D_\infty(\mathbf{R}^n) \subseteq D_t(\mathbf{R}^n)$$

if $p \leqslant r$ (e.g. [Schwartz, 5, pp. 199–203 and pp. 237–243]) and that

$$L^1(\mathbf{R}^n) \subseteq D_1(\mathbf{R}^n) \qquad \text{and} \qquad L^\infty(\mathbf{R}^n) \subseteq D_\infty(\mathbf{R}^n).$$

c) Because of (E2.2.2) and the remarks on the Fourier transform in Exercise 1.3.6, we see that for each $\Phi \in D_\infty(\mathbf{R}^n)$ there is $T \in D_t(\mathbf{Я}^n)$ whose Fourier transform is Φ. Prove the following form of Theorem 2.2.1 (taking Proposition 2.2.2 into account): given $\hat{T} = \Phi \in D_\infty(\mathbf{R}^n)$; $\gamma \in \operatorname{supp} T \Leftrightarrow$ there is a sequence $\{f_n : n = 1, \ldots\} \subseteq C_{rd}(\mathbf{R}^n)$ such that $f_n * \Phi(x)$ converges to $e^{ix\gamma}$ in the T_α topology.

d) Prove the following form of Wiener's Tauberian theorem [Benedetto, 1]: given $f \in L^1(\mathbf{R})$ with a non-vanishing Fourier transform and $\Phi \in C_\infty(\mathbf{R})$, and assume that

$$\lim_{x \to \infty} f * \Phi(x) = r \int_{\mathbf{R}} f(y)\, dy;$$

then, for each $g \in D_1(\mathbf{R})$,

$$\lim_{x \to \infty} g * \Phi(x) = r\langle g, 1 \rangle.$$

2.2.3 *Bernstein's inequality for $A'(\mathbf{Я})$*

We give this exercise because of the use of Proposition 1.4.7 in the proof of Theorem 2.2.4.

a) Prove that if $T \in D_c(\mathbf{Я})$ then its Fourier transform is the restriction to $\mathbf{R}$ of an entire function.

b) $\Phi : \mathbf{C} \to \mathbf{C}$ is a function of *exponential type* λ if

$$\forall\, \varepsilon > 0\ \exists\, K_\varepsilon \qquad \text{such that} \qquad \forall\, z \in \mathbf{C},\ |\Phi(z)| \leqslant K_\varepsilon \exp[(\lambda + \varepsilon)|z|].$$

Prove that if $T \in D_c(\mathbf{Я})$ and $\operatorname{supp} T \subseteq [-\lambda, \lambda]$ then $L(T)(z)$ is of exponential type λ (the Laplace transform L was defined in Exercise 1.3.6). (*Hint*: Taking $\varepsilon > 0$ it is possible to write $T = \varphi^{(n)} + \psi$, where φ and ψ are continuous functions supported by $[-\lambda, \lambda + \varepsilon]$ and $\varphi^{(n)}$ represents distributional differentiation).

c) Prove *Bernstein's inequality*: if $T \in A'(\mathbf{Я})$ and $\operatorname{supp} T \subseteq [-\lambda, \lambda]$ then

$$\|\hat{T}'\|_\infty \leqslant \lambda \|T\|_{A'}$$

(cf. [Benedetto, 6, Section A.10; Donoghue, 1, pp. 227–229; Meyer, 5, pp. 149–151]). Gårding and Hörmander have noted that it is easy to prove Wiener's Tauberian theorem using Bernstein's inequality, e.g. [Donoghue, 1, pp. 230–231].

d) Prove that if $T \in A'(\mathbf{Я})$ is supported by $[-\lambda, \lambda]$, $\|\hat{T}'\|_\infty = \lambda \|T\|_{A'}$, and $\|\hat{T}'\|_\infty = \hat{T}'(y)$ for some y, then $\hat{T}(x) = K \sin(\lambda x - y)$.

2.2.4 The harmonic spectrum and $\operatorname{supp}_P T$

a) $\Phi \in L^1_+(\mathbf{R})$ if

$$\forall\, \sigma < 0, \int_{-\infty}^{\infty} |\Phi(x)| e^{\sigma|x|}\, dx < \infty.$$

For such an element Φ and for $\sigma + i\gamma \in \mathbf{C}$,

$$H_\Phi(\sigma, \gamma) = \int_{-\infty}^{\infty} \Phi(x)\, e^{\sigma|x| + i\gamma x}\, dx$$

is an harmonic function in the half-plane $\sigma < 0$. $\gamma \in \operatorname{sp}_H \Phi$, the *harmonic spectrum* of $\Phi \in L^1_+(\mathbf{R})$, if

$$\forall\, \varepsilon > 0, \lim_{\sigma \to 0^-} \int_{\gamma-\epsilon}^{\gamma+\epsilon} |H_\Phi(\sigma, \lambda)|\, d\lambda > 0.$$

Prove that $\operatorname{sp}_H \Phi$ is closed for each $\Phi \in L^1_+(\mathbf{R})$; and that there is $\Phi \in L^1_+(\mathbf{R}) \setminus \{0\}$ such that $\operatorname{sp}_H \Phi = \emptyset$ (cf. [Vretblad, 1]). [Beurling, 3] has determined a satisfactory spectral analysis for a certain subspace of $L^1_+(\mathbf{R})$ as well as having studied a large class of related spectral analysis problems. À propos this material, Hille and Tamarkin have contributed to the ideal theory of the algebra of Laplace–Stieltjes transforms absolutely convergent in a fixed half-plane, e.g. [Hille and Phillips, 1] for references.

b) With regard to the β topology and priorities we now comment on some work of [Povzner, 1] (1947). Let $P(\mathbf{R})$ be the set of even, positive, continuous functions p which increase to infinity at infinity and which satisfy the condition that $p(x+y) \leqslant M(y)p(x)$ for $x, y \in \mathbf{R}$. For each $p \in P(\mathbf{R})$, $\|\Phi\|^p = \|\Phi/p\|_\infty$ defines a norm on $C_b(\mathbf{R})$ and $V_\Phi(p)$ will be the $\|\ \ \|^p$-closed variety generated by $\Phi \in C_b(\mathbf{R})$. For $\hat{T} = \Phi \in C_b(\mathbf{R})$, define

$$\operatorname{supp}_p T = \{\gamma \in \mathbf{Я} : (\gamma, \cdot) \in V_\Phi(p)\}$$

and

$$\operatorname{supp}_P T = \bigcap_{p \in P(\mathbf{R})} \operatorname{supp}_p T.$$

If $\hat{T} = \Phi \in C_b(\mathbf{R}) \setminus \{0\}$ prove that $\operatorname{supp}_P T = \operatorname{supp}_p T = \emptyset$ for some $p \in P(\mathbf{R})$. Povzner announces (there are essentially no proofs in the paper) that

$$\operatorname{supp}_P T = \operatorname{supp} T$$

for $\hat{T} \in C_b(\mathbf{R})$ (cf. (E2.2.1)). A related "L^1" theory is given in Section 1.4.3. [Beurling, 2] (1945) proved that

$$T \neq 0 \quad \Rightarrow \quad \operatorname{supp}_\alpha T \neq \emptyset$$

if $\hat{T}$ is uniformly continuous, but he does not verify (2.2.6); on the other hand he states that (2.2.6) is easily proved in a footnote in [Beurling, 6] (1949).

2.2.5 The uniform closure of $B(G)$

From Theorem 2.2.3 we know that the uniform closure of $M_f(\Gamma)^\wedge$ is $AP(G)$ and, of course, the uniform closure of $L^1(\Gamma)^\wedge$ is $C_0(G)$. An internal characterization of the uniform closure of the set of Fourier–Stieltjes transforms of continuous singular measures is not so easy. In this exercise, $\overline{B(G)}$ will denote the uniform closure of $B(G)$.

a) Prove that $\Phi \in C_b(G)$ is an element of $AP(G)$ if and only if for each $\| \quad \|_1$-bounded *directed system* $\{\mu_\alpha\} \subseteq M(G)$, for which $\lim_\alpha \hat{\mu}_\alpha = 0$ pointwise, we can conclude that

$$\lim_\alpha \int_G \Phi \, d\mu_\alpha = 0$$

[Edwards, 4, p. 254; Ramirez, 2].

b) Let G be σ-compact. Prove that $\Phi \in C_b(G)$ is an element of $\overline{B(G)}$ if and only if for each $\| \quad \|_1$-bounded *sequence* $\{\mu_n : n = 1, \ldots\} \subseteq M(G)$, for which $\lim_{n\to\infty} \hat{\mu}_n = 0$ pointwise, we can conclude that

$$\lim_{n\to\infty} \int_G \Phi \, d\mu_n = 0$$

(cf. [Edwards, 4, Theorem 1.7]). This result was first proved by Beurling and Hewitt (unpublished). The extension to arbitrary G as well as characterizations of $\overline{B(G)}$ in terms of the β and weak operator topologies are due to [Ramirez, 1] (cf. Section 3.2.17). Surveys on the properties of $M(G)$ are found in [Dunkl and Ramirez, 1; Hewitt, 2; Rudin, 5].

c) Prove that $C_0(G) \subseteq \overline{B(G)} \subseteq WAP(G)$ and that each element of $WAP(G)$ is uniformly continuous. [Rudin, 3], e.g. [Dunkl and Ramirez, 1], proved that the inclusion $\overline{B(G)} \subseteq WAP(G)$ is proper. The theory of weakly almost periodic functions was developed by Eberlein via his research in ergodic theory; in the process he was quite aware that synthesizable pseudo-measures have transforms with generalized almost periodic properties [Eberlein, 1, Section 7–8]. [Burckel, 1] is an exposition on $WAP(G)$ and [Argabright, 1] shows that the invariant means on $WAP(G)$ provide a simple way to compute the point values of pseudo-measures T for which $\hat{T} \in WAP(G)$.

2.2.6 Almost periodic functions

Prove Theorem 2.2.3.

2.3 Classical Tauberian theorems

2.3.1 Abel's and Tauber's theorems

Proposition 2.3.1 (Abel). *If* $f(x) = \sum_0^\infty a_n x^n$ *converges on* $[0, 1)$ *and* $\sum_0^\infty a_n = S$ *then* $f(1-) = S$.

Proof. It is obviously sufficient to prove that $\sum a_k x^k$ converges uniformly on $[0, 1]$. Assume each $a_k \in \mathbf{R}$ so that by hypothesis

$$\forall\, \varepsilon > 0\ \exists\, N \qquad \text{such that } \forall k > m \geqslant N,$$
$$-\varepsilon < a_{m+1} + \ldots + a_k < \varepsilon.$$

Now, by partial summation (Abel's technique),

$$\sum_{m+1}^{n} a_k x^k = \sum_{m+1}^{n} \left(\sum_{m+1}^{k} a_j \right)(x^k - x^{k+1}) + x^{n+1} \sum_{m+1}^{n} a_k$$
$$\leqslant \varepsilon \left(\sum_{m+1}^{n} (x^k - x^{k+1}) + x^{n+1} \right) = \varepsilon x^{m+1} \leqslant \varepsilon.$$

q.e.d.

The first Tauberian theorem (by Tauber in 1897) came as a response to the problem of finding some sort of converse to Proposition 2.3.1 besides the trivial converse one obtains by assuming each $a_k \geqslant 0$ (this "$a_k \geqslant 0$" case was observed by Pringsheim in 1900 although Abel himself had practically done the calculation). Tauber proved: *if* $f(x) = \sum a_n x^n$ *converges on* $[0, 1)$, $f(x-) = S$, *and* $a_n = o(1/n)$, $n \to \infty$, *then* $\sum a_n = S$. The boundedness condition "$a_n = o(1/n)$" is the "Tauberian condition" necessary to effect the converse. Thus, the hypothesis "$\Phi \in L^\infty(G)$" is the Tauberian condition in Theorem 2.1.5 [Hardy, 1, p. 285] (cf. [Rudin, 6, p. 209]).

2.3.2 A technical lemma

Proposition 2.3.2. *Define* $f(x) = \sum_0^\infty a_n x^n$ *and*

$$(2.3.1) \qquad \Phi(x) = \begin{cases} \sum_0^{[x]} a_n & \text{for } x \geqslant 0 \\ 0 & \text{for } x = 0. \end{cases}$$

If $x = e^{-\varepsilon}$ *then*

$$(2.3.2) \qquad \forall\, x \in (0, 1), \qquad f(x) = (1 - x) \sum_{n=0}^{\infty} \Phi(n)\, x^n = \varepsilon \int_0^\infty \Phi(t)\, e^{-\varepsilon t}\, dt,$$

where $\varepsilon \in (0, \infty)$, *in the sense that if any of the quantities in* (2.3.2) *exists then the other two also exist and all are equal.*

Proof. a) Fix $x < 1$ and $r \in (x, 1)$.
Assume that $f(y)$ converges on $[0, 1)$. Thus, there is a constant K such that for each k, $|a_k| < Kr^k$.
Consequently, $\lim_{n\to\infty} |\Phi(n)x^n| = 0$ since $x/r < 1$ and

$$|\Phi(n)x^n| \leqslant Kx^n\left(1 + \frac{1}{r} + \ldots + \frac{1}{r^n}\right) = \frac{Kx^n}{r^n(1-r)}.$$

Therefore $(1-x)\sum_{n=0}^{\infty}\Phi(n)x^n$ converges to $f(x)$ since

$$\sum_{k=0}^{n} a_k x^k = \sum_{k=0}^{n-1}\Phi(k)(x^k - x^{k+1}) + \Phi(n)x^n = (1-x)\sum_{k=0}^{n-1}\Phi(k)x^k + \Phi(x)x^n.$$

A similar argument works to prove that $f(x)$ converges on $[0, 1)$ if $(1-y)\sum_{n=0}^{\infty}\Phi(n)y^n$ converges on $[0, 1)$.

b) Note that $\varepsilon\int_n^{n+1} e^{-\varepsilon t}\,dt = (1-x)x^n$, where $x = e^{-\varepsilon}$. Then if $\varepsilon\int_0^{\infty}\Phi(t)e^{-\varepsilon t}dt$ exists,

$$\varepsilon\int_0^{\infty}\Phi(t)e^{-\varepsilon t}\,dt = \varepsilon\sum_{n=0}^{\infty}\int_n^{n+1}\Phi(t)e^{-\varepsilon t}\,dt = \varepsilon\sum_{n=0}^{\infty}\Phi(n)\int_n^{n+1}e^{-\varepsilon t}\,dt.$$

q.e.d.

2.3.3 Frobenius' theorem: Cesàro summable implies Abel summable. Observe that

$$\frac{1}{n+1}\sum_{j=0}^{n} s_j = \sum_{k=0}^{n}\left(1 - \frac{k}{n+1}\right)a_k, \qquad \text{where } s_j = \sum_{k=0}^{j} a_k$$

(cf. (2.1.31) and (2.1.32)). We write

$$\sum_{n=0}^{\infty} a_n = S, \qquad (C, 1),$$

if

$$\lim_{n\to\infty}\sum_{k=0}^{n}\left(1 - \frac{k}{n+1}\right)a_k = S$$

(cf. (E1.3.5) for the case $x = 0$); and in this case we say that $\sum_0^{\infty} a_n$ is *Cesàro summable* to S. Similarly, in light of Proposition 2.3.1, we write

$$\sum_{n=0}^{\infty} a_n = S, \qquad (A),$$

and say that $\sum_0^\infty a_n$ is *Abel summable* to S, if

$$\lim_{x\to 1-} \sum_{n=0}^{\infty} a_n x^n = S.$$

We give the following improvement of Proposition 2.3.1.

Proposition 2.3.3. (Frobenius) *If* $\sum a_n = S$, $(C, 1)$, *then* $\sum a_n = S$, (A).

Proof. Let $\varepsilon = 1/R$ in (2.3.2). With the notation of Proposition 2.3.2, we set

$$S_R = \frac{1}{R}\int_0^\infty \Phi(t)\mathrm{e}^{-t/R}\,\mathrm{d}t \qquad \text{and} \qquad s(x) = \int_0^x \Phi(t)\,\mathrm{d}t,$$

and prove that if

(2.3.3) $$\lim_{R\to\infty} s(R)/R = S$$

then

(2.3.4) $$\lim_{R\to\infty} S_R = S.$$

By hypothesis, there is $K > 0$ such that for all $R > 0$, $|s(R)| \leqslant KR$.
Because of this boundedness and since

$$\frac{1}{R}\int_0^x \mathrm{e}^{-t/R}\,\mathrm{d}s(t) = \frac{s(x)}{R}\mathrm{e}^{-x/R} + \int_0^x \frac{s(t)}{R^2}\mathrm{e}^{-t/R}\,\mathrm{d}t,$$

we compute

(2.3.5) $$S_R = \frac{1}{R}\int_0^\infty s(t)\mathrm{e}^{-t/R}\,\mathrm{d}(t/R) = \frac{1}{R}\int_0^\infty s(tR)\mathrm{e}^{-t}\,\mathrm{d}t.$$

Now, $|s(tR)\mathrm{e}^{-t}/R| \leqslant Kt\,\mathrm{e}^{-t} \in L^1(0,\infty)$. Consequently, we use (2.3.3) and the dominated convergence theorem on (2.3.5) to compute (2.3.4).

q.e.d.

2.3.4 Littlewood's Tauberian theorem. The first deep Tauberian theorem is due to [Littlewood, 1] (1910). We shall deduce it from Wiener's Tauberian theorem in the form of Theorem 2.1.5. [Wiener, 7, pp. 105–106] first showed that if $f(x) = \sum a_n x^n$ converges on $(0, 1)$ and $f(1-) = S$ then $S = \lim_{x\to\infty} \frac{1}{x}\int_0^x \Phi(y)\,\mathrm{d}y$ ("Φ" as in (2.3.1)); he used his general Tauberian theorem to establish this fact. We present this part of his proof, up to but not including the application of his Tauberian theorem. The second part of his proof is Hardy's Tauberian theorem (1909) which states that *if* $a_n = 0(1/n)$, $n \to \infty$, *and* $\sum a_n = S$, $(C, 1)$, *then* $\sum a_n = S$, e.g. [Hardy, 1, Theorem 63]. At this point our proof follows the more direct route observed by [Glasser, 1]. On the other hand it is no surprise that Wiener proceeded as he did. In fact, Hardy had noticed that, in view of Proposition 2.3.3, the easiest way of finding non-convergent series which are Abel summable is to consider Cesàro summable series (that are not convergent); this, coupled with his (Hardy's) Tauberian theorem, led him to the problem which Littlewood solved by means of the following result. Clearly, Theorem 2.3.1 includes Hardy's Tauberian theorem as well as Tauber's.

Theorem 2.3.1. (Littlewood) *If* $f(x) = \sum a_n x^n$ *converges on* $[0, 1)$, $f(1-) = S$, *and*

(2.3.6) $$a_n = 0(1/n), \qquad n \to \infty,$$

then $\sum a_n = S$.

Proof. a) We use (2.3.6) to calculate

(2.3.7) $$\left|\sum_0^N a_n - \sum_0^\infty a_n e^{-n/N}\right| \leqslant \sum_0^N |a_n|(1 - e^{-n/N}) + \sum_{N+1}^\infty |a_n| e^{-n/N}$$
$$\leqslant K\left(\sum_0^N \frac{1 - e^{-n/N}}{n} + \sum_{N+1}^\infty \frac{e^{-n/N}}{n}\right).$$

Clearly, by expanding $e^{-n/N}$ as a power series,

(2.3.8) $$\sum_{n=1}^N \frac{1 - e^{-n/N}}{n} \leqslant \sum_{n=1}^N \left(\frac{1}{N} + \frac{n}{2!N^2} + \frac{n^2}{3!N^3} + \ldots\right) \leqslant \sum_{j=1}^\infty \frac{1}{j!}.$$

Since $\{e^{-n/N}/n : n = N+1, \ldots\}$ is a decreasing sequence we can apply the integral test to obtain a bound for $\sum_{N+1}^\infty \frac{1}{n} e^{-n/N}$ independent of N.

Thus, from (2.3.7) and (2.3.8),

(2.3.9) $$\exists\, J > 0 \text{ such that } \forall\, N > 1, \qquad \left|\sum_0^N a_n - \sum_0^\infty a_n e^{-n/N}\right| \leqslant J.$$

Setting $x = e^{-1/N}$ we apply our hypothesis, $f(1-) = S$, to (2.3.9) and obtain a constant M such that

(2.3.10) $$\forall\, x \in [0, \infty), \qquad \left|\sum_0^{[x]} a_n\right| \leqslant M.$$

b) Define $\Phi(x)$ as in (2.3.1). Then from Proposition 2.3.2,

$$\forall\, \varepsilon \in (0, \infty), \qquad f(e^{-\varepsilon}) = \varepsilon \int_0^\infty \Phi(t) e^{-\varepsilon t}\, dt;$$

and so if we let $\varepsilon = e^{-y}$ we have

$$\forall\, y \in \mathbf{R}, \qquad f(x) = e^{-y} \int_0^\infty \Phi(t) e^{-te^{-y}} dt = \int_{-\infty}^\infty \Phi(e^z) e^{-e^{(z-y)}} e^{(z-y)} dz,$$

where $x = e^{-\varepsilon}$.

Thus, by Exercise 1.1.1 a.i),

$$S = \lim_{y \to \infty} \int_{-\infty}^\infty \Phi(e^z) g(z-y)\, dz = \lim_{y \to \infty} \Psi * g_-(y), \tag{2.3.11}$$

where $\hat{g}(\gamma) = \Gamma(1 + i\gamma)$ and $\Psi(z) = \Phi(e^z)$. Recall that the gamma function doesn't vanish.

c) We see that

$$\lim_{n \to \infty} \int_{\mathbf{R}} g_n(x)\, dx = 1 \tag{2.3.12}$$

for

$$g_n(x) = \sqrt{\frac{n}{\pi}}\, e^{-nx^2}$$

(e.g. Exercise 1.1.1 a.ii); and note that $\Psi \in L^\infty(\mathbf{R})$ by (2.3.10). Finally we show that

$$\lim_{n \to \infty} \Psi * g_{n-}(z) = \Psi(z), \qquad \text{pointwise a.e.} \tag{2.3.13}$$

To do this, write

$$|\Psi(z) - \Psi * g_{n-}(z)| \leqslant \int_{\mathbf{R}} |\Psi(z)|\, |g_-(y) - g_{n-}(y)|\, dy + \int_{\mathbf{R}} |g_{n-}(y)|\, |\Psi(z) - \Psi(y - z)|\, dy.$$

The first integral tends to 0 by (2.3.12) and the fact that $\Gamma(1) = 1$. The second integral tends to 0 for almost all z since Ψ is continuous a.e. and since modulo some trivialities, we only have to worry about small neighborhoods of the origin.

Thus, in the terminology of Theorem 2.1.5, $L(x) = \Psi(x)$ and $L = 1$; and obviously (2.3.11), (2.3.12), and (2.3.13) are the hypotheses required to apply Theorem 2.1.5. Consequently, $\lim_{z \to \infty} \Psi(z) = S$ (recalling again that $\hat{g}_-(0) = \Gamma(1) = 1$); and so by the definition of Ψ, $\sum a_n = S$.

q.e.d.

Remark 1. Theorem 2.3.1 is sharp in the sense that if b_n tends to infinity then there is a non-convergent series which is Abel summable and which satisfies the condition, $|na_n| \leqslant b_n$.

2. [Karamata, 1] (1930) gave a direct proof of Theorem 2.3.1 which depended on Weierstrass' approximation theorem and polynomials with variable, $e^x e^{-e^x}$ (cf. Remark 2 after Theorem 2.1.4). As in Wiener's proof (that we referenced above) of Littlewood's theorem, Karamata was able to complete his proof by an application of Hardy's Tauberian theorem. [Wielandt, 1] (1952) was able to carry out Karamata's approximation technique internally without using Hardy's result. [Delange, 1] and [Korevaar, 1] have also given an elementary proof of Littlewood's theorem; and [Ingham, 1] has made the interesting suggestion that Littlewood's original and technically difficult repeated differentiation procedure and Karamata's approximation procedure do indeed have a non-trivial relation.

2.3.5 Slowly oscillating functions and Pitt's pointwise conclusion to Wiener's Tauberian theorem. In 1913, [Landau, 1] and Hardy and Littlewood generalized Theorem 2.3.1 to read: *if* $f(x) = \sum a_n x^n$ *converges on* $[0,1)$, $f(1-) = S$, *and*

(2.3.14) $\quad \exists\, K$ *such that* $\forall\, n, \quad na_n \geqslant -K,$

then $\sum a_n = S$, e.g. [Wiener, 7, pp. 108–111]. Actually, Landau's condition was even weaker than (2.3.14) and perhaps led to R. Schmidt's (1925) notion of a slowly oscillating function. $\Phi: \mathbf{R} \to \mathbf{C}$ is *slowly oscillating* if

$$\forall\, \varepsilon > 0\ \exists\, N \text{ and } \exists\, \delta > 0 \text{ such that} \qquad \forall\, |x| > N \text{ and } \forall\, |y| > N,$$

$$|x - y| < \delta \quad \Rightarrow \quad |\Phi(x) - \Phi(y)| < \varepsilon.$$

If $\Phi \in C_b(\mathbf{R})$ is uniformly continuous then Φ is slowly oscillating, and there are discontinuous slowly oscillating functions.

If (2.3.6) is satisfied and Φ is defined by (2.3.1) then Φ is slowly oscillating. In fact, if $y > x$ and we write $y = rx$ where $r > 1$ then

$$|\Phi(y) - \Phi(x)| \leqslant \sum_{n=[x]}^{[y]} |a_n| \leqslant K \sum_{n=[x]}^{[rx]} \frac{1}{n};$$

and the right-hand side is estimated by $K(\log rx - \log x) = K \log r$ when x is large. Also observe that if $f(x) = \sum a_n x^n$ then the condition, $f(1-) = 0$, is written as

$$\lim_{y \to \infty} \int_0^\infty \Phi(xy)\, e^{-x}\, dx = 0 \tag{2.3.15}$$

because of (2.3.2) (where Φ is defined in (2.3.1)). With this background we state R. Schmidt's (1925) generalization of Theorem 2.3.1, e.g. [Wiener, 6, pp. 36–39]: *let* $\Phi \in L^\infty(\mathbf{R})$ *be slowly oscillating and assume that* (2.3.15) *holds*; *then*

$$\lim_{x \to \infty} \Phi(x) = 0. \tag{2.3.16}$$

It was natural to try to generalize Schmidt's result by replacing the function $g(x) = e^{-x}$ in (2.3.15) by an arbitrary element $f \in L^1(0, \infty)$. This attempt fails. In fact, if

$$\varphi(\gamma) = \int_0^\infty x^{i\gamma} f(x)\, dx = 0$$

and we define $\Phi(x) = x^{i\gamma}$, then

$$\lim_{y \to \infty} \int_0^\infty f(x)(yx)^{i\gamma} dx = 0,$$

whereas $\lim_{x\to\infty} \Phi(x) \neq 0$. Note that if $x = e^y$ then

$$\int_0^\infty x^{i\gamma} f(x)\, dx = \hat{g}(\gamma),$$

where $g(y) = f(e^y)e^y$. We are now in a position to realize, because of Theorem 1.1.3 and Theorem 2.1.5, that we really couldn't expect this generalization to hold. What is true is the following Tauberian theorem which is a corollary to Theorem 1.1.3. The "pointwise" conclusion is due to [Pitt, 1].

Theorem 2.3.2. *Let $\Phi \in L^\infty(\mathbf{R})$ be slowly oscillating and assume that $\hat{f} = \varphi \in A(Я)$ never vanishes. If*

(2.3.17) $$\lim_{x \to \infty} f * \Phi(x) = r \int_{\mathbf{R}} f(t)\, dt$$

then

(2.3.18) $$\lim_{x \to \infty} \Phi(x) = r.$$

Proof. Given $\varepsilon > 0$, and choose N and δ as in the definition of a slowly oscillating function. We see that

$$\forall\, x, \qquad \Phi(x) - g * \Phi(x) = \frac{1}{\delta} \int_{-\delta/2}^{\delta/2} [\Phi(x) - \Phi(x-y)]\, dy,$$

where $g = \frac{1}{\delta} \chi_{[-\delta/2, \delta/2]}$.

Thus,

$$\forall\, x \in [-2N, 2N]^\sim, \qquad |\Phi(x) - g * \Phi(x)| < \varepsilon.$$

Consequently, since $\lim_{x \to \infty} g * \Phi(x) = r \int g = r$, by Theorem 1.1.3 (or Theorem 2.1.5), we conclude with (2.3.18).

q.e.d.

Since the above proof requires no special properties of **R**, and because our definition of a slowly oscillating function can obviously be made on G, we see that the statement of Theorem 2.3.2 is valid when "**R**" and "**Я**" are replaced by "G" and "Γ".

The conclusion to Theorem 2.3.2 is what one would expect in a generalization of Theorem 2.3.1. The hypothesis that Φ is slowly oscillating is the Tauberian condition; and (2.3.17), as a convolution, represents an hypothesis concerning information in the average, and, so, corresponds to the hypothesis, $f(1-) = S$, in Theorem 2.3.1. Recall Exercise 2.1.3b which is the Abelian theorem to which Theorem 2.3.2 addresses itself.

2.3.6 A remark on Tauberian theorems. Wiener's Tauberian theorem, Theorem 2.1.5, relates summability methods instead of relating a summability method such as (2.3.17) with pointwise convergence such as (2.3.18). Thus Theorem 2.1.4 (or 2.1.5) has the form: if a function Φ has a mean of a certain type at infinity and if the growth of the function is properly restricted then the function has a large class of means at infinity (cf. Section 2.1.10). There is a large literature on Tauberian theorems and we list the following classics: [Bochner and Chandrasekharan, 1, Section 17 of Chapter 1 and Chapter 6; Hardy, 1, Chapters 6, 7, and 12; Pitt, 2; Widder, 1, Chapter 5; Wiener, 6; 7, Chapters 2 and 3].

2.3.7 $\zeta(s)$ and the fundamental theorem of arithmetic. We shall devote the remainder of Paragraph 2.3 to the study of the distribution of primes; our approach shall be one-sidedly Tauberian (sic). Since analytic number theory is so widely advertised we shall show more restraint than usual with historical and motivational notes for this topic.

Proposition 2.3.4 is an analytical statement of the *fundamental theorem of arithmetic*. Let P denote the set $\{2, 3, 5, \ldots, 2^{127}-1, \ldots, 2^{1279}-1, \ldots\}$ of primes.

Proposition 2.3.4. (Euler) *For each* $s = \sigma + i\gamma \in \mathbf{C}$, $\sigma > 1$,

$$\zeta(s) = \sum_{n=1}^{\infty} \frac{1}{n^s} = \prod_{p \in P} (1 - p^{-})^{-1}. \tag{2.3.19}$$

Proof. Fix s. Setting

$$Q(x) = \prod_{p \in P,\, p \leqslant x} (1 + p^{-s} + p^{-2s} + \ldots),$$

and expanding the product, we obtain

$$Q(x) = \sum_{n \in S_x} \frac{1}{n^s},$$

where S_x consists of those positive integers which have no prime factor greater than x.

Thus, if $S = \sum_{n=1}^{\infty} 1/n^s$,

$$Q(x) - S = - \sum_{n \in R_x} \frac{1}{n^s},$$

where R_x consists of those positive integers which have at least one prime factor greater than x.

Since $n > x$ if $n \in R_x$ we have

$$|Q(x) - S| \leqslant \sum_{n > x} \frac{1}{n^s},$$

and so

$$\sum_{n=1}^{\infty} \frac{1}{n^s} = \prod_{p \in P} (1 + p^{-s} + p^{-2s} + \ldots).$$

Consequently, we obtain (2.3.19) by properties of the geometric series.

q.e.d.

2.3.8 Number theoretic functions. Define

$$\forall\, x \geqslant 0,\ \pi(x) = \operatorname{card}\{p \in P : p \leqslant x\},$$

$$\forall\, n \in \mathbf{Z}^+, \qquad \Lambda(n) = \begin{cases} \log p & \text{for } \ n = p^k \text{ for some } p \in P \text{ and positive integer } k, \\ 0 & \text{for } \ n \neq p^k \text{ for any } p \in P \text{ and positive integer } k, \end{cases}$$

and

$$\forall\, x \geqslant 0, \qquad \psi(x) = \sum_{n \leqslant x} \Lambda(n).$$

Proposition 2.3.5.

$$\text{(2.3.20)} \qquad \underline{\lim}_{x \to \infty} \frac{\psi(x)}{x} \leqslant \underline{\lim}_{x \to \infty} \frac{\pi(x) \log x}{x} \leqslant \overline{\lim}_{x \to \infty} \frac{\pi(x) \log x}{x} \leqslant \overline{\lim}_{x \to \infty} \frac{\psi(x)}{x}.$$

Proof. i) We first prove that

$$\text{(2.3.21)} \qquad \psi(x) = \sum_{p \in P,\, p \leqslant x} \left[\frac{\log x}{\log p}\right] \log p.$$

To do this, write

$$\psi(x) = \sum_{p^m \leqslant x} \log p.$$

Now, for a given $p \leqslant x$, $\log p$ is added to itself k times where $p < p^2 < \ldots < p^k \leqslant x < p^{k+1}$. Taking the log of "$p^k \leqslant x < p^{k+1}$" we obtain $k \leqslant \dfrac{\log x}{\log p} \leqslant k+1$, i.e. $k = \left[\dfrac{\log x}{\log p}\right]$. This gives (2.3.21).

ii) To prove the first inequality of (2.3.20) we have from (2.3.21) that

$$\psi(x) \leqslant \sum_{p \leqslant x} \log x = \pi(x) \log x.$$

Dividing by x yields the result.

iii) If $x > e$, define $y = x/(\log x)^2$ (and so $x > y$).

We compute

$$\pi(x) = \pi(y) + \sum_{y<p\leqslant x} \frac{\log p}{\log p} \leqslant \pi(y) + \frac{1}{\log y} \sum_{y<p\leqslant x} \log p \leqslant y + \frac{\psi(x)}{\log y}.$$

Therefore,

$$\frac{\pi(x)\log x}{x} \leqslant \frac{y \log x}{x} + \frac{\psi(x)\log x}{x \log y} = \frac{1}{\log x} + \frac{\psi(x)\log x}{x} \cdot \frac{1}{\log x - 2\log\log x}$$

$$\leqslant \frac{1}{\log x} + \frac{\psi(x)}{x} \frac{1}{1 - (2\log\log x)/\log x};$$

and the last inequality in (2.3.20) follows.

q.e.d.

2.3.9 The prime number theorem. The *prime number theorem* asserts that

$$\text{(PNT)} \qquad \pi(x) = \frac{x}{\log x} + o\left(\frac{x}{\log x}\right), \qquad x \to \infty.$$

Because of Proposition 2.3.5, (PNT) will follow from

Theorem 2.3.3 (*prime number theorem*)

(2.3.22) $\quad \psi(x) = x + o(x), \qquad x \to \infty.$

In order to prove (2.3.22) we need Proposition 2.3.6, Proposition 2.3.7, and Proposition 2.3.8. We refer to [Donoghue, 1, p. 243] for another and interesting proof of Proposition 2.3.6b. Also, we balance the devious trick in the proof of Proposition 2.3.6b by referring to [Titchmarsh, 1, p. 39] for an exposition of Hadamard's basic idea.

2.3.10 Fundamental properties of $\zeta(s)$.

Proposition 2.3.6. a) (Riemann) *$\zeta(s)$ is analytic for $\sigma > 0$ except for a pole of order* 1 *and residue* 1 *at* $s = 1$.

b) (Hadamard *and* de la Vallée-Poussin)

$$\forall \gamma, \qquad \zeta(1+i\gamma) \neq 0.$$

Proof. a) Clearly, if $\sigma > 1$,

$$\zeta(s) = s\int_1^\infty \frac{[x]\,dx}{x^{s+1}} = s\int_1^\infty \frac{[x]-x}{x^{s+1}}\,dx + s\int_1^\infty \frac{dx}{x^s} = s\int_1^\infty \frac{[x]-x}{x^{s+1}}\,dx + \frac{s}{1-s}. \tag{2.3.23}$$

The expression (2.3.23) converges for $\sigma > 0$, $s \neq 1$.

If $\sigma_0 > 0$ is fixed, then

$$\forall\, \sigma \geqslant \sigma_0 \text{ and } \forall\, x \in [1,\infty), \qquad \frac{x-[x]}{|x^{s+1}|} \leqslant \frac{1}{x^{\sigma_0}};$$

and so the integral in (2.3.23) converges uniformly, and this yields the analyticity. The pole-residue information is obvious from (2.3.23).

b. i) By the Taylor series expansion of $\log(1+z)$ we have

$$\log\left(1-\frac{1}{p^s}\right) = -\sum_{n=1}^\infty \frac{1}{n}\frac{1}{p^{ns}}.$$

Thus, from (2.3.19),

$$\log\zeta(s) = \sum_{p\in P}\sum_{n=1}^\infty \frac{1}{n}\frac{1}{p^{ns}}, \quad \sigma > 1;$$

and so

$$\log|\zeta(s)| = \operatorname{Re}(\log\zeta(s)) = \operatorname{Re}\left(\sum_{p\in P}\sum_{n=1}^\infty \frac{1}{n}\frac{1}{p^{ns}}\right) \tag{2.3.24}$$

$$= \sum_{p\in P}\sum_{n=1}^\infty \frac{1}{n}\frac{1}{p^{n\sigma}}\cos(n\gamma\log p), \qquad s = \sigma + i\gamma, \sigma > 1.$$

ii) Set

$$f(s) = |\zeta(\sigma)|^{3/4}|\zeta(\sigma+i\gamma)|\,|\zeta(\sigma+2i\gamma)|^{1/4}, \; s = \sigma+i\gamma, \; \sigma > 1.$$

By (2.3.24)

$$\log f(s) = \sum_{p\in P}\sum_{n=1}^\infty \frac{1}{n}\frac{1}{p^{n\sigma}}\left\{\frac{3}{4} + \cos(n\gamma\log p) + \frac{1}{4}\cos(2n\gamma\log p)\right\}.$$

Hark!

$$\frac{3}{4} + \cos\lambda + \frac{1}{4}\cos 2\lambda = \frac{1}{2}(1+\cos\lambda)^2 \geqslant 0.$$

Thus, $\log f(s) \geqslant 0$, and so

$$|(\sigma-1)\,\zeta(\sigma)|^3 \left|\frac{\zeta(\sigma+i\gamma)}{\sigma-1}\right|^4 |\zeta(\sigma+2i\gamma)| \geqslant \frac{1}{\sigma-1},\ \sigma>1. \tag{2.3.25}$$

iii) We assume $\zeta(1+i\gamma_0)=0$ and obtain a contradiction by proving that the left-hand side of (2.3.25) converges to a finite limit as σ tends to 1. Note that $\gamma_0 \neq 0$ because of part a).

From (2.3.23) we compute

$$\lim_{\sigma\to 1+} |(\sigma-1)\,\zeta(\sigma)\,|^3 = 1 \quad\text{and}\quad \lim_{\sigma\to 1+} |\zeta(\sigma+2i\gamma_0)| = |\zeta(1+2i\gamma_0)|.$$

Also

$$\lim_{\sigma\to 1+} \left|\frac{\zeta(\sigma+i\gamma_0)}{\sigma-1}\right|^4 = |\zeta'(1+i\gamma_0)|^4 < \infty,$$

since $\gamma_0 \neq 0$.

q.e.d.

2.3.11 Lambert series. One would not expect a result about prime numbers to be proved completely on an $\varepsilon-\delta$ trip. It is time to count. A *Lambert series* has the form

$$\sum_{n=1}^{\infty} a_n \frac{y^n}{1-y^n}. \tag{2.3.26}$$

Proposition 2.3.7 a) *If* $\sum_1^{\infty} a_n y^n$ *converges for* $|y|<1$ *then*

$$\forall\ |y|<1, \quad \sum_{n=1}^{\infty} a_n \frac{y^n}{1-y^n} = \sum_{n=1}^{\infty} y^n \left(\sum_{m|n} a_m\right). \tag{2.3.27}$$

b) $\forall\ x>0$,

$$\sum_{n=1}^{\infty} (\Lambda(n)-1)\frac{e^{-nx}}{1-e^{-nx}} = \sum_{n=1}^{\infty} (\log n - d(n))\, e^{-nx}$$

where

$$d(n) = \sum_{j|n} 1.$$

c) (Dirichlet)

$$\sum_{n \leqslant y} \mathrm{d}(n) = y \log y + 2(\gamma - 1) y + O(y^{1/2}), \quad y \to \infty,$$

where γ *is Euler's constant.*

d)
$$\sum_{n=1}^{\infty} \frac{\Lambda(n) - 1}{n} \frac{n e^{-nx}}{1 - e^{-nx}} = -\frac{2\gamma}{x} + O(x^{-1/2}), \quad x \to 0+.$$

Proof. a) Formally, by cross multiplying,

$$\frac{y^n}{1 - y^n} = y^n + y^{2n} + y^{3n} + \dots.$$

Thus, (2.3.26) is $\sum_{n=1}^{\infty} a_n(y^n + y^{2n} + \dots)$; and by regrouping we obtain (2.3.27).

These formal operations are legitimate if $\sum a_n y^n$ converges for $|y| < 1$; in fact, this hypothesis obviously entails the absolute convergence of (2.3.26) which, in turn, allows for the rearrangement.

b) If $a_n = 1$ for each n, part a) immediately yields (for $y = e^{-x}$)

$$\text{(2.3.28)} \qquad \forall\, x > 0, \qquad \sum_{n=1}^{\infty} \frac{e^{-nx}}{1 - e^{-nx}} = \sum_{n=1}^{\infty} \mathrm{d}(n)\, e^{-nx}.$$

Now let $a_m = \Lambda(m)$. Fix a positive integer $n = \prod p^k$, where the product is the unique factorization of n. If $m | n$ and $\Lambda(m) \neq 0$ then $m = p^k$; consequently, by the definition of Λ,

$$\text{(2.3.29)} \qquad \sum_{m|n} \Lambda(m) = \log n.$$

b) follows when we combine (2.3.29) with (2.3.27) and (2.3.28).

c) This is Exercise 2.3.4a.

d) Because of b) we use partial summation on $\sum (\log n - \mathrm{d}(n))\, e^{-nx}$. Thus, using *Stirling's formula*, $n! \sim n^n e^{-n} \sqrt{2\pi n}$, $n \to \infty$ (e.g. [Widder, 2, pp. 139–140]), and c) we compute

$$\text{(2.3.30)} \qquad \sum_{n=1}^{\infty} (\log n - \mathrm{d}(n))\, e^{-nx} = (1 - e^{-x}) \sum_{n=1}^{\infty} e^{-nx}(-2\gamma n + O(n^{1/2})).$$

Set $y = e^{-x}$ and calculate $\sum_{n=1}^{\infty} n y^n = \sum_{0}^{\infty} (n+1) y^n - 1 - \sum_{1}^{\infty} y^n = y/(1-y)^2$.

Thus,

$$\text{(2.3.31)} \qquad -2\gamma(1 - e^{-x}) \sum_{n=1}^{\infty} n e^{-nx} = -2\gamma \frac{y}{1-y} = \frac{-2\gamma}{e^x - 1}.$$

Since $\lim_{x\to\infty}(e^x - 1)/x = 1$ we shall replace $(e^x - 1)$ by x in our final calculation.

In Exercise 2.3.4b we prove that

$$\text{(2.3.32)} \qquad \exists\, K \text{ such that } \forall\, y \in [0,1), \qquad (1-y)\left(\log\frac{1}{y}\right)^{1/2} \sum_{n=1}^{\infty} n^{1/2} y^n \leqslant K.$$

Combining (2.3.31) and (2.3.32) with the right-hand side of (2.3.30), we obtain d).

q.e.d.

2.3.12 Theorems of Chebyshev and Mertens. Define

$$\forall\, x \geqslant 1, \qquad H(x) = \sum_{n \leqslant x} \frac{\Lambda(n) - 1}{n}$$

and

$$\Phi_H(x) = \begin{cases} H(e^x) & \text{for } x \geqslant 0, \\ 0 & \text{for } x < 0. \end{cases}$$

Proposition 2.3.8 a) (Chebyshev) $\psi(x) = O(x), \quad x \to \infty.$
b) (Mertens) $\Phi_H \in L^\infty(\mathbf{R})$.
c) $f * \Phi_H(x) = -2\gamma + O(e^{-(1/2)x}), \quad x \to \infty$, *where*

$$f(x) = \frac{d}{dx}\frac{e^{-x}\exp(-e^{-x})}{1 - \exp(-e^{-x})}.$$

Proof. a) This is Exercise 2.3.4c.
b.i) We first compute how many times $t(p,m)$ a prime p divides $m!$ (i.e., $p^{t(p,m)} | m!$ but $p^{t(p,m)+1} \nmid m!$).
In the set $\{1,\ldots, m\}$ there are exactly $[m/p]$ integers which are divisible by p, viz.

$$\text{(2.3.33)} \qquad \left\{p, 2p, 3p, \ldots, \left[\frac{m}{p}\right]p\right\}.$$

In the set $\{1,\ldots,m\}$ there are exactly $[m/p^2]$ integers which are divisible by p^2, viz.

$$\text{(2.3.34)} \qquad \left\{p^2, 2p^2, \ldots, \left[\frac{m}{p^2}\right]p^2\right\}.$$

The set in (2.3.34) is a subset of that in (2.3.33).
Let S_n be the subset of $\{1,\ldots,m\}$ each of whose elements is divisible by p^n but not by p^{n+1}. Then card S_n is exactly

$$\left[\frac{m}{p^n}\right] - \left[\frac{m}{p^{n+1}}\right];$$

and so p divides S_n exactly $n\left(\left[\frac{m}{p^n}\right]-\left[\frac{m}{p^{n+1}}\right]\right)$ times.

Clearly, $\{1,\ldots,m\} = \bigcup_{n\geq 1} S_n$ and $\{S_n : n \geq 1\}$ is a disjoint collection.

Thus

$$t(p,m) = \sum_{n=1}^{\infty} n\left(\left[\frac{m}{p^n}\right]-\left[\frac{m}{p^{n+1}}\right]\right) = \sum_{n=1}^{\infty}\left[\frac{m}{p^n}\right], \tag{2.3.35}$$

a finite sum.

ii) The unique factorization of m! is

$$m! = \prod_{p\in P,\, p\leq m} p^{t(p,m)}.$$

We compute

$$\log m! = \sum_{p\in P,\, p\leq m} t(p,m)\log p = \sum_{p\in P,\, p^n\leq m}\left[\frac{m}{p^n}\right]\log p = \sum_{n=1}^{m}\left[\frac{m}{n}\right]\Lambda(n). \tag{2.3.36}$$

Setting $\frac{m}{n} = \left[\frac{m}{n}\right] + \varepsilon_n$, $0 \leq \varepsilon_n < 1$, we have from (2.3.36) that

$$m\sum_{n=1}^{m}\frac{\Lambda(n)}{n} = \sum_{n=1}^{m}\varepsilon_n\Lambda(n) + \log m!. \tag{2.3.37}$$

Applying Stirling's formula to the right-hand side of (2.3.37) yields

$$\sum_{n=1}^{m}\frac{\Lambda(n)}{n} = \frac{1}{m}\sum_{n=1}^{m}\varepsilon_n\Lambda(n) + \log m + O(1), \qquad m \to \infty.$$

From the definition of $\psi(x)$,

$$\frac{1}{m}\sum_{n=1}^{m}\varepsilon_n\Lambda(n) \leq \frac{\psi(m)}{m};$$

and so by part a),

$$\sum_{n=1}^{m}\frac{\Lambda(n)}{n} = \log m + O(1), \qquad m \to \infty. \tag{2.3.38}$$

On the other hand, it is a result from calculus that

$$\sum_{n=1}^{m}\frac{1}{n} = \log m + O(1), \qquad m \to \infty. \tag{2.3.39}$$

Subtracting (2.3.39) from (2.3.38) entails the desired boundedness.

c) We first use Proposition 2.3.7d to compute

$$(2.3.40)\qquad -\int_1^\infty H(t)\frac{\mathrm{d}}{\mathrm{d}t}\left(\frac{u t\,\mathrm{e}^{-tu}}{1-\mathrm{e}^{-tu}}\right)\mathrm{d}t = u\int_1^\infty \frac{t\,\mathrm{e}^{-tu}}{1-\mathrm{e}^{-tu}}\,\mathrm{d}H(t) = u\sum_{n=1}^\infty \frac{\Lambda(n)-1}{n}\,\frac{n\,\mathrm{e}^{-nu}}{1-\mathrm{e}^{-nu}}$$

$$= -2\gamma + O(u^{1/2}),\qquad u \to 0+.$$

Let $t=\mathrm{e}^y$ and $u=\mathrm{e}^{-x}$ in (2.3.40). This yields part c).

q.e.d.

2.3.13 Wiener's Tauberian theorem and the proof of the prime number theorem.

Proof. (cf. Theorem 2.3.3) i) Take f as in Proposition 2.3.8c.

We evaluate

$$(2.3.41)\qquad \int_{-\infty}^\infty f(x)\,\mathrm{d}x = \lim_{x\to+\infty}\frac{\mathrm{e}^{-x}\mathrm{e}^{-\mathrm{e}^{-x}}}{1-\mathrm{e}^{-\mathrm{e}^{-x}}} - \lim_{x\to-\infty}\frac{\mathrm{e}^{-x}\mathrm{e}^{-\mathrm{e}^{-x}}}{1-\mathrm{e}^{-\mathrm{e}^{-x}}}$$

$$= \lim_{x\to\infty}\frac{-\mathrm{e}^{-x}\mathrm{e}^{-\mathrm{e}^{-x}}+\mathrm{e}^{-2x}\mathrm{e}^{-\mathrm{e}^{-x}}}{-\mathrm{e}^{-x}\mathrm{e}^{-\mathrm{e}^{-x}}} = \lim_{x\to\infty}(1+\mathrm{e}^{-x}) = 1.$$

ii) Take $\varepsilon>0$. Integrating by parts we compute

$$\hat f(-\gamma+i\varepsilon) = \int_{-\infty}^\infty f(x)\,\mathrm{e}^{-x(\mathrm{i}\gamma+\varepsilon)}\,\mathrm{d}x = (\mathrm{i}\gamma+\varepsilon)\int_{-\infty}^\infty \frac{\mathrm{e}^{-x}\mathrm{e}^{-\mathrm{e}^{-x}}}{1-\mathrm{e}^{-\mathrm{e}^{-x}}}\,\mathrm{e}^{-x(\mathrm{i}\gamma+\varepsilon)}\,\mathrm{d}x;$$

so that by the change of variable $y=\mathrm{e}^{-x}$ we obtain

$$\hat f(-\gamma+\mathrm{i}\varepsilon) = (\mathrm{i}\gamma+\varepsilon)\int_0^\infty \frac{\mathrm{e}^{-y}}{1-\mathrm{e}^{-y}}\,y^{(\mathrm{i}\gamma+\varepsilon)}\,\mathrm{d}y$$

$$= (\mathrm{i}\gamma+\varepsilon)\int_0^\infty y^{\mathrm{i}\gamma+\varepsilon}(\mathrm{e}^{-y}+\mathrm{e}^{-2y}+\ldots)\,\mathrm{d}y$$

$$= (\mathrm{i}\gamma+\varepsilon)\sum_{n=1}^\infty\int_0^\infty \mathrm{e}^{-ny}y^{\mathrm{i}\gamma+\varepsilon}\,\mathrm{d}y$$

$$= (\mathrm{i}\gamma+\varepsilon)\sum_{n=1}^\infty \frac{1}{n^{\mathrm{i}\gamma+\varepsilon+1}}\int_0^\infty \mathrm{e}^{-y}y^{\mathrm{i}\gamma+\varepsilon}\,\mathrm{d}y$$

$$= (\mathrm{i}\gamma+\varepsilon)\Gamma(1+\varepsilon+\mathrm{i}\gamma)\zeta(1+\varepsilon+\mathrm{i}\gamma).$$

iii) We let ε tend to 0.

There is no problem about switching limits because of the smoothness of f.

Thus,

$$\hat{f}(-\gamma) = \mathrm{i}\gamma\Gamma(1+\mathrm{i}\gamma)\zeta(1+\mathrm{i}\gamma). \tag{2.3.42}$$

We note that Γ never vanishes, the factor "iγ" cancels the pole of $\zeta(1+\mathrm{i}\gamma)$ at $\gamma = 0$ (this sort of thing would have to happen since $f \in L^1(\mathbf{R})$), and $\zeta(1+\mathrm{i}\gamma)$ never vanishes (Proposition 2.3.6).

Hence, $\hat{f}$ never vanishes.

iv) Also,

$$\lim_{x\to\infty} f * \Phi_H(x) = -2\gamma \int_{-\infty}^{\infty} f(y)\,\mathrm{d}y.$$

v) Because of parts iii) and iv), we can apply Wiener's Tauberian theorem. We could apply it in the form of Theorem 2.1.4 since (2.1.33) is obviously satisfied in our case. For technical convenience we apply it in the form of Theorem 2.3.2.

We must check that Φ_H is slowly oscillating. This is easy.

Take $\delta < 1$ so that if $0 \leqslant y - x < \delta$ then $\Phi_H(y) - \Phi_H(x) = 0$ or

$$-\frac{1}{n} \leqslant \Phi_H(y) - \Phi_H(x) \leqslant \frac{\Lambda(n)}{n} < \frac{\log p}{p}$$

in the case that $n \in [x, y]$.

By Theorem 2.3.2,

$$\lim_{x\to\infty} H(x) = \sum_{n=1}^{\infty} \frac{\Lambda(n) - 1}{n} = -2\gamma. \tag{2.3.43}$$

vi) Clearly,

$$H(x) = \int_{1/2}^{x} \frac{\mathrm{d}(\psi(t) - [t])}{t} \tag{2.3.44}$$

and, generally,

$$\int_0^t u\,\mathrm{d}(H(u) + 2\gamma) = t(H(t) + 2\gamma) - \int_0^t (H(u) + 2\gamma)\,\mathrm{d}u. \tag{2.3.45}$$

Since $\lim_{u\to\infty}(H(u) + 2\gamma) = 0$, the mean, $\frac{1}{t}\int_0^t (H(u) + 2\gamma)\,\mathrm{d}u$, also tends to 0. Thus, (2.3.45) becomes

(2.3.46) $$\lim_{t \to \infty} \frac{1}{t}(\psi(t) - [t]) = 0$$

with the substitution of (2.3.44).

We are done.

q.e.d.

We have paid our homage to Lambert series. Our proof of (2.3.22) showed that convergence of the "Lambert mean" implied the convergence of the Cesàro mean, $\frac{1}{N}\sum_{1}^{N} \Lambda(n)$.

2.3.14 Perspective

Remark 1. As we've seen, Wiener's proof of (PNT) required knowledge of the behavior of ζ only on the line $\sigma = 1$. The first proofs of (PNT) required that ζ have no zeros in a region to the *left* of the line $\sigma = 1$. Our proof generally follows the guidelines set in [Levinson, 3; Widder, 1, Chapter V, Section 16; Wiener, 7, pp. 112 ff.]. As we've indicated in Paragraph 2.1, Wiener was led to his method of proof because of Hardy and Littlewood's result that (PNT) is equivalent to a Tauberian theorem concerning Lambert series, and because of their failure to give an independent proof of this latter result and thus obtain a proof of (PNT). Wiener first published his proof in [Wiener, 4], although it is smoother going in [Wiener, 7].

2. One of the most efficient means of proving (PNT) (and also the prime number theorem for primes in an arithmetic progression) is by means of Ikehara's complex variable Tauberian theorem [Chandrasekharan, 1; Ikehara, 1; Widder, 1; Wiener, 7]. A deep extension of Ikehara's work in complex variable Tauberian theorems is due to [Agmon, 1]. Essentially, with the added analytic structure, a complex variable Tauberian theorem will describe the asymptotic growth of the coefficients a_n (in (2.3.1), say), whereas the Tauberian theorems that we have proved only give information on the asymptotic growth of the partial sums, $\sum_{0}^{k} a_n$.

2.3.15 The prime number theorem and the Riemann hypothesis. The *Riemann hypothesis*, (RH), asserts that if $\zeta(s) = 0$, for $\sigma > 0$, then $\sigma = 1/2$. It is not known whether or not (RH) is true. (RH) can be considered as a strong form of (PNT) because of the following well-known result, e.g. [Blanchard, 1, pp. 89–93].

Theorem 2.3.4. (RH) *is valid* $\Leftrightarrow$ *for each* $\varepsilon > 0$

(2.3.47) $$\psi(x) = x + O(x^{\varepsilon + (1/2)}), \quad x \to \infty$$

(cf. (2.3.22)).

Actually, the original Hadamard and de la Vallée–Poussin (PNT) is equivalent to

(2.3.48) $\quad \forall\, m > 0, \quad \psi(x) = x + O(x/\log^m x), \quad x \to \infty$

[Bombieri, 1]. There are even more refined estimates than (2.3.48) and we mention it only as a prototype in which to compare (PNT) with (RH).

2.3.16 Tauberian theorems and the Riemann hypothesis. [Wiener, 6] was aware that (RH) is valid if and only if

(2.3.49) $\quad \forall\, \sigma \in (\tfrac{1}{2}, 1), \quad V_{f_\sigma} = L^1(\mathbf{R}),$

where

$$f_\sigma(x) = e^{(\sigma-1)x} \frac{d}{dx}\left(\frac{-e^x}{1 - \exp e^x}\right).$$

The prognoses in [Levinson, 1, p. 840; Wiener, 6, pp. 49–50; 7, pp. 136–137] on the usefulness of Tauberian theorems to prove the validity of (RH) are negative, although nonetheless there have been several interesting results related to (2.3.49) that have been proved since Wiener's work; in particular, we give Theorem 2.3.5 and Theorem 2.3.6 below. These "prognoses" are associated with the *numerical Tauberian problem* of determining how fast one summability method converges in the conclusion of a Tauberian theorem if in the hypotheses (of the Tauberian theorem) we know how fast the given summability method converges. This problem is also called the *Tauberian remainder problem* and there is an extensive literature associated with it; we refer to [Beurling, 1, pp. 364–365; Ganelius, 2; Korevaar, 1, Section 7 and Section 8; Lyttkens, 1; 2] (e.g. Exercise 2.3.5) and move on.

2.3.17 Salem's Tauberian characterization of the Riemann hypothesis.

Theorem 2.3.5. ([Salem, 3]) *Define*

$$\forall\, \sigma \in (\tfrac{1}{2}, 1), \quad f_\sigma(x) = \frac{e^{\sigma x}}{1 + \exp e^x}.$$

The following are equivalent:

a) (RH) *is valid*;

b) $\forall\, \sigma \in (\tfrac{1}{2}, 1)$, $V_{f_\sigma} = L^1(\mathbf{R})$;

c) $\forall\, \sigma \in (\tfrac{1}{2}, 1)$ and $\forall\, \Phi \in L^\infty(\mathbf{R})$,

$$f_\sigma * \Phi = 0 \quad \Rightarrow \quad \Phi = 0.$$

Proof. In light of Theorem 1.1.3 and Theorem 1.3.1c, it is sufficient to prove that for $\sigma \in (\tfrac{1}{2}, 1)$,

(2.3.50) $\quad \forall\, \gamma \in \hat{\mathbf{R}}, \quad \hat{f_\sigma}(\gamma) = \Gamma(s)(1 - 2^{1-s})\zeta(s)$

(recalling that Γ never vanishes).

By the definition of Γ,

$$\forall\, \sigma > 0, \qquad \frac{\Gamma(s)}{n^s} = \int_0^\infty x^{s-1}\,\mathrm{e}^{-nx}\,\mathrm{d}x$$

(the result being true if the positive integer n is replaced by any $z \in \mathbf{C}$ for which $\mathrm{Re}\, z > 0$) (cf. Exercise 1.4.4c).

Direct multiplication yields

$$(2.3.51) \qquad \forall\, \sigma > 1, \qquad (1 - 2^{1-s})\,\zeta(s) = \sum_{n=1}^{\infty} \frac{(-1)^{n+1}}{n^s}.$$

Now for each $\sigma > 0$, $\{1/n^\sigma : n = 1, \ldots\}$ decreases to 0 and so the alternating series $\sum (-1)^{n+1}/n^\sigma$ converges for $\sigma > 0$. Consequently, by the fundamental convergence property of Dirichlet series, the right-hand side of (2.3.51) converges for $\sigma > 0$. Therefore, because of Proposition 2.3.6a,

$$(2.3.52) \qquad \forall\, \sigma > 0, \qquad \Gamma(s)(1 - 2^{1-s})\,\zeta(s) = \sum_1^\infty (-1)^{n+1} \left(\int_0^\infty x^{s-1}\,\mathrm{e}^{-nx}\,\mathrm{d}x \right).$$

Now, $x^{\sigma-1}\mathrm{e}^{-x} \in L^1(0, \infty)$ if $\sigma > 0$ and

$$0 \leqslant 1 - \frac{1}{\mathrm{e}^x} + \frac{1}{\mathrm{e}^{2x}} - \frac{1}{\mathrm{e}^{3x}} + \ldots (-1)^{n+1} \frac{1}{\mathrm{e}^{nx}} \leqslant 1;$$

and so we can switch operations in (2.3.52) by the dominated convergence theorem. Thus, since $\sum_1 (-1)^{n+1} z^n = z/(1+z)$, (2.3.52) becomes

$$\forall\, s = \sigma + \mathrm{i}\gamma, \sigma > 0, \qquad \Gamma(s)(1 - 2^{1-s})\,\zeta(s) = \int_0^\infty \frac{x^{s-1}}{1 + \mathrm{e}^x}\,\mathrm{d}x = \hat{f}_\sigma(\gamma),$$

by the substitution $x = \mathrm{e}^y$; this is (2.3.50).

q.e.d.

2.3.18 Beurling's functions and a spectral analysis problem. Analogous, although more complicated, results have been proved by [Beurling, 10; Levinson, 1]. Because of our concern with spectral analysis problems we shall now comment on Beurling's result (cf. Remark 3 after Exercise 2.1.6).

Let S be the multiplicative semigroup $\{x \in \mathbf{R} : 0 < x \leqslant 1\}$. If $\Phi \in L^2(S)$ then V_Φ^r, $1 \leqslant r \leqslant 2$, will denote the $L^r(S)$-closed variety generated by $\{\Phi(xy) : y \in S\}$. As is well known, each (continuous and normalized) character on S has the form $\chi(x) = x^z, z \in \mathbf{C}$. It is not difficult to check that a character $\chi \in L^2(S)$ if and only if $\mathrm{Re}\, z > -\frac{1}{2}$ (cf. [Paley and Wiener, 1, Chapter II, e.g. p. 30]). The spectral analysis question of whether sets having the form V_Φ^r contain characters is very complex and generally has a negative answer [Beurling, 7; Nyman, 1; Vretblad, 1]. [Beurling, 9] proved the following spectral analysis result with the restrictive condition (2.3.53):

If $\Phi \in L^2(S)$ has the property

$$(2.3.53) \qquad \forall\, x \in S, \qquad \int_0^x |\Phi(t)|\, dt > 0$$

then there is a character $\chi \in L^2(S) \cap \bigcap_{1 \leq r < 2} V_\Phi^r$ which is not necessarily in V_Φ^2.

In order to give Beurling's (RH) theorem it is necessary to define a certain class $\mathscr{B} \subseteq L^\infty(S)$. Define

$$\begin{aligned} a:(0,\infty) &\to [0,1) \\ x &\mapsto x - [x]. \end{aligned}$$

Clearly,

$\forall\, y \in S$ and $\forall\, x \in (0,\infty)$,

$$x > y \;\Rightarrow\; a\left(\frac{y}{x}\right) = \frac{y}{x};$$

and for a fixed $y \in S$ we set

$$\begin{aligned} b_y:(0,\infty) &\to [0,1) \\ x &\mapsto a(y/x). \end{aligned}$$

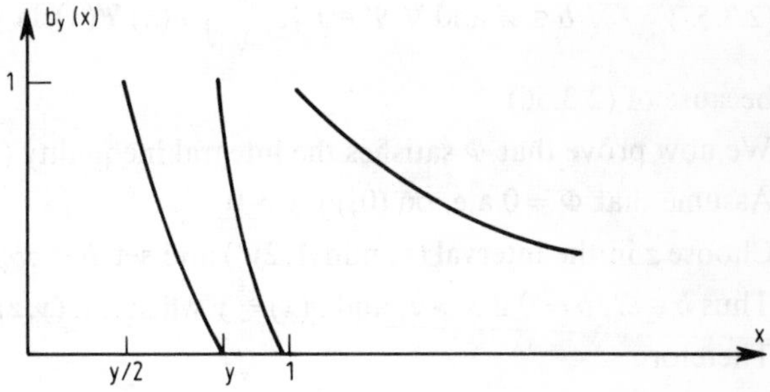

Fig. 1

b_y has a countable set $\{1, \frac{y}{n}: n = 1, \ldots\}$ of jump discontinuities.

Define

$$(2.3.54) \quad \forall\, x \in (0,\infty),\; b(x) = \sum_{y \in F} a_y b_y(x), \quad \text{where card } F < \infty,\; F \subseteq S, \text{ and } a_y \in \mathbf{C};$$

and assume that

$$(2.3.55) \qquad \sum_{y \in F} a_y y = 0.$$

Consequently, if $x > \max\{y : y \in F\}$, then $b(x) = 0$. Therefore, we can finally define $\mathscr{B}$ as the vector space of functions b having the form (2.3.54) and satisfying (2.3.55).

2.3.19 Beurling's Tauberian characterization of the Riemann hypothesis. With these preliminaries, we are in a position to state Beurling's theorem.

Theorem 2.3.6 ([Beurling, 10]). (RH) *is valid* $\Leftrightarrow \overline{\mathscr{B}} = L^2(S)$.

It is not too difficult to prove that (RH) is valid if $\overline{\mathscr{B}} = L^2(S)$; we refer to [Beurling, 10; Donoghue, 1, pp. 252–254] for a proof. We outline the proof of the other and more difficult direction to show the use of (2.3.53).

Outline of Proof. We must prove that if $\overline{\mathscr{B}} \subsetneq L^2(S)$ then $\zeta(\sigma_0 + i\gamma_0) = 0$ for some $\sigma_0 \in (\frac{1}{2}, 1)$, $\gamma_0 \in \mathbb{R}$.

By the Hahn–Banach theorem there is $\Phi \in L^2(S) \setminus \{0\}$ such that

$$\forall f \in L^2(S), \qquad \int_S f(x)\,\Phi(x)\,dx = 0.$$

It is then easy to check that

$$\text{(2.3.56)} \qquad \forall\, b \in \mathscr{B} \text{ and } \forall\, y \in S, \qquad y \int_S b(x)\,\Phi(xy)\,dx = 0.$$

Define V_Φ^r, $1 \leqslant r \leqslant 2$, as above. Since $\mathscr{B} \subseteq L^\infty(S)$, we see that for any $r \in [1,2]$,

$$\text{(2.3.57)} \qquad \forall\, b \in \mathscr{B} \text{ and } \forall\, \Psi \in V_\Phi^r, \qquad \int b(x)\,\Psi(x)\,dx = 0,$$

because of (2.3.56).

We now prove that Φ satisfies the integral inequality (2.3.53).

Assume that $\Phi = 0$ a.e. on $(0, y)$, $y > 0$.

Choose z in the interval $(y, \min\{1, 2y\})$ and set $b = zb_y - yb_z$.

Thus $b \in \mathscr{B}$, $b = 0$ if $x > z$, and $b(x) = y$ when $x \in (y, z)$.

Therefore

$$0 = \int_S b(x)\,\Phi(x)\,dx = y \int_y^z \Phi(x)\,dx,$$

and so $\Phi = 0$ a.e. in the interval $(y, \min\{1, 2y\})$.

We repeat the argument a finite number of times and see that $\Phi = 0$ a.e. on S, a contradiction.

Consequently, we can apply **Beurling's** result, (2.3.53), which, when coupled with (2.3.57), yields that

$$\text{(2.3.58)} \qquad \exists\, z = (\sigma_0 - 1) + i\gamma_0, \qquad \operatorname{Re}(\sigma_0 - 1) > -\frac{1}{2}, \qquad \text{such that } \forall\, b \in \mathscr{B},$$

$$\int_S b(x)\,x^z\,dx = 0.$$

Define $b = b_1 - \dfrac{1}{y} b_y$, and, by direct calculation, we obtain from (2.3.58) (for $z = s_0 - 1$) that

$$\forall\, y \in (0,1), \qquad \frac{\zeta(s_0)}{s_0}(1 - y^{s_0 - 1}) = 0.$$

If $s_0 = 1$ then $\int_S b(x)\,dx = 0$, which is not true. Consequently, we choose $y \in (0,1)$ such that $y^{s_0-1} \neq 1$, and so $\zeta(s_0) = 0$. Hence, (RH) is not valid since $\sigma_0 > 1/2$.

"q.e.d."

Exercises 2.3

2.3.1 Tauber's theorem

Prove Tauber's Tauberian theorem directly. (*Hint*: Note that for each $\varepsilon > 0$ there is N such that if $n \geqslant N$ we have $|a_n n| < \varepsilon/3$,

$$\frac{1}{n}\sum_1^n |ka_k| < \varepsilon/3,$$

and

$$\forall\, x \in \left(1 - \frac{1}{N}, 1\right), \quad \left|S - \sum_0^\infty a_k x^k\right| < \varepsilon/3;$$

then for $n \geqslant N$ and the appropriate choice of x,

$$\left|S - \sum_0^n a_k\right| \leqslant \left|S - \sum_0^\infty a_k x^k\right| + \sum_1^n |a_k|\,|x^k - 1| + \sum_{n+1}^\infty |a_k|\, x^k < \varepsilon).$$

2.3.2 Hardy's Tauberian theorem

Prove Hardy's Tauberian theorem directly, e.g. [Edwards, 5, I, p. 86].

2.3.3 $\zeta(1 + i\gamma) \neq 0$

Use (2.3.22) and *not* the razzle-dazzle of Proposition 2.3.6b to prove that $\zeta(1 + i\gamma) \neq 0$ for all $\gamma \in \mathbb{R}$. (*Hint*: Assume $\zeta(1 + \gamma_0)$ has a zero of order m. Then

$$\lim_{s \to 1 + i\gamma_0} \frac{\zeta'(s)}{\zeta(s)} (s - (1 + i\gamma_0)) = m.$$

Compute

$$-\frac{\zeta'(s)}{\zeta(s)} = \frac{s}{s-1} + s\int_1^\infty \frac{\psi(x) - x}{x^{s+1}}\,dx, \qquad \sigma > 1$$

(cf. (2.3.23)), and obtain a contradiction by estimating $|\zeta'(s)(\sigma - 1)/\zeta(s)|$).

2.3.4 Number theoretic estimates

a) Prove Proposition 2.3.7c. This is a well-known elementary and ingenious number theoretic estimate. We refer to [Chandrasekharan, 1, pp. 53–54].

b) Prove (2.3.32). Note that [Widder, 1, p. 231] proves Proposition 2.3.7d in a similar way but with a final appeal to an Abelian theorem.

c) Prove Proposition 2.3.8a. (*Hint*: In light of the proof of Proposition 2.3.5b it is sufficient to find a bound for $\frac{\pi(x)\log x}{x}$. This is done in [Chandrasekharan, 1, pp. 67–68]).

2.3.5 A Tauberian remainder theorem

Let $f \in L^1(\mathbf{R})$ have the property that $L(f)^{-1}$ is analytic in a region $R = \{s = \sigma + i\gamma: |\gamma| < a, a > 0\}$ ("L" is the Laplace transform operator), and let $\Phi \in L^\infty(\mathbf{R})$. Assume that

$$\exists\, K_1, k > 0 \text{ such that } \forall\, s \in R,\ |L(f)(s)^{-1}| \leqslant K_1 e^{k|s|}$$

(cf. Exercise 2.2.3b) and that

$$\exists\, K_2 > 0 \quad \text{and} \quad \exists\, x_0 \in \mathbf{R} \text{ such that } \forall\, x > x_0$$
$$\Phi(x) + K_2 x \text{ is non-decreasing.}$$

If there is $a >$ ɔr which $f * \Phi(x) = O(e^{-ax})$, $x \to \infty$, prove that

$$\Phi(x) = O\left(\frac{1}{x}\right), \quad x \to \infty,$$

e.g. [Ganelius, 1].

2.4 Wiener's inversion of Fourier series

2.4.1 Remarks on Wiener's inversion of Fourier series. We shall now give Wiener's original proof of his theorem on the inversion of Fourier series, viz. Proposition 1.1.5b. Wiener proved Proposition 1.1.5b in [Wiener, 6, p. 14; 7, p. 91]. Theorem 2.4.1b, below, played a major role in this proof; and the property of "local membership" which is used in Theorem 2.4.1 is a valuable concept in harmonic analysis.

The approximate identity results, Proposition 1.1.4 and Theorem 1.2.1, are needed in parts b) and c), respectively, of Theorem 2.4.1; and so it is important to note that the proof of Theorem 1.2.1 does not require Proposition 1.1.5b.

We also mention Wiener's extension, Proposition 1.2.5a, of Proposition 1.1.5b which was essential in the proof of the generalization of Wiener's Tauberian theorem that we gave in Theorem 1.2.3. Proposition 1.2.5a tells us that *if* $Z\psi \subseteq \operatorname{int} Z\varphi$, *where* $\psi, \varphi \in A(\Gamma)$, *then* $\varphi/\psi \in A(\Gamma)$, *where* $\varphi(\gamma)/\psi(\gamma)$ *is defined to be* 0 *if* $\varphi(\gamma) = 0$. ([Wiener, 7, p. 92] and [Mandelbrojt, 1, p. 50] have given a strange acknowledgement of this result to Denjoy and Luzin; in fact, the papers by Denjoy and Luzin that they cite deal with the Denjoy–Luzin theorem, e.g. [Kahane and Salem, 4, Chapitre VII], and [Wiener, 7, pp. 91–92] actually gives a proof of the Denjoy–Luzin theorem after his proof of Proposition 1.2.5a).

Recall from Paragraph 2.1.11 that Wiener's original proof of his Tauberian theorem, Theorem 1.1.3, in 1928, did not use Proposition 1.1.5b and Proposition 1.2.5a, which are intrinsic properties of $A(\Gamma)$, but did assume added integrability conditions such as (2.1.33).

2.4.2 Local membership. Let $I \subseteq A(\Gamma)$ be an ideal. A function $\varphi: \Gamma \to \mathbf{C}$ *belongs to* I *locally at* $\gamma \in \Gamma$ (resp. *at infinity*) if

(2.4.1) $\exists\, \psi_\gamma \in I$ and $\exists\, N_\gamma$, an open neighborhood of γ, such that

$$\forall\, \lambda \in N_\gamma, \qquad \psi_\gamma(\lambda) = \varphi(\lambda)$$

(resp. $\exists\, \psi_\infty \in I$ and $\exists\, K \subseteq \Gamma$, compact, such that

$$\forall\, \lambda \in K^\sim, \qquad \psi_\infty(\lambda) = \varphi(\lambda));$$

in this case we write $\varphi \in I_{\text{loc}}(\gamma)$ (resp. $\varphi \in I_{\text{loc}}(\infty)$). Thus, if G is discrete and $I \subseteq A(\Gamma)$ is an ideal then

$$\forall\, \varphi \in A(\Gamma), \qquad \varphi \in I_{\text{loc}}(\infty).$$

Theorem 2.4.1. *Let* $I \subseteq A(\Gamma)$ *be an ideal and let* $\varphi: \Gamma \to \mathbf{C}$ *be a function.*

a) *If* φ *has compact support then* $\varphi \in I_{\text{loc}}(\infty)$.

b) *If* $\varphi \in I_{\text{loc}}(\gamma)$ *for each* $\gamma \in \Gamma$ *and if* $\varphi \in I_{\text{loc}}(\infty)$ *then* $\varphi \in I$.

c) *Given* $\varphi \in A(\Gamma)$. *If* $\varphi \in I_{\text{loc}}(\gamma)$ *for each* $\gamma \in \Gamma$ *and if* I *is closed then* $\varphi \in I$.

Proof. a) Take $K = \operatorname{supp}\varphi$ and $\psi_\infty = 0$ in (2.4.1).

b) Since $\varphi \in I_{\text{loc}}(\infty)$ there is a compact set L and an element $\psi \in I$ such that $\varphi = \psi$ on $L^\sim$.

Setting $\theta = \varphi - \psi$ we see that $K = \operatorname{supp}\theta$ is compact, that $\theta \in I_{\text{loc}}(\infty)$, by part a), and that $\theta \in I_{\text{loc}}(\gamma)$ for each $\gamma \in \Gamma$. Using this information we'll show that $\theta \in I$; and, thus, $\varphi \in I$.

For each $\gamma \in K$ choose $\psi_\gamma \in I$ and an open neighborhood N_γ of γ for which $\psi_\gamma = \theta$ on N_γ.

By the compactness, let $\{N_1, \ldots, N_n\} \subseteq \{N_\gamma : \gamma \in K\}$ cover K and, for each $j = 1, \ldots, n$, let $K_j \subseteq N_j$ be a compact neighborhood such that $K \subseteq \bigcup_{j=1}^{n} K_j$. If $N_j = N_\gamma$ we write $\psi_j = \psi_\gamma$.

Because of Proposition 1.1.4 we choose $\{\varphi_j : j = 1, \ldots, n\} \subseteq A(\Gamma)$ for which $\varphi_j = 1$ on K_j and $\varphi_j = 0$ on $N_j^\sim$.

Since I is an ideal, $\varphi_j \psi_j \in I$.

We now note that

$$\text{(2.4.2)} \qquad \forall\, \gamma \in \Gamma \text{ and } \forall\, j = 1, \ldots, n, \qquad \varphi_j(\gamma)\psi_j(\gamma) = \varphi_j(\gamma)\theta(\gamma).$$

This follows easily by checking the various possibilities.

Define

(2.4.3) $\theta_1 = \theta[1 - (1-\varphi_1)(1-\varphi_2)\dots(1-\varphi_n)]$,

and observe that $\theta_1 \in I$ because of (2.4.2) (the "1's" cancel when we compute the right-hand side of (2.4.3)).

Finally we show that $\theta = \theta_1$. In fact, if $\gamma \notin K$ then $\theta(\gamma) = 0$ and so $\theta_1(\gamma) = 0$ by (2.4.3); and if $\gamma \in K$ then $1 - \varphi_j(\gamma) = 0$ for some j, and so the value of the expression inside the bracket (of (2.4.3)) is 1.

c) Choose the directed system $\{\varphi_\alpha\} \subseteq A_c(\Gamma)$ of Theorem 1.2.1.

Then it is clear that $\varphi\varphi_\alpha \in I_{\text{loc}}(\gamma)$ for each $\gamma \in \Gamma$ and for each α.

Because of a) and b), we have $\varphi\varphi_\alpha \in I$; and so $\varphi \in I$ by Theorem 1.2.1.

q.e.d.

There are generalizations of Theorem 2.4.1, e.g. [Katznelson, 5, Chapter VIII, Section 5.2; Loomis, 1, Section 25E] and Exercise 2.4.5a.

2.4.3 Wiener's proof to invert Fourier series. Wiener used the following example to prove Proposition 1.1.5b.

Example 2.4.1. *Given* $\hat{f} = \varphi \in A(\mathbf{T})$ *and assume that*

(2.4.4) $$2|f(0)| = 2\left|\int_{\mathbf{T}} \varphi(\gamma)\,\mathrm{d}\gamma\right| > \|\varphi\|_A.$$

We shall show that $\frac{1}{\varphi} \in A(\mathbf{T})$. We combine the series expansion, $1/(1+\psi) = 1 - \psi + \psi^2 - \psi^3 + \dots$, and (2.4.4) to compute for $\psi(\gamma) = \sum' \frac{f(n)}{f(0)}\,\mathrm{e}^{in\gamma}$ that

$$\frac{1}{\varphi(\gamma)} = \frac{1}{f(0)}(1 - \psi + \psi^2 - \psi^3 + \dots).$$

Thus, by (2.4.4) again,

$$\left\|\frac{1}{\varphi}\right\|_A \leqslant \frac{1}{|f(0)|}\sum_{n=0}^{\infty} \|\psi\|_A^n = \frac{1}{|f(0)|}\,\frac{1}{1 - \|\psi\|_A} = \frac{1}{2|f(0)| - \|\varphi\|_A} < \infty.$$

This example is obviously valid when $\mathbf{T}$ is replaced by any compact Γ.

We can now give Wiener's proof of Proposition 1.1.5b. It should be noted that we use Theorem 1.2.2a, a fact which Wiener proved [Wiener, 7, pp. 88–90], and that the proof of Theorem 1.2.2a did not in any way employ Proposition 1.1.5b. Wiener realized the importance of Theorem 1.2.2a in [Wiener, 6, pp. 12–15]; also, compare the proof of Theorem 1.2.2a contained in [Zygmund, 1, pp. 141–143].

Theorem 2.4.2. *Let Γ be compact and take $\varphi \in A(\Gamma)$.*
a) If $\varphi(\gamma_0) \neq 0$ then there is $\psi \in A(\Gamma)$ such that $\varphi = \psi$ on a neighborhood of γ_0 and $1/\psi \in A(\Gamma)$.
b) If φ never vanishes then $1/\varphi \in A(\Gamma)$.

Proof. a) Let $\gamma_0 = 0$ and define

$$\theta_\alpha(\gamma) = \varphi(0) + \psi_\alpha(\gamma)(\varphi(\gamma) - \varphi(0)), \tag{2.4.5}$$

where $\{\psi_\alpha\} \subseteq A_c(\Gamma)$ is the directed system of Theorem 1.2.2.
Choose $\eta = |\varphi(0)|/3$; and apply Theorem 1.2.2a to find α such that $\|\psi_\alpha(\varphi - \varphi(0))\|_A < \eta$. For this α set $\psi = \theta_\alpha$.
Clearly, since the modulus of the "0th Fourier coefficient", $\left|\int_\Gamma \psi_\alpha(\gamma)(\varphi(\gamma) - \varphi(0))\,d\gamma\right|$, is bounded by $\|\psi_\alpha(\varphi - \varphi(0))\|_A$, we have

$$\left|\int_\Gamma \psi(\gamma)\,d\gamma\right| > |\varphi(0)| - \eta. \tag{2.4.6}$$

On the other hand it is immediate from (2.4.5) that

$$\|\psi\|_A < |\varphi(0)| + \eta. \tag{2.4.7}$$

Combining (2.4.6) and (2.4.7) we obtain

$$2\left|\int_\Gamma \psi(\gamma)\,d\gamma\right| > \|\psi\|_A. \tag{2.4.8}$$

From the definition of ψ_α, $\psi = \varphi$ on a neighborhood of 0; and because of (2.4.8) and Example 2.4.1, $\frac{1}{\psi} \in A(\Gamma)$.

b) For each $\gamma \in \Gamma$ choose $\psi_\gamma \in A(\Gamma)$ such that $\psi_\gamma = \varphi$ on a neighborhood of γ and $\frac{1}{\psi_\gamma} \in A(\Gamma)$. Thus, $\frac{1}{\varphi} \in A(\Gamma)_{\text{loc}}(\gamma)$ for each $\gamma \in \Gamma$ and so $\frac{1}{\varphi} \in A(\Gamma)$ by Theorem 2.4.1a, b.

q.e.d.

2.4.4 Wiener–Lévy theorem. In an important, and in some sense elementary, paper, [Lévy, 1] proved the following generalization of Theorem 2.4.2b for the case $\Gamma = \mathbf{T}$. Theorem 2.4.3 is the *Wiener–Lévy theorem.*

Theorem 2.4.3. *Let $N \subseteq \mathbf{C}$ be an open neighborhood and let $F: N \to \mathbf{C}$ be an analytic function.*
Take $\varphi \in A(\Gamma)$.
a) If $K \subseteq \Gamma$ is compact and $\varphi(K) \subseteq N$ then $F \circ \varphi \in A(\Gamma)$.

b) *If Γ is compact and $\varphi(\Gamma) \subseteq N$ then $F \circ \varphi \in A(\Gamma)$.*

c) *If $F(0) = 0$ and $\overline{\varphi(\Gamma)} \subseteq N$ then $F \circ \varphi \in A(\Gamma)$.*

Proof. i) We shall prove that *for each $\lambda \in K$ there is $\psi_\lambda \in A(\Gamma)$ such that $F \circ \varphi = \psi_\lambda$ in a neighborhood of λ*, i.e., $F \circ \varphi \in A(\Gamma)_{\text{loc}}(\lambda)$.

Let $z_\lambda = \varphi(\lambda)$. By hypothesis,

$$(2.4.9) \qquad F(z) = F(z_\lambda) + \sum_{n=1}^{\infty} c_n (z - z_\lambda)^n$$

in some neighborhood of z_λ.

By Hadamard's formula for the radius of convergence of an analytic function we have that

$$\exists\, r \in \mathbf{R} \text{ and } \exists\, K > 0 \qquad \text{such that} \qquad \forall\, n \geqslant 0,\ |c_n| \leqslant K e^{nr}.$$

Take $\varepsilon > 0$ such that $r + \log \varepsilon < 0$.

Choose the de la Vallée–Poussin kernel $\{\psi_\alpha\}$ about λ, as in Theorem 1.2.2, so that for the appropriate α,

$$(2.4.10) \qquad \|(\varphi - \varphi(\lambda))\, \psi_\alpha\|_A < \varepsilon.$$

Thus, from (2.4.10),

$$\psi_\lambda = F(z_\lambda)\, \psi_\alpha + \sum_{n=1}^{\infty} c_n \left[(\varphi - \varphi(\lambda))\, \psi_\alpha\right]^n \in A(\Gamma);$$

and because of (2.4.9) and the fact that $\psi_\alpha = 1$ in a neighborhood of λ, we have that $\psi_\lambda = F \circ \varphi$ in a neighborhood of λ.

ii) Part a) follows from part i) and Theorem 2.4.1a, b.

iii) Parts b) and c) are now easy.

q.e.d.

2.4.5 Other proofs of the Wiener–Lévy theorem. Calderón has given a slick classical proof of Theorem 2.4.3 for the case of $\Gamma = \mathbf{T}$ which is independent of Theorem 2.4.1 [Zygmund, 2, I, pp. 245–246]. A soft proof of Theorem 2.4.2b, which tries to skirt around classical considerations and maximal ideal spaces is found in [Eisen and Gindler, 1]; this proof should be compared with the proof of Proposition 1.1.6. There is also the proof by [Carleman, 1, pp. 67 ff.] which dates back to 1935 (cf. Mandelbrojt's introductory remarks in [Akutowicz, 1]).

2.4.6 Generalizations of the Wiener-Lévy theorem.

Remark. In 1939, [Gelfand, 1] proved (among other things) the analogue of Theorem 2.4.2b for a commutative Banach algebra with unit (cf., the relevant introductory remarks in [Cohen 4]). Then, in [Gelfand, 2] (1941), he proved the analogue of the

Wiener–Lévy theorem for commutative Banach algebras with unit at the end of what is a fundamental exposition on Banach algebra theory (an English translation of this paper appears as the appendix in [Mackey, 1]). Segal [Katznelson, 5, Chapter VIII, Section 3; Mirkil, 2, pp. 13–14; Segal, 3] extended Gelfand's Wiener–Lévy result to commutative Banach algebras without unit. The Wiener–Lévy theorem has since been generalized to locally analytic operations on commutative Banach algebras; this latter situation includes analytic functions of several complex variables and was devised to deal with "the square root" and "the logarithm" operations (cf. Exercise 2.4.4c). The main work in this program is due to Shilov (1953 and 1960) and [Arens and Calderón, 1]; and expositions are found in [Gelfand, Raikov, and Shilov, 1, Section 13 and Chapter IX; Hoffman, 2].

2.4.7 The maximal ideal space of $M(G)$ and the extension of the Wiener–Lévy theorem to $B(\Gamma)$. With Theorem 2.4.3 and Section 2.4.6, with its list of generalizations (of Theorem 2.4.3), it is natural to ask if the analogue of Theorem 2.4.3 is true for $B(\Gamma)$ instead of $A(\Gamma)$. In particular, since $\frac{1}{\varphi} \in A(\Gamma)$ if $\varphi \in A(\Gamma)$ never vanishes and Γ is compact, is it true that $\frac{1}{\varphi} \in B(\Gamma)$ if $\varphi \in B(\Gamma)$ satisfies the condition $\inf\{|\varphi(\gamma)| : \gamma \in \Gamma\} > 0$ and Γ is non-compact?

In general the question is answered in the negative and this is due to the complicated structure of the maximal ideal space, $M(G)^m$, of $M(G)$, when G is non-discrete; there are general characterizations of $M(G)^m$ dating from Šreider (1948–1951), e.g. [Dunkl and Ramirez, 1, Chapter 1; Hewitt, 2, Chapter 3, Section 3–Section 5], but a good deal of mystery still surrounds such aspects of $M(G)$. [Pitt and Wiener, 1] gave the first example of a function $\varphi \in B(\mathbf{Я})$ for which $\inf\{|\varphi(\gamma)| : \gamma \in \mathbf{Я}\} > 0$ and $\frac{1}{\varphi} \notin B(\mathbf{Я})$; there was some question as to the details of their proof, but this is settled in [Pitt, 2, pp. 104 ff.]. The problem of finding such examples was pursued by Šreider (1950), [Hewitt, 1], and [Williamson, 1]; and is closely related to the fact that $M(G)$, G non-discrete, is not self-adjoint, e.g. [Katznelson, 5, Chapter VIII, Section 9; Rudin, 5, Section 5.3].

2.4.8 The Wiener–Pitt theorem. There are some positive results of the Wiener–Lévy type for $B(\Gamma)$. Essentially, the analogue of the Wiener–Lévy result for $B(\Gamma)$ is true for each $\mu \in M(G)$ that has a trivial continuous singular part, e.g. [Segal, 3, pp. 99–100]. Classical contributions to this problem are due to [Beurling, 1, pp. 356–358; Cameron and Wiener, 1; Pitt and Wiener, 1]; and the following result was proved in [Pitt and Wiener, 1] with the hypothesis that $\inf\{|\hat{\mu}(\gamma)| : \gamma \in \Gamma\} > 0$.

Theorem 2.4.4. *Given $\mu = \mu_a + \mu_s + \mu_d \in M(G)$, where μ_a is the absolutely continuous part, μ_s is the singular continuous part, and μ_d is the discontinuous part. Assume that $\hat{\mu}$*

never vanishes and that

$$\|\mu_s\|_1 < \inf\{|\hat{\mu}_d(\gamma)| : \gamma \in \Gamma\}.$$

Then $\frac{1}{\hat{\mu}} \in B(\Gamma)$.

Proof. i) Since $M(G)$ is a commutative Banach algebra with unit, we know that μ is invertible if and only if

$$\forall \mu' \in M(G)^m, \qquad \langle \mu', \mu \rangle \neq 0.$$

In ii) we consider the case that $\langle \mu', f \rangle \neq 0$ for some $f \in L^1(G) \subseteq M(G)$, and in iii) we suppose $\langle \mu', f \rangle = 0$ for each $f \in L^1(G)$; consequently, in this latter case,

$$\langle \mu', \mu \rangle = \langle \mu', \mu_s \rangle + \langle \mu', \mu_d \rangle.$$

We observe, from elementary considerations, that

$$|\langle \mu', \mu_s \rangle| \leqslant \|\mu_s\|_1$$

and

$$\langle \mu', \mu_d \rangle = \sum_{j=1}^{\infty} a_j(\chi, x_j), \qquad x_j \in G,$$

where $\chi : G \to \mathbf{C}$ is a character, $\mu_d = \sum_1^{\infty} a_j \delta_{x_j}$, and $\sum_1^{\infty} |a_j| < \infty$.

ii) Assume $\langle \mu', f \rangle \neq 0$ for some $f \in L^1(G)$. We shall find $\gamma \in \Gamma$ such that

$$\forall \nu \in M(G), \qquad \langle \mu', \nu \rangle = \hat{\nu}(\gamma),$$

and thus $\langle \mu', \mu \rangle \neq 0$.

By the L^1-theory, there is $\gamma \in \Gamma$ for which $\langle \mu', g \rangle = \hat{g}(\gamma)$ for each $g \in L^1(G)$. Since $L^1(G)$ is an ideal in $M(G)$, $\nu * f \in L^1(G)$, and so

$$\hat{f}(\gamma)\langle \mu', \nu \rangle = \langle \mu', f \rangle \langle \mu', \nu \rangle = \langle \mu', f * \nu \rangle = (f * \nu)\hat{\ }(\gamma) = \hat{f}(\gamma)\,\hat{\nu}(\gamma).$$

iii) From part ii) and the discussion in part i) it is sufficient to prove that $\langle \mu', \mu_s \rangle + \langle \mu', \mu_d \rangle \neq 0$, and this follows, again from the discussion in part i), once we verify that

$$|\langle \mu', \mu_d \rangle| \geqslant \inf\{|\hat{\mu}_d(\gamma)| : \gamma \in \Gamma\}.$$

Kronecker's theorem tells us that for each $\varepsilon > 0$ and each n there is $\lambda \in \Gamma$ such that

$$\forall j = 1, \ldots, n, \qquad |(\chi, x_j) - (\lambda, x_j)| < \varepsilon/(4 \sum |a_k|).$$

(We shall prove Kronecker's theorem in Section 3.2.12 after recording some necessary facts about the Bohr compactification in Section 3.1.16 which will allow for a very quick proof.)

Choosing n for which $\sum_{n+1}^{\infty} |a_k| < \varepsilon/4$, we obtain

$$\left|\langle \mu', \mu_d \rangle - \sum_{j=1}^{\infty} a_j(\lambda, x_j)\right| < \varepsilon.$$

Since this inequality is valid for all $\varepsilon > 0$ (where λ varies with ε), we obtain the desired inequality.

q.e.d.

2.4.9 The homomorphism problem. Note that if $\hat{f} = \varphi \in A(\mathbf{R})$ then $\varphi_{ab}(\gamma) = \varphi(a\gamma + b)$ is of the form $\varphi \circ \alpha \in A(\mathbf{R})$, where $a, b \in \mathbf{R}$, $a \neq 0$, and $\alpha(\gamma) = a\gamma + b$; in fact, $\hat{g} = \varphi_{ab}$ where $g(x) = \frac{1}{|a|} e^{ixb/a} f\left(\frac{x}{a}\right)$. Besides proving Theorem 2.4.3, [Lévy, 1] also posed the following problem (for the case $\Gamma_1 = \Gamma_2 = \mathbf{T}$): find all functions $\alpha : \Gamma_2 \to \Gamma_1$ such that

$$\text{(2.4.11)} \qquad \forall\, \varphi \in A(\Gamma_1), \qquad \varphi \circ \alpha \in A(\Gamma_2).$$

Clearly, (2.4.11) defines an algebra homomorphism, and we refer to Lévy's problem as the *homomorphism problem*. This problem has been solved in the following works: for $\Gamma_1 = \Gamma_2 = \mathbf{R}$ by [Beurling and Helson, 1] (1953); for $\Gamma_1 = \Gamma_2 = \mathbf{T}$ by [Leibenson, 1] (1954) with an assist by [Kahane, 1; 2] (1955); for $\Gamma_1 = \Gamma_2 = \mathbf{Z}$ by [Rudin, 1] (1956); for the general case, after further contributions by Helson and Rudin, by [Cohen, 2; 3] (1960). Expositions of the homomorphism problem are found in [Cohen, 5; Kahane, 7; Katznelson, 5, Chapter VIII, Section 4; Rudin, 5, Chapters 3 and 4].

Naturally, there is an homomorphism problem for quotient algebras, $A(E)$, and spectral synthesis can play a role in this case, e.g. [Kahane, 13, Chapitre IX]. With quotient algebras in mind we give the following result in a fairly general setting.

Proposition 2.4.1. *Let X_i, $i = 1, 2$, be complex semi-simple commutative Banach algebras with maximal ideal spaces $X_i^m = \{x' \in X_i' \setminus \{0\} : \forall\, x, y \in X_i,\ \langle x', xy \rangle = \langle x', x \rangle \langle x', y \rangle\}$, $i = 1, 2$.*

a) *Let h: $X_1 \to X_2$ be an algebra homomorphism. There is a function α: $X_2^m \to X_1^m \cup \{0\} \subseteq X_1'$ such that if $x_2' \in X_2^m$ then*

$$\text{(2.4.12)} \qquad \forall\, x_1 \in X_1, \qquad \langle \alpha(x_2'), x_1' \rangle = \langle x_2', h(x_1) \rangle$$

(i.e., the right-hand side of (2.4.12) defines α).

b) *Let $X_1 = X_2 = X$ and let $h : X \to X$ be an algebra automorphism. Then α, defined by (2.4.12), is an homeomorphism $X^m \to X^m$, where X^m has the induced $\sigma(X', X)$ topology.*

c) *Let $X_1 = X_2 = X$ and let $\alpha : X^m \to X^m$ be an homeomorphism. α is defined in terms of an endomorphism $h : X \to X \Leftrightarrow$*

$$\forall\, x \in X, \qquad x \circ \alpha \in X,$$

where $x \circ \alpha : X^m \to \mathbf{C}$ is defined as $x \circ \alpha(x') = \langle \alpha(x'), x \rangle$.

The proof of Proposition 2.4.1 is left as Exercise 2.4.5b.

Remark 1. If α: $\mathbf{T} \to \mathbf{T}$ is permitted in the sense that (2.4.11) is valid, then $\hbar(\varphi) = \varphi \circ \alpha$ defines an endomorphism $\hbar: A(\mathbf{T}) \to A(\mathbf{T})$, and so $\hbar$ is continuous (e.g. the remarks at the end of Exercise 2.4.5b). In particular,

$$(2.4.13) \qquad \forall\, n, \quad \|e^{in\alpha}\|_A \leqslant \|\hbar\| < \infty.$$

It turns out that the boundedness of (2.4.13) is the condition required on any function $\alpha: \mathbf{T} \to \mathbf{T}$ to ensure that

$$(2.4.14) \qquad \alpha(\gamma) = a\gamma + b, \qquad a \in \mathbf{Z}$$

[Beurling and Helson, 1; Leibenson, 1]. This solves the homomorphism problem. Thus, in terms of Proposition 2.4.1, *if $\hbar: A(\mathbf{T}) \to A(\mathbf{T})$ is an algebra automorphism then there is an homeomorphism $\alpha: \mathbf{T} \to \mathbf{T}$ of the form* (2.4.14) *such that $\hbar(\varphi) = \varphi \circ \alpha$ for each $\varphi \in A(\mathbf{T})$*. This result can actually be strengthened by assuming only that $\hbar: A(\mathbf{T}) \to A(\mathbf{T})$ is an algebra endomorphism, e.g. [Katznelson, 5, pp. 220–221]. $A(\Gamma)$ and $A'(\Gamma)$ norm boundedness of "exponentials" (as in (2.4.13)) play an important role in spectral synthesis and we shall return to this matter in Chapter 3.

2. Cohen's solution to the homomorphism problem is based on the following result: *let $\hbar: A(\Gamma_1) \to A(\Gamma_2)$ be an algebra homomorphism, let α be defined as in* (2.4.12), assume that Γ_1 and Γ_2 are discrete, and set

$$S = \{(\gamma_2, \alpha(\gamma_2)) : \gamma_2 \in \Gamma_2 \text{ and } \alpha(\gamma_2) \in \Gamma_1\} \subseteq \Gamma_2 \times \Gamma_1;$$

then $\hat{\mu} = \chi_S$, where μ is an idempotent measure. Thus, a solution of the homomorphism problem is more or less reduced to the problem of characterizing idempotent measures, cf. Section 1.2.5.

Cohen used his technique for idempotent measures to prove that there is a constant K such that for any set $\{n_1, .., n_k\} \subseteq \mathbf{Z}$,

$$(2.4.15) \qquad \int_{\mathbf{T}} |e^{in_1\gamma} + \ldots + e^{in_k\gamma}|\, d\gamma \geqslant K\left(\frac{\log k}{\log\log k}\right)^{1/8};$$

[Davenport, 1] modified one of Cohen's combinatorial lemmata and was able to replace the exponent, 1/8, by 1/4 and to take $K \geqslant 1/8$ for large k; Pichorides has gone a step further and been able to replace 1/4 by 1/2. In his famous list of research problems (actually there were three such lists), Littlewood conjectured that the right-hand side of (2.4.15) can be replaced by $K \log k$; if the conjecture is true, an upper bound K has been computed by Karanikolov.

2.4.10 Beurling's criterion for $A(\mathbf{T})$. Getting back to the spirit of Theorem 2.4.3 we shall now prove a theorem, which stems from [Beurling, 6], from which it is possible to find conditions on $\varphi \in A(\mathbf{T})$ so that if $F(z) = |z|$ then $F \circ \varphi = |\varphi| \in A(\mathbf{T})$. Note that if $\hat{f} = \varphi \in L^2(\mathbf{T})$ and $\hat{g} = \psi = |\varphi|$ then

(2.4.16) $$\forall\, \gamma, \lambda \in \mathbf{T}, |\psi(\gamma + \lambda) - \psi(\gamma)| \leqslant |\varphi(\gamma + \lambda) - \varphi(\gamma)|,$$

$$\int_{\mathbf{T}} |\varphi(\gamma + \lambda) - \varphi(\gamma)|^2 \,\mathrm{d}\gamma = \sum_n |f(n)|^2 \sin^2 \frac{\lambda n}{2},$$

and

$$\int_{\mathbf{T}} |\psi(\gamma + \lambda) - \psi(\gamma)|^2 \,\mathrm{d}\gamma = \sum |g(n)|^2 \sin^2 \frac{\lambda n}{2}.$$

As such, we define

(2.4.17) $$A(\varphi) = \frac{1}{2\pi} \int_0^{2\pi} \frac{1}{\lambda^{3/2}} \left[\int_{\mathbf{T}} |\varphi(\gamma + \lambda) - \varphi(\gamma)|^2 \,\mathrm{d}\gamma \right]^{1/2} \mathrm{d}\lambda,$$

(2.4.18) $$B(\varphi) = \sum_{n=1}^{\infty} \frac{1}{n^{1/2}} \left[\sum_{|k|=n}^{\infty} |f(k)|^2 \right]^{1/2},$$

and

(2.4.19) $$C(\varphi) = \sum_{n=1}^{\infty} \frac{1}{n^{3/2}} \left[\sum_{|k|=1}^{n} |kf(k)|^2 \right]^{1/2},$$

where $\hat{f} = \varphi \in L^2(\mathbf{T})$. We say *A (resp. B or C) is finite for the sequence* $\{d_n : n = 1, \ldots\} \subseteq \mathbf{R}^+$ if $A(\varphi) < \infty$ (resp. $B(\varphi) < \infty$ or $C(\varphi) < \infty$) where $\hat{f} = \varphi$, $f(n) = d_{|n|}$, and $f(0) = 0$.

Theorem 2.4.5. a) *If $\hat{f} = \varphi \in L^2(\mathbf{T})$ and $B(\varphi) < \infty$ then $\varphi \in A(\mathbf{T})$.*

(b) If $\{d_n : n = 1, \ldots\} \subseteq \mathbf{R}^+$ is a decreasing sequence and $\sum d_n < \infty$ then B and C are finite for $\{d_n : n = 1, \ldots\}$.

c) *If $\{d_n : n = 1, \ldots\} \subseteq \mathbf{R}^+$ and $\sum_1 d_n \sqrt{n} < \infty$ then B and C are finite for $\{d_n : n = 1, \ldots\}$.*

d) *If $\hat{f} = \varphi \in L^2(\mathbf{T})$ and $B(\varphi) + C(\varphi) < \infty$ then $A(\varphi) < \infty$.*

Proof. a) First note, by drawing the triangular arrays, that formally

(2.4.20) $$\sum_{k=1}^{\infty} \sum_{n=1}^{k} \frac{|f(k)|}{k} = \sum_{n=1}^{\infty} \sum_{k=n}^{\infty} \frac{|f(k)|}{k}.$$

Also, formally,

(2.4.21) $$\sum_{k=1}^{\infty} |f(k)| = \sum_{k=1}^{\infty} \sum_{n=1}^{k} \frac{|f(k)|}{k}.$$

Combining (2.4.20) and (2.4.21) yields

$$\text{(2.4.22)} \qquad \sum_{1}^{\infty} |f(k)| = \sum_{n=1}^{\infty} \sum_{k=n}^{\infty} \frac{|f(k)|}{k};$$

and, using Hölder's inequality, the right-hand side of (2.4.22) is bounded by

$$\sum_{n=1}^{\infty} \left[\sum_{k=n}^{\infty} \frac{1}{k^2} \right]^{1/2} \left[\sum_{k=n}^{\infty} |f(k)|^2 \right]^{1/2} \leqslant \sum_{n=1}^{\infty} n^{-1/2} \left(\sum_{k=n}^{\infty} |f(k)|^2 \right)^{1/2} \leqslant B(\varphi) < \infty.$$

b) We shall use the Cauchy "2^n-criterion" for convergence: *if* $\{c_n : n = 1, \ldots\}$ *decreases to* 0 *then* $\sum c_n$ *converges if and only if* $\sum 2^n c_{2^n}$ *converges*, e.g. [Bary, 2, I, pp. 5–6].

i) There is a constant $a > 0$ such that

$$\text{(2.4.23)} \qquad \sum_{n=1}^{\infty} n^{-3/2} \left(\sum_{k=1}^{n} k^2 d_k^2 \right)^{1/2} = \sum_{m=0}^{\infty} \sum_{n=2^m}^{2^{m+1}-1} n^{-3/2} \left(\sum_{k=1}^{n} k^2 d_k^2 \right)^{1/2}$$

$$\leqslant \sum_{m=0}^{\infty} \left(\sum_{k=1}^{2^{m+1}} k^2 d_k^2 \right)^{1/2} \sum_{n=2^m}^{2^{m+1}} n^{-3/2}$$

$$\leqslant a \sum_{m=0}^{\infty} 2^{-m/2} \left(\sum_{k=1}^{2^{m+1}} k^2 d_k^2 \right)^{1/2},$$

by an integral estimate of

$$\int_{2^m}^{2^{m+1}-1} x^{-3/2}\, dx.$$

Since $\{d_k : k = 1, \ldots\}$ is decreasing, the right-hand side of (2.4.23) is

$$\text{(2.4.24)} \qquad a \sum_{m=0}^{\infty} 2^{-m/2} \left(\sum_{j=1}^{m} \sum_{k=2^j}^{2^{j+1}-1} k^2 d_k^2 \right)^{1/2} \leqslant a \sum_{m=0}^{\infty} 2^{-m/2} \left(\sum_{j=1}^{m} 2^{3j} d_{2^j}^2 \right)^{1/2},$$

by an integral estimate of

$$\int_{2^j}^{2^{j+1}-1} x^2\, dx.$$

The right-hand side of (2.4.24) is bounded by

$$\text{(2.4.25)} \qquad a \sum_{m=1}^{\infty} 2^{-m/2} \left(\sum_{j=1}^{m} 2^{3j/2} d_{2^j} \right) = a \sum_{j=1}^{\infty} 2^{3j/2} d_{2^j} \left(\sum_{m=j}^{\infty} 2^{-m/2} \right)$$

because $\left(\sum_1^m a_j^2\right)^{1/2} \leqslant \sum_1^m a_j$ if each $a_j \geqslant 0$ and because

$$\sum_{m=1}^{\infty} a_m\left(\sum_{j=1}^{m} b_j\right) = \sum_{j=1}^{\infty} b_j\left(\sum_{m=j}^{\infty} a_m\right).$$

Now, $\sum_{m=j}^{\infty} 2^{-m/2}$ is estimated by an integral estimate to be $2^{-j/2}$, and so (2.4.25) becomes $\sum_{m=1}^{\infty} 2^m d_{2^m}$.

Thus the left-hand side of (2.4.23) is bounded.

ii) To show that B is finite for $\{d_n : n = 1, \ldots\}$ it is sufficient to prove that

$$\sum_{n=0}^{\infty} 2^{-n/2}\left(\sum_{k=2^n}^{\infty} d_k^2\right)^{1/2} < \infty. \tag{2.4.26}$$

Since $\{d_k : k = 1, \ldots\}$ is decreasing and $2^{j+1} - 2^j = 2^j$, the left-hand side of (2.4.26) is

$$\sum_{n=0}^{\infty} 2^{-n/2}\left(\sum_{j=n}^{\infty}\sum_{k=2^j}^{2^{j+1}-1} d_k^2\right)^{1/2} \leqslant \sum_{n=0}^{\infty} 2^{-n/2}\left(\sum_{j=n}^{\infty} 2^j d_{2^j}^2\right)^{1/2}, \tag{2.4.27}$$

and, as in the analogous calculation in part i), the right-hand side of (2.4.27) is bounded by

$$\sum_{n=0}^{\infty} 2^{-n/2}\left(\sum_{j=n}^{\infty} 2^{j/2} d_{2^j}\right) \leqslant \sum_{j=0}^{\infty} 2^{j/2} d_{2^j} \sum_{n=0}^{\infty} 2^{-n/2}.$$

Thus, we apply the Cauchy "2^n-criterion" to obtain (2.4.26).

c) If $\sum_1^{\infty} d_n\sqrt{n} < \infty$, then, by using integral estimates in the obvious way, we see that

$$\sum_{n=1}^{\infty} n^{-3/2}\left(\sum_{k=1}^{n} k d_k\right) + \sum_{n=1}^{\infty} n^{-1/2}\left(\sum_{k=n+1}^{\infty} d_k\right)$$

$$= \sum_{k=1}^{\infty} k d_k\left(\sum_{n=k}^{\infty} n^{-3/2}\right) + \sum_{k=1}^{\infty} d_k\left(\sum_{n=1}^{k-1} n^{-1/2}\right) \leqslant a\sum_{k=1}^{\infty} d_k\sqrt{k} < \infty.$$

d) From the definition of $A(\varphi)$ and Parseval's formula it is sufficient to estimate

$$\int_0^1 \lambda^{-3/2}\left[\sum_{k=1}^{\infty} |f(k)|^2 \sin^2 k\lambda\right]^{1/2} d\lambda \leqslant \sum_{n=2}^{\infty}\int_{n^{-1}}^{(n-1)^{-1}} \lambda^{-3/2}\left\{\left[\sum_{k=1}^{n} |f(k)|^2 k^2\lambda^2\right]^{1/2}\right.$$

$$\left.+\left[\sum_{k=n+1}^{\infty} |f(k)|^2\right]^{1/2}\right\} d\lambda = \sum_{n=2}^{\infty}\left(\sum_{k=1}^{n} |kf(k)|^2\right)^{1/2}\int_{n^{-1}}^{(n-1)^{-1}} \lambda^{-1/2}\, d\lambda$$

$$+\sum_{n=2}^{\infty}\left(\sum_{k=n+1}^{\infty} |f(k)|^2\right)^{1/2}\int_{n^{-1}}^{(n-1)^{-1}} \lambda^{-3/2}\, d\lambda \leqslant C(\varphi) + B(\varphi) < \infty.$$

q.e.d.

The simple proof of Theorem 2.4.5a is due to Stečkin; there are more complicated proofs.

Actually, for $\hat{f} = \varphi$, Leindler has shown that $A(\varphi) < \infty$ if and only if $B(\varphi) < \infty$; and [Sunouchi, 1] has used Jensen's inequality to show that $B(\varphi) < \infty$ if and only if $C(\varphi) < \infty$. On this matter we also mention the contributions of Konyushkov (1958) and Boas (1960).

Using (2.4.16), Theorem 2.4.5a,b,d, and the direction,

$$A(\varphi) < \infty \quad \Rightarrow \quad B(\varphi) < \infty,$$

of Leindler's theorem, we have: *given* $\hat{f} = \varphi \in A(\mathbf{T})$; *if there is a decreasing sequence* $\{d_n : n = 1, \ldots\} \subseteq \mathbf{R}^+$ *for which* $\sum d_n < \infty$ *and if*

$$\forall n > 0, \qquad |f(\pm n)| \leqslant d_n,$$

then $|\varphi| \in A(\mathbf{T})$.

Contraction conditions such as (2.4.16) have also been developed by Beurling into a very important tool in spectral synthesis. In fact, not only does $A(\varphi) < \infty$ imply that $\varphi \in A(\mathbf{T})$, but, as we'll prove in Section 3.2.4, it implies that φ is synthesizable.

2.4.11 Katznelson's converse to the Wiener–Lévy theorem. As we note in Exercise 2.4.4b, $F(z) = |z|$ does not have the property that $F(A(\mathbf{T})) \subseteq A(\mathbf{T})$. In 1958 [Kahane, 3] proved that there are infinitely differentiable functions F such that $F \circ \varphi \notin A(\mathbf{T})$ for some $\varphi \in A(\mathbf{T})$ (cf. Theorem 2.4.3). Later in 1958, [Katznelson, 1] made the breakthrough that establishes the converse to Theorem 2.4.3; the general statement of this result as developed in [Helson and Kahane, 1; Helson, Kahane, Katznelson, and Rudin, 1; Katznelson, 1] and exposited in [Katznelson, 5, Chapter VIII, Section 8; Rudin, 5, Chapter 6] essentially says that if a function F satisfies the condition, $F(A(\Gamma)) \subseteq A(\Gamma)$, then F is analytic (an exact statement is given in Exercise 2.4.2).

Just as the Wiener (-Lévy) theorem is fundamental in establishing a solution to the spectral analysis problem, so the Katznelson theorem is intimately related to the solution of spectral synthesis problems.

[Varopoulos, 6] has given a tensor algebra proof of Katznelson's theorem. The Katznelson conjecture [Katznelson, 4] associated with this theorem is discussed in Exercise 2.4.2. There is a natural analogue of Katznelson's theorem when $A(\Gamma)$ is replaced by $B(\Gamma)$ [Graham, 1; Rudin, 5, pp. 135 ff.]; also see [Katznelson, 2].

2.4.12 Tensor products. We shall now give an application of Theorem 2.4.1 to tensor products; the introduction of tensor products allows us to give the intrinsic definition of convolution that we promised after Proposition 1.3.1. As we shall later see, tensor products play a surprising and interesting role in harmonic analysis.

Let X_1, X_2 be complex Hausdorff locally convex topological vector spaces, let Y and Z be complex vector spaces, and let Z^* be the algebraic dual of Z. We denote the space of bilinear maps,

$$X_1 \times X_2 \;\to\; Y,$$

by $\mathscr{B}_Y$. The *tensor product* of X_1 and X_2 is a vector space $X_1 \otimes X_2$ and a map $\alpha \in \mathscr{B}_{X_1 \otimes X_2}$ such that for any vector space Y and any map $\beta \in \mathscr{B}_Y$ there is a unique linear map

$$L: X_1 \otimes X_2 \;\to\; Y$$

for which the following diagram commutes-

$$\begin{array}{ccc} & X_1 \times X_2 & \\ {}^{\alpha}\swarrow & & \searrow^{\beta} \\ X_1 \otimes X_2 & \xrightarrow[L]{} & Y. \end{array}$$

It is a standard algebraic fact that $X_1 \otimes X_2$ exists and is uniquely defined for given X_1, X_2. For each $(s,t) \in X_1 \times X_2$ we designate

$$\alpha(s,t) = s \otimes t;$$

and it is easy to check that

(2.4.29) $\quad \forall\, u \in X_1 \otimes X_2, \qquad u = \sum s \otimes t$, a finite sum,

$$(s_1 + s_2) \otimes t = s_1 \otimes t + s_2 \otimes t, \qquad s \otimes t_1 + s \otimes t_2 = s \otimes (t_1 + t_2),$$

and

$$s \otimes (ct) = c(s \otimes t) = (cs) \otimes t,$$

where $s, s_j \in X_1$, $t, t_j \in X_2$, and $c \in \mathbf{C}$. The representation in (2.4.29) is not necessarily unique; on the other hand it can be shown that each $u \in X_1 \otimes X_2$ has a unique representation

(2.4.30) $$u = \sum_{i \in I} s_i \otimes h_i,$$

$\in$

where I is an index set for an Hamel basis $\{h_i : i \in I\}$ of X_2 and where $s_i = 0$ except for finitely many i, e.g. [Benedetto, 2, pp. 128–130].

An important property of algebraic tensor products is that

$$(X_1 \otimes X_2)^* = \mathscr{B}_{\mathbf{C}}\,.$$

This "equation" is (one of) the raison d'être for tensor products since, because of it, vector spaces of multilinear maps can be considered as vector spaces of intrinsically linear maps. More generally, if $\mathscr{L}_Y$ is the vector space of linear maps $X_1 \otimes X_2 \to Y$

then

$$\mathscr{L}_Y = \mathscr{B}_Y .$$

2.4.13 An intrinsic definition of supp T. Note that the dual group of $G \times G$ is $\Gamma \times \Gamma$.

Proposition 2.4.2. a) *There is a natural injection*

$$A(\Gamma) \otimes A(\Gamma) \quad \to \quad A(\Gamma \times \Gamma).$$

b) $\overline{A(\Gamma) \otimes A(\Gamma)} = A(\Gamma \times \Gamma)$.

Proof. a) If $f,g \in L^1(G)$ then $fg \in L^1(G \times G)$ and $\widehat{fg}(\lambda,\gamma) = \varphi(\lambda)\psi(\gamma)$, where $\hat{f} = \varphi$ and $\hat{g} = \psi \in A(\Gamma)$. Thus $A(\Gamma) \otimes A(\Gamma) \subseteq A(\Gamma \times \Gamma)$ and so we have a linear map $a: A(\Gamma) \otimes A(\Gamma) \to A(\Gamma \times \Gamma)$.

To prove that this map is injective first note that if $\{\theta_i : i \in I\}$ is an Hamel basis of $A(\Gamma)$ then $\{\theta_i \otimes \theta_j : (i,j) \in I \times I\}$ is an Hamel basis for $A(\Gamma) \otimes A(\Gamma)$.

Let $\Phi = \sum \varphi \otimes \psi \in A(\Gamma) \otimes A(\Gamma)$, and, using (2.4.30), expand each φ (i.e., each s_i) as a finite linear combination of elements from $\{\theta_i : i \in I\}$. We then regroup this sum and obtain the unique representation

$$\Phi = \sum c_{jk} \theta_j \otimes \theta_k, \qquad \text{a finite sum,}$$

where $\{\theta_j\}$ and $\{\theta_k\}$ are two linearly independent sets.

Now if $a(\Phi) = 0$ we have

$$\forall\, \lambda, \gamma \in \Gamma, \qquad \sum c_{jk} \theta_j(\lambda)\, \theta_k(\gamma) = 0.$$

Thus, for each fixed $\lambda \in \Gamma$,

$$\forall\, k, \qquad \sum_j c_{jk} \theta_j(\lambda) = 0$$

by the linear independence; and by a second application of the linear independence, we see that $c_{jk} = 0$ for each j and k.

Consequently, $\Phi = 0$ and so a is injective.

b) Write $I = \overline{A(\Gamma) \otimes A(\Gamma)} \subseteq A(\Gamma \times \Gamma)$. Clearly I is a closed ideal by Proposition 1.2.6. Let $\Phi \in A(\Gamma \times \Gamma)$, and take $(\lambda, \gamma) \in \Gamma \times \Gamma$. Choose $\varphi \in A(\Gamma)$ (resp. $\psi \in A(\Gamma)$) which is 1 in a neighborhood of λ (resp. γ).

Then, $\varphi\psi \in A(\Gamma) \otimes A(\Gamma)$, and so $\Phi\varphi\psi \in I$ since I is an ideal. Also, $\Phi\varphi\psi = \Phi$ in a neighborhood of (λ, γ).

Consequently, by Theorem 2.4.1c, $\Phi \in I$.

q.e.d.

The problem solved in Proposition 2.4.2b has an analogue in distribution theory where it is necessary to prove that $\overline{C_c^\infty(\mathbf{R}^n) \otimes C_c^\infty(\mathbf{R}^m)} = C_c^\infty(\mathbf{R}^n \times \mathbf{R}^m)$; this latter result is proved by means of a lemma due to H. Cartan [Horváth, 1, pp. 368–369; Schwartz, 5].

Proposition 2.4.3. *Given $S, T \in A'(\Gamma)$.*

a) *There is a unique element $S \otimes T \in A'(\Gamma \times \Gamma)$ such that*

$$\forall\, \varphi, \psi \in A(\Gamma), \qquad \langle S \otimes T, \varphi\psi\rangle = \langle S, \varphi\rangle \langle T, \psi\rangle.$$

b) *For any $\Phi \in A(\Gamma \times \Gamma)$,*

$$\langle S \otimes T, \Phi\rangle = \langle S_\gamma, \langle T_\lambda, \Phi(\lambda, \gamma)\rangle\rangle = \langle T_\lambda, \langle S_\gamma, \Phi(\lambda, \gamma)\rangle\rangle.$$

Both parts of Proposition 2.4.3 are proved by verifying the result on $A(\Gamma) \otimes A(\Gamma)$, and then extending to $A(\Gamma \times \Gamma)$ via Proposition 2.4.2b and an easy calculation. $S \otimes T$ is the *tensor product* of S and T; and it is now immediate that

(2.4.31) $\qquad \forall\, S, T \in A'(\Gamma)$ and $\forall\, \varphi \in A(\Gamma), \qquad \langle S * T, \varphi\rangle = \langle S \otimes T, \Phi\rangle,$

where $\Phi(\lambda, \gamma) = \varphi(\lambda + \gamma)$. (2.4.31) is the promised intrinsic (in $A'(\Gamma)$) definition of convolution.

The proof of the following is left as Exercise 2.4.6.

Proposition 2.4.4. *Given $S, T \in A'(\Gamma)$ and $\varphi \in A(\Gamma)$.*

a) $\operatorname{supp} S \otimes T = (\operatorname{supp} S) \times (\operatorname{supp} T)$.

b) *Assume that for each compact set $K \subseteq \Gamma$,*

$$(\operatorname{supp} S \times \operatorname{supp} T) \cap \{(\lambda, \gamma) \in \Gamma \times \Gamma : \lambda + \gamma \in K\} \subseteq \Gamma \times \Gamma$$

is compact. Then $\operatorname{supp} S + \operatorname{supp} T$ *is closed and*

(2.4.32) $\qquad \operatorname{supp} S * T \subseteq \operatorname{supp} S + \operatorname{supp} T.$

c) *If* $\operatorname{supp} S$ *is compact then* (2.4.32) *is valid.*

d) $\operatorname{supp} S * \varphi \subseteq \operatorname{supp} S + \operatorname{supp} \varphi$.

Exercises 2.4

2.4.1 Local membership and a property of $A_c(\Gamma)$

a) Let $E \subseteq \Gamma$ be closed and let $\varphi \in k(E)$.

Prove that

$$\forall\, \gamma \in \Gamma \setminus (\partial Z\varphi \cap \partial E), \qquad \varphi \in j(E)_{\mathrm{loc}}(\gamma).$$

b) We noted in Proposition 1.2.4a that if $I \subseteq A(\Gamma)$ is an ideal then $I \cap A_c(\Gamma)$ is norm dense in I. Prove that

$$A_c(\Gamma) = \cap\, \{I : I \subseteq A(\Gamma) \text{ is a norm dense ideal in } A(\Gamma)\}.$$

(*Hint*: Use Proposition 1.1.5b).

2.4.2 The conjecture of dichotomy

Let $E \subseteq \mathbf{T}$ be a closed set. A function $F: I \to \mathbf{C}$, where $I \subseteq \mathbf{R}$ is an open interval, *operates* in $A(E)$ if $F \circ \varphi \in A(E)$ whenever $\varphi \in A(E)$ has its range contained in I. Katznelson's theorem mentioned after Theorem 2.4.5 tells us that if E contains an interval and $F: I \to \mathbf{C}$ operates in $A(E)$ then F is analytic in a neighborhood of I; the same result holds if E has the property that for each n E contains an arithmetic progression having n terms [Kahane and Katznelson, 1; Katznelson, 3]. [Kahane and Katznelson, 2] have also proved that if E is a set of strict multiplicity (in particular, if $mE > 0$) and F operates in $A(E)$ then F is analytic; more generally, they obtained the same result for closed sets E with the property that

$$\forall\, \varepsilon > 0 \;\; \exists\, \mu \in M(E) \qquad \text{such that}$$
$$\overline{\lim_{|n| \to \infty}} \, (|\hat{\mu}(n)| / \|\mu\|_{A'}) < \varepsilon.$$

a) Let $E \subseteq \mathbf{T}$ be closed and infinite, and let $I \subseteq \mathbf{R}$ be an open interval. Prove that if $F: I \to \mathbf{C}$ operates in $A(E)$ then F is continuous.

b) Let $E \subseteq \mathbf{T}$ be an Helson set. Prove that if $F: I \to \mathbf{C}$ is a continuous function then F operates in $A(E)$. Helson sets and sets of strict multiplicity are mutually exclusive classes, e.g. [Benedetto, 6, Chapter 7].

We say that a closed set $E \subseteq \mathbf{T}$ is a *set of analyticity* if F is analytic whenever $F: I \to \mathbf{C}$ operates in $A(E)$. The *Katznelson conjecture*, or *conjecture of dichotomy*, mentioned after Theorem 2.4.5, is that any closed set $E \subseteq \mathbf{T}$ is either an Helson set or a set of analyticity; an interesting recent result on the Katznelson conjecture is found in [Katznelson and McGehee, 2]. A criterion for E to be a set of analyticity, proved by means of a combinatoric argument, is given in [Salinger and Varopoulos, 1]; and a probabilistic contribution to the Katznelson conjecture, whose presentation is a shockastic process, is given in [Katznelson and Malliavin, 1; 2] (cf. [Katznelson and Malliavin, 3]). Körner's example of a pseudo-function supported by an Helson set provides motivation to pursue the techniques in [Kahane and Katznelson, 2].

2.4.3 Norm estimates for $1/\varphi$

Assume that $\hat{f} = \varphi \in A(\mathbf{T})$ doesn't vanish.

a) Set $m = \inf\{|\varphi(\gamma)| : \gamma \in \mathbf{T}\}$. Provide an estimate for $\|1/\varphi\|_\infty$ in terms of $\sum_{|n|>N} |nf(n)|$ and m, where N is chosen so that $\sum_{|n|>N} |f(n)| < m/4$, e.g. [Cohen, 4].

b) Can you provide an estimate for $\|1/\varphi\|_A$ in terms of $\|\varphi\|_A$?

2.4.4 $F \circ \varphi \notin A(\mathbf{T})$ for $\varphi \in A(\mathbf{T})$

a) Prove that $\varphi \circ \varphi \notin A(\mathbf{T})$, where $\varphi \in A(\mathbf{T})$ is defined by

$$\varphi(\gamma) = \begin{cases} 0 & \text{for } \gamma = \pi \quad \text{or } \gamma \in (-\pi, 0], \\ \dfrac{1}{(\log \gamma)^2} & \text{for } \gamma \in (0, \tfrac{1}{2}), \\ \text{linear} & \text{otherwise.} \end{cases}$$

This example is due to [Marcinkiewicz, 2]. Thus, $\varphi^n \in A(\mathbf{T})$ if $\varphi \in A(\mathbf{T})$, whereas it is not necessarily true that $\varphi \circ \ldots \circ \varphi \in A(\mathbf{T})$ if $\varphi \in A(\mathbf{T})$. Also, if φ is a uniformly convergent Fourier series then φ^2 is not necessarily uniformly convergent. (*Hint*: If $\hat{f} = \varphi$ then

$$|f(n)| = \left| \frac{1}{2\pi n} \int_0^{2\pi} \varphi'(\gamma)\, e^{-in\gamma}\, d\gamma \right| \leqslant \frac{1}{2\pi|n|} \int_0^{2\pi} \left| \varphi'(\gamma) - \varphi'\left(\gamma + \frac{\pi}{n}\right) \right| d\gamma$$

$$= O(1/|n \log^2 n|),\ |n| \to \infty.$$

By using an argument with conjugate functions it is not difficult to prove that if $\psi \in A(\mathbf{T})$ then

$$\lim_{\varepsilon \to 0} \int_\varepsilon^\pi (\psi(\gamma) - \psi(-\gamma)) \cot \frac{\gamma}{2}\, d\gamma$$

exists and is finite; whereas it is easy to check that this limit is infinite for $\psi = \varphi \circ \varphi$. Cf. the example in [Bary, 2, II, pp. 194–196] which uses Shilov's criterion for the absolute convergence of Fourier series [Bary, 2, II, pp. 181–183]).

Marcinkiewicz's main theorem associated with this example is: let $\hat{f} = \varphi \in A(\mathbf{T})$ be real-valued and assume that $\sum |f(n)|^s < \infty$ for some $s \in [0,1]$; if

$$\forall\, z \in (a,b) \text{ and } \forall\, n, \quad |F^{(n)}(z)| \leqslant n^{n/s},$$

where $(\inf \varphi(\gamma), \sup \varphi(\gamma)) \subseteq (a,b)$, then $F \circ \varphi \in A(\mathbf{T})$ (cf. [Kahane, 11] which is interesting in this regard for several other related problems).

b) Define

$$\varphi_1(\gamma) = K \frac{\sin \gamma}{|\sin \gamma|} (1 - \log|\sin \gamma|)^{-3/2}$$

and

$$\varphi_2(\gamma) = \sum_{n=1}^{\infty} \sum_{j=1}^{2^n} \frac{1}{n^2\, 2^n} \sin(4^n + j2^n)\gamma,$$

where $K > 0$ is chosen large enough so that

$$\varphi = \varphi_1 - \varphi_2$$

is non-negative on $[0, 2\pi]$. Check that $\varphi_1, \varphi_2 \in A(\mathbf{T})$ (and so $\varphi \in A(\mathbf{T})$), and prove that $|\varphi| \notin A(\mathbf{T})$, e.g. [Kahane, 1; 2, pp. 255–259].

c) Given $\varphi \in A(\mathbf{T})$ and write $\varphi(\gamma) = r(\gamma)(\cos r_\gamma + i \sin r_\gamma)$, where $r(\gamma) \geqslant 0$ and $r_\gamma \in \mathbf{R}$. Prove that $\sqrt{\varphi} \in A(\mathbf{T})$ if and only if

$$\forall\, \gamma \in \mathbf{R},\ \exists\, n_\gamma \in \mathbf{Z} \text{ such that } r_\gamma = 4n_\gamma \pi + r_{\gamma + 2\pi}$$

(cf. the proof of the Wiener–Lévy theorem in [Zygmund, 2, I, pp. 245–246]).

2.4.5 Algebra homomorphisms

a) Let X be a commutative semi-simple regular Banach algebra with unit and let $Y \subseteq X$ be a subalgebra which separates the points of the maximal ideal space of X. Prove that $X = Y$ if each element of X is locally in Y, e.g. [Rainwater, 1].

b) Prove Proposition 2.4.1. With regard to Proposition 2.4.1a it is well-known and easy to prove (by the closed graph theorem) that every algebra homomorphism $h: X_1 \to X_2$ of Banach algebras X_1 and X_2 is continuous as long as X_1 is semi-simple. The analogous problem for an algebra homomorphism $h: X \to \mathbf{C}$, where X is a Fréchet algebra, is unsolved; a spectral synthesis technique is used with regard to this problem in [Benedetto, 7].

2.4.6 Tensor products and support

Prove Proposition 2.4.4. (*Hint*: Draw a planar picture of the sets for parts b) and c). Part d) is much easier).

2.4.7 Potential theory

A *potential kernel* $K: \mathbf{Я} \to \mathbf{R}$ is an even positive function which is convex on $(0, \infty) \subseteq \mathbf{Я}$. K_α is the potential kernel which in a neighborhood of $0 \in \mathbf{Я}$ equals $1/|\gamma|^\alpha$, for $\alpha > 0$, or $\log(1/|\gamma|)$, for $\alpha = 0$, and which is infinitely differentiable for each $\gamma \neq 0$. The *α-potential* of $T \in D(\mathbf{Я})$ is $U_T^\alpha = T * K_\alpha$, $\alpha \in [0, \infty)$. A compact set $E \subseteq \mathbf{Я}$ has *zero α-capacity* if

$$\forall\, T \in D(\mathbf{Я}) \setminus \{0\} \text{ for which } \operatorname{supp} T \subseteq E,\ U_T^\alpha \notin L^\infty(\mathbf{Я})$$

[Hedberg, 1]; the usual definition of zero α-capacity replaces "$D(\mathbf{Я})$" by "$M(\mathbf{Я})$". If $\alpha = 0$ the α-capacity is called the *logarithmic capacity*.

a) Let $E \subseteq \mathbf{Я}$ be a compact set having zero logarithmic capacity and assume that φ takes constant values on the countable collection of open intervals contiguous to E. If

$$\forall\, \gamma \in \mathbf{Я}, \qquad \varphi \in A(\mathbf{Я})_{\mathrm{loc}}(\gamma),$$

prove that φ is a constant function. (*Hint*: Let $T = \varphi'$, the distributional derivative of φ. By the local membership hypothesis there is $\psi \in A(\mathbf{Я})$ and a bounded open interval $I \supseteq E$ such that $\varphi = \psi$ on I, $\psi = 0$ on $\bar{I}^\sim$, and ψ is infinitely differentiable on $I^\sim$. Define the distributional derivative $S = \psi'$ of ψ; and note that

$$T = S + \theta,$$

and

$$U_T^0 = U_S^0 + U_\theta^0,$$

where $\theta \in C_c^\infty(\mathbf{Я})$. By taking Fourier transforms we see that $U_T^0 \in L^\infty(\mathbf{Я})$. Consequently, $T = 0$ because of our hypothesis, and so we can conclude that φ is a constant). This result is due to [Hedberg, 1]; and he also showed that the result is false for each K^α, $\alpha > 0$.

b) Let $E \subseteq \mathbb{R}$ be the perfect symmetric set characterized by the condition that $\gamma \in E$ if and only if $\gamma = \sum \varepsilon_j r_j$, $\varepsilon_j = 0, 1$, e.g. (E1.1.3). Prove that if

$$\underline{\lim}_{j \to \infty} \left(\frac{1}{2^j} \log \frac{r_j}{r_{j+1}} \right) > 0,$$

then E has zero logarithmic capacity. The conclusion of part a) was first proved for sets E with this $\underline{\lim}$ property in [Kahane and Katznelson, 3].

2.4.8 Inverses in $M(G)$, multipliers, and the Tauberian theorem

This exercise should be compared with Theorem 2.4.4.

a) Prove that

(E2.4.1) $\quad \forall\, \mu \in M(G), \qquad \|\mu_d\|_{A'} \leqslant \|\mu\|_{A'},$

where μ_d is the discontinuous part of μ. (*Hint*: Use Wiener's characterization of continuous measures to prove that for $\varepsilon > 0$ and $\gamma \in \Gamma$ there is a sequence $\{\gamma_m : m = 1, \ldots\} \subseteq \Gamma$ such that if $j < n$ then

$$|(\mu - \mu_d)\hat{\ }(\gamma + \gamma_n - \gamma_j)| < \varepsilon/2.$$

A subsequence of $\{\gamma_m : m = 1, \ldots\}$ exists for which

$$|\hat{\mu}_d(\gamma) - \hat{\mu}_d(\gamma + \gamma_n - \gamma_j)| < \varepsilon/2$$

for j large and $n > j$. Thus, for such j and n,

$$|\hat{\mu}_d(\gamma) - \mu(\gamma + \gamma_n - \gamma_j)| < \varepsilon.).$$

(E2.4.1) is due to [Eberlein, 3] who proved that projections $WAP(\Gamma) \to AP(\Gamma)$ of norm 1 exist (noting, of course, that $B(\Gamma) \subseteq WAP(\Gamma)$); the above proof is due to [Glicksberg and Wik, 1].

b) Recall that $L^1(G)$ is a closed ideal in $M(G)$. Prove the following generalization of Wiener's Tauberian theorem: given $\mu \in M(G)$; $\hat{\mu}$ is never 0 if and only if

(E2.4.2) $\quad \overline{\hat{\mu} A(\Gamma)} = A(\Gamma).$

c) Given $\mu \in M(G)$. Prove that $\mu^{-1} \in M(G)$ if and only if

(E2.4.3) $\quad \hat{\mu} A(\Gamma) = A(\Gamma).$

(*Hint*: If $\mu^{-1} \in M(G)$ then (E2.4.3) is evident. The converse requires the following well-known multiplier theorem: *if $\hbar \in L(L^1(G), L^1(G))$ has the property that*

(E2.4.4) $\quad \forall\, f \in L^1(G)$ and $\forall\, x \in G, \;\; \hbar(\tau_x f) = \tau_x(\hbar f),$

then $\hbar(f) = f * \mu$ for some $\mu \in M(G)$ (cf. Exercise 3.1.3). If (E2.4.3) holds then the map $A(\Gamma) \to A(\Gamma)$, $\varphi \mapsto \hat{\mu}\varphi$, is a continuous bijection; consequently, the inverse map $\hbar$ is continuous by the open mapping theorem. $\hbar$ satisfies (E2.4.4) and so there is an elemen $\nu \in M(G)$ such that

$$\forall\ \varphi \in A(\Gamma),\ \varphi = \hat{\mu}\hat{\nu}\varphi).$$

Thus, *if* $\overline{\hat{\mu}A(\Gamma)} = \hat{\mu}A(\Gamma)$ *and* $\hat{\mu}$ *never vanishes, for* $\mu \in M(G)$, *then* $\mu^{-1} \in M(G)$. [Beurling 1] initially posed the problem of finding when $\hat{\mu}A(\Gamma)$ is closed in $A(\Gamma)$. A recent contribution is due to [Glicksberg, 1]: *assume that* $\hat{\mu} \in B(\Gamma)\backslash\{0\}$ *and that* Γ *is connected* $\hat{\mu}A(\Gamma)$ *is closed in* $A(\Gamma)$ *if and only if* $\mu^{-1} \in M(G)$.

With regard to (E2.4.3), [Dieudonné, 1] (1960) has shown that if Γ is not compact the

$$\forall\ \varphi \in A(\Gamma), \qquad \varphi A(\Gamma) \neq A(\Gamma)$$

(cf. Proposition 1.1.6b and the Cohen factorization theorem).

2.5 The Tauberian theorem in spectral synthesis

2.5.1 Historical commentary on S-sets and Tauberian theorems. We shall now trace the development from Wiener's Tauberian theorem to a generalization (Theorem 2.5.1 and Theorem 2.5.2) which is most important in spectral synthesis. As we shall see, it evolved in the 1940's that such a generalization was possible in light of the fact, which Wiener proved in [Wiener, 7, Lemma 6_{13}, p. 88] (1933) and which we discussed in Paragraph 1.2, that one-point sets have very good spectral synthesis properties.

For perspective, note that Wiener's Tauberian theorem, or, at a more basic level, Proposition 1.1.5b, can be viewed as a theorem in both spectral analysis *and* synthesis; the spectral analysis result was given in Section 1.4.1 and the spectral synthesis result is that $\emptyset$ *is an S-set* (i.e., choose $\varphi \in j(\emptyset)$ which never vanishes and conclude that $\overline{j(\emptyset)} = k(\emptyset) = A(\text{Я})$ by Theorem 1.1.3).

[Carleman, 1, p. 78] (1935) showed that *finite sets are S-sets* by proving that if $\operatorname{card} \operatorname{sp} \mathscr{T}_{\Phi} < \infty$, for $\Phi \in L^{\infty}(\mathbf{R})$, then $\Phi(x) = \sum a_{\gamma} e^{i\gamma x}$, a finite sum. As we indicated earlier, this result was essentially given by Wiener. Carleman's technique involved the function-theoretic "Carleman transform", e.g. [Katznelson, 5, Chapter VI, Section 8].

Next came the remarkable paper by [Ditkin, 1] (1939). Ditkin proved, in our terminology, that *if* $I \subseteq A(\text{Я})$ *is a closed ideal,* $ZI = E \subseteq \text{Я}$, *and* $\operatorname{card} \partial E \leqslant \aleph_0$, *then* $I = k(E)$, i.e., E *is an S-set*. His technique is based on the $D(\varphi, I)$ lemma: if $I \subseteq A(\text{Я})$ is a closed ideal, $\varphi \in A(\text{Я})$, $ZI \subseteq Z\varphi$, and

$$D(\varphi, I) = \{\gamma \in \Gamma : \varphi \notin I_{\mathrm{loc}}(\gamma)\}$$

(for $\Gamma = \text{Я}$ in this case), then $D(\varphi, I)$ is perfect. The proof of the $D(\varphi, I)$ lemma depends on the likes of Theorem 1.2.2a which was then available for $\Gamma = \text{Я}$, i.e., Wiener's "Lemma 6_{13}" mentioned above. The proof also uses Theorem 2.4.1. Shortly after itkin's work, and unaware of it, [Segal, 1; 3, p. 88] (1940) essentially proved the bove Ditkin S-set theorem for $E \subseteq \mathbf{R}$ and card $\partial E \leqslant \aleph_0$. His theorem depended on 'iener's "Lemma 6_{13}" [Segal, 3, Lemma 2.7.1] coupled with his general results on the spectral resolution of ideals" [Segal, 3, Theorem 2.2]. [Fukamiya, 1] (1942) and egal, 3] also proved Theorem 1.1.3 for G and Γ; Fukamiya was aware of [Segal, 1] ut unaware of Ditkin's paper.

he importance of proving a result for G and Γ corresponding to Wiener's "Lemma $_{13}$" was established beyond a doubt in [Shilov, 1] which dates from his thesis in 941. His major theorem in this regard [Loomis, 1, p. 86; Shilov, 1, Theorem 13, . 110] was proved for regular semi-simple Banach algebras X and has the following orm for $X = A(\Gamma)$: *assume that one-point sets in Γ and the point at infinity (of Γ) are -sets, and let $I \subseteq A(\Gamma)$ be a closed ideal with $E = ZI$; if $\varphi \in k(E)$ then $D(\varphi, I)$ is perfect*; ith this result it is easy to check that *if $E = E_1 \cup E_2$ is closed, where E_1 is scattered and $_2$ is open, then E is an S-set*, cf. *Exercise* 2.5.8.

he next stage of development then was to ascertain the spectral synthesis properties f finite sets in an arbitrary Γ. The fact that *finite sets are synthesizable for any Γ* was stablished in 1949 by [Kaplansky, 1] and Riss, e.g. Exercise 1.3.7. Kaplansky eemed unaware of the above Shilov paper (which appeared in 1947) but was aware of itkin's and Segal's work. His technique used structure theory and Segal's ethods; and as such he was able to prove also the above Ditkin–Segal S-set heorem for a large family of LCAG Γ. In 1951, [Helson, 1] proved that *finite sets are -sets in any Γ without using structure theory*; and then, using a modification of an npublished Beurling result for $\Gamma = \mathbf{R}$ (from Beurling's Harvard lectures in 1949), elson proved that *if $T \in A'(\Gamma)$ and ∂ supp T is scattered then $T \in A'_S(\Gamma)$*. Independent f Helson, [Reiter, 1] (1952) proved that *if $I \subseteq A(\Gamma)$ is a closed ideal, $\varphi \in A(\Gamma)$, $I \subseteq Z\varphi$, and $\partial Z\varphi \cap \partial ZI$ is denumerable, then $\varphi \in I$.* The proof of this result works if denumerable" is replaced by "scattered"; and, as Reiter points out, Helson's roof actually gives the full generality of Reiter's statement. Reiter was aware of itkin's and Segal's work but was unaware of Shilov's work; his methods, lthough abstract, are influenced by [Agmon and Mandelbrojt, 1] (1950) which tilized the Carleman transform. Agmon and Mandelbrojt's theorem is the same s Reiter's but is proved for the case that $\Gamma = \mathbf{R}$; as such, the Agmon–Mandelbrojt esult also follows by Shilov's theorem and Wiener's "Lemma 6_{13}". The use of the arleman transform as well as some corollaries special for $\Gamma = \mathbf{R}$ still make [Agmon nd Mandelbrojt, 1] a valuable contribution.

he other major contributions during this period are due to [Beurling, 2] (1945) (cf. aragraph 2.2), [Godement, 1; 2] (1946), and [Mackey, 1] (1951 lectures). Later, Cotlar, 1] (1954) provided a technical clarification.

.5.2 Ditkin's lemma, C-sets, and the major generalization of Wiener's Tauberian heorem. We shall now prove the most appropriate general form of the Tauberian theorem for spectral synthesis, viz. Theorem 2.5.2. The idea of proof stems fro Wiener's Theorem 2.4.1, Ditkin's $D(\varphi, I)$ lemma, Shilov's work, and the fa (which we'll prove) that scattered sets are C-sets; the result is due to [Warner, 1]. W begin with Ditkin's $D(\varphi, I)$ lemma.

Proposition 2.5.1. *Let $I \subseteq A(\Gamma)$ be an ideal and take $\varphi \in A(\Gamma)$.*
a) If $\gamma \notin ZI$ then $\varphi \in I_{\text{loc}}(\gamma)$.
b) If I is closed and $ZI \subseteq Z\varphi$ then $D(\varphi, I)$ is a perfect set.

Proof. a) Take $\gamma \notin ZI$.
Let $\psi \in I$ be 1 at γ and choose $\theta \in A_c(\Gamma)$ such that $\|\theta\|_A < \varepsilon < 1$ and

(2.5.1) $\theta = \psi(\gamma) - \psi = 1 - \psi$

in some neighborhood V of γ; this can be accomplished via Exercise 2.5.1b, which, a we prove there, is an easy consequence of Theorem 1.2.2a.
Since $\|\varphi \sum_0 \theta^n\|_A \leqslant \|\varphi\|_A \sum_0 \|\theta\|_A^n < \|\varphi\|_A \sum \varepsilon^n < \infty$, we compute, by means of th geometric series, that

$$\frac{\varphi}{1-\theta} \in A(\Gamma).$$

Since $\psi \in I$ we have $\dfrac{\varphi\psi}{1-\theta} \in I$; and because of (2.5.1)

$$\frac{\varphi\psi}{1-\theta} = \varphi \qquad \text{on } V.$$

b) For each $\gamma \notin D(\varphi, I)$ we can choose an open neighborhood $N_\gamma \subseteq D(\varphi, I)^\sim$ of γ becau of (2.4.1). Thus, $D(\varphi, I)^\sim$ is open and so $D(\varphi, I)$ is closed.
Let $\gamma \in D(\varphi, I)$ be an isolated point; we shall obtain a contradiction and thus conclu that $D(\varphi, I)$ is perfect.
Because of part a), $\gamma \in ZI$, and therefore $\varphi(\gamma) = 0$ by hypothesis.
Since γ is an isolated point of $D(\varphi, I)$ we can find a compact neighborhood U of γ suc that

(2.5.2) $(U \setminus \{\gamma\}) \cap D(\varphi, I) = \emptyset.$

Let $\psi \in A(\Gamma) \setminus \{0\}$ vanish outside of U.
Since one-point sets are C-sets, there is a sequence $\{\varphi_n : n = 1, \ldots\} \subseteq j(\{\gamma\})$ for which

$$\lim \|\varphi - \varphi\varphi_n\|_A = 0.$$

Using (2.5.2) we easily calculate that

$$\forall \lambda \in \Gamma \text{ and } \forall n, \qquad \varphi\psi\varphi_n \in I_{\text{loc}}(\lambda).$$

We apply Theorem 2.4.1 and conclude that $\varphi\psi \in I$ since I is closed and

$$\|\varphi\psi\varphi_n - \varphi\psi\|_A \leqslant \|\psi\|_A \|\varphi\varphi_n - \varphi\|_A.$$

We can specify ψ to be 1 on a neighborhood $V \subseteq U$ of γ, and so $\varphi \in I_{\text{loc}}(\gamma)$. This is the desired contradiction.

q.e.d.

Thus, *if $E \subseteq \Gamma$ is closed and $\varphi \in k(E)$ then $D(\varphi, \overline{j(E)}) \subseteq \partial E$.* For $E \subseteq \Gamma$ closed we define

$$D(k,j) = \cup \{D(\varphi, \overline{j(E)}) : \varphi \in k(E)\}.$$

Thus, *if E is an S-set then $D(k,j) = \emptyset$.* The proof of Theorem 2.5.1c, especially part (iii), shows that *the converse is also true.* $D(k,j)$ is closed if Γ is metric.

Example 2.5.1 a) Let $n \geqslant 2$ be an integer and let $\mathbf{Z}(n)$ be the cyclic subgroup of $\mathbf{T}$ consisting of the n elements $\gamma = 2\pi j/n, j = 0, \ldots, n-1$. $\widehat{\mathbf{Z}(n)} = \mathbf{Z}(n)$. Define the *complete direct sum*

$$\Gamma_{n,I} = \sum_{i \in I} \mathbf{Z}_i(n),$$

where I is an index set and for each $i \in I$, $\mathbf{Z}_i(n) = \mathbf{Z}(n)$. Clearly, $\Gamma_{n,I}$ is a compact, abelian, perfect, and totally disconnected group. If $I = \mathbf{Z}^+$ then $\Gamma_{n,I}$ is metrizable with metric

$$d(\lambda, \gamma) = \sum_{i=1}^{\infty} \frac{d_i(\lambda_i, \gamma_i)}{2^i},$$

where $\lambda = (\lambda_1, \ldots)$, $\gamma = (\gamma_1, \ldots) \in \Gamma_{n,I}$ and d_i is the natural metric on $\mathbf{Z}_i(n)$. Thus, $\Gamma_{n,I}$ is homeomorphic to the Cantor set when $\operatorname{card} I = \aleph_0$. We designate $\Gamma_{2,I}$, $\operatorname{card} I = \aleph_0$, by D_∞.

b) Let $I = [0, 1]$ and set

$$E = \{\gamma \in \Gamma_{2,I} : \operatorname{card}\{i \in I : \gamma_i = \pi\} \leqslant 1\}.$$

The group identity $0 \in \Gamma_{2,I}$ is the only limit point of E and E is closed. E provides an example of an uncountable scattered set. Scattered sets are countable in separable groups.

c) A closed interval contained in $\mathbb{R}$ is an uncountable set with a scattered boundary. There are no uncountable closed subsets of $\mathbb{R}^n$, $n \geqslant 2$, which have scattered boundaries.

Theorem 2.5.1. *Let $E \subseteq \Gamma$ be closed.*

a) *Assume that for each $\varphi \in k(E)$, $\partial E \cap \partial Z\varphi$ is scattered. Then E is a C-set.*

b) *Assume that for each $\varphi \in k(E)$, $\partial E \cap \partial Z\varphi$ is a C-set. Then E is a C-set.*

c) *If ∂E is scattered then E is a C-set.*

Proof. a.i) *Assuming that* $\varphi \in j(E)_{\mathrm{loc}}(\gamma)$ *we'll prove that*

$$\varphi \in \varphi j(E)_{\mathrm{loc}}(\gamma); \tag{2.5.3}$$

it is not necessary that ∂E be scattered to verify (2.5.3).
Take $\psi_\gamma \in j(E) \cap A_c(\Gamma)$ and N_γ (as in (2.4.1)) such that

$$\forall\, \lambda \in N_\gamma, \qquad \psi_\gamma(\lambda) = \varphi(\lambda).$$

Since $E \cap \operatorname{supp} \psi_\gamma = \emptyset$ and $\operatorname{supp} \psi_\gamma$ is compact, there is $\psi \in j(E) \cap A_c(\Gamma)$ for which $\psi =$ on $\operatorname{supp} \psi_\gamma$.
Thus,

$$\forall\, \lambda \in N_\gamma, \qquad \varphi(\lambda) = \psi_\gamma(\lambda) = \psi(\lambda)\, \psi_\gamma(\lambda) = \psi(\lambda)\, \varphi(\lambda),$$

and we have (2.5.3).
ii) *Let* $\varphi \in k(E)$. It is clear that

$$\forall\, \gamma \in \Gamma \setminus (\partial Z\varphi \cap \partial E), \qquad \varphi \in j(E)_{\mathrm{loc}}(\gamma),$$

e.g. Exercise 2.4.1a.
By part i),

$$\forall\, \gamma \in \Gamma \setminus (\partial Z\varphi \cap \partial E), \qquad \varphi \in \varphi j(E)_{\mathrm{loc}}(\gamma); \tag{2.5.4}$$

it is not necessary that ∂E be scattered to verify (2.5.4).
iii) Let $I = \overline{\varphi j(E)}$ where $\varphi \in k(E)$. Then, by Proposition 2.5.1 and the fact that $ZI \subseteq Z\varphi$ we see that $D(\varphi, I)$ is perfect.
Because of (2.5.4), we have

$$D(\varphi, I) \subseteq \partial Z\varphi \cap \partial E,$$

and, consequently, $D(\varphi, I) = \emptyset$ since $\partial Z\varphi \cap \partial E$ is scattered.
Thus, we apply Theorem 2.4.1, and conclude that $\varphi \in \overline{\varphi j(E)}$.
b) Take $\varphi \in k(E)$, set $F = \partial Z\varphi \cap \partial E$, and choose a sequence $\{\varphi_n : n = 1, \ldots\} \subseteq j(F) \cap A_c(\Gamma)$ for which

$$\lim_{n \to \infty} \|\varphi - \varphi\varphi_n\|_A = 0,$$

since F is a C-set.
Note that $(E \cap \operatorname{supp} \varphi) \cap \operatorname{supp} \varphi_n = F \cap \operatorname{supp} \varphi_n = \emptyset$; and, therefore, since $\varphi_n \in A_c(\Gamma)$, we choose $\psi_n \in j(E)$ which is equal to 1 on $\operatorname{supp} \varphi \cap \operatorname{supp} \varphi_n$.
Clearly, $\varphi\varphi_n = \varphi\varphi_n\psi_n$. Thus

$$\lim_{n \to \infty} \|\varphi - \varphi(\varphi_n\psi_n)\|_A = 0,$$

nd, because $\varphi_n \psi_n \in j(E)$, we conclude that E is a C-set.

) is clear from part a).

q.e.d.

n light of Theorem 2.5.1b, c the following is tempting to state and, alas, is true (Exercise .5.3a).

roposition 2.5.2. *Let $E \subseteq \Gamma$ be closed. If ∂E is a C-set then E is a C-set.*

Theorem 2.5.1c is interesting in light of the above historical development and Theorem .5.2. Note that if $ZI \subseteq Z\varphi$ then

$$\partial ZI \cap \partial Z\varphi = ZI \cap \operatorname{supp} \varphi.$$

Theorem 2.5.2. a) *Let $I \subseteq A(\Gamma)$ be a closed ideal and assume that*

$$ZI \subseteq Z\varphi, \qquad \text{where } \varphi \in A(\Gamma).$$

If $\partial ZI \cap \partial Z\varphi$ is a C-set then $\varphi \in I$.

b) *Given $T \in A'(\Gamma)$ and set $E = \operatorname{supp} T$. If $\varphi \in k(E)$ and $\partial E \cap \partial Z\varphi$ is a C-set then $\langle T, \varphi \rangle = 0$.*

Proof. a) Let $E = ZI$ so that $\varphi \in k(E)$. By hypothesis we choose $\{\psi_n : n = 1, \ldots\} \subseteq$ $i(\partial E \cap \partial Z\varphi) \cap A_c(\Gamma)$ such that

$$(2.5.5) \qquad \lim_{n \to \infty} \| \varphi - \varphi\psi_n \|_A = 0.$$

From Exercise 2.4.1 and the fact that $\overline{j(E)} \subseteq I$ we see that

$$\forall \gamma \in \Gamma \text{ and } \forall n, \qquad \varphi\psi_n \in I_{\text{loc}}(\gamma).$$

Thus, $\varphi\psi_n \in I$ and hence $\varphi \in I$ because of (2.5.5).

b) is clear from a).

q.e.d.

Theorem 1.2.3 is an immediate corollary of Theorem 2.5.2.

2.5.3 Singular support. Given $T \in A'(\Gamma)$. Let $U \subseteq \Gamma$ consist of all points $\gamma \in \Gamma$ for which there is an open neighborhood U_γ of γ and $\mu_\gamma \in M(\Gamma)$ such that $T - \mu_\gamma = 0$ on U_γ. U is open and the *singular support of* T is defined as

$$\operatorname{sing supp} T = \Gamma \setminus U.$$

[Hörmander, 2] used a similar notion with the measure μ_γ replaced by a C^∞-function (on $\mathbb{R}^n$). [Edwards, 3] proved that if $E \subseteq \mathbb{R}^n$ is a closed set then there is $T \in A'(\mathbb{R}^n)$ for which $E = \operatorname{sing supp} T$; this result is clear when "singsupp" is replaced by "supp". Clearly, $\operatorname{sing supp} T \subseteq \operatorname{supp} T$, and [Edwards, 3] gives conditions for equality in terms of Helson sets. In light of Theorem 2.5.2 (properly translated) and the definition of the singular support the following comes as no surprise: *given* $T \in A'(\Gamma)$, $\operatorname{supp} T = E$, *and* $\varphi \in k(E)$; *if* $\varphi \in j(\operatorname{sing\,supp} T)$ *then* $T\varphi = 0$. [Edwards, 4, p. 273] makes an interesting observation on this point and the β topology.

2.5.4 Herz sets. As we mentioned before Theorem 1.4.4, it is not known if every *S*-set is a *C*-set; this is the *C-set-S-set problem.*

Let $E \subseteq \mathbf{T}$ *be an S-set* and let $\varphi \in A(\mathbf{T})$ have the property that $Z\varphi = E$ (not, $Z\varphi \supseteq E$). If $\|\varphi - \varphi_n\|_A < \frac{1}{n}$ for some $\varphi_n \in j(E)$, then, by Proposition 1.1.5b, $\varphi\psi_n = 1$ on $(\operatorname{int} Z\varphi_n)^\sim$ for some $\psi_n \in A(\mathbf{T})$. Thus, $\varphi_n = \varphi(\varphi_n \psi_n)$ on $\mathbf{T}$, and so *we have the C-set property for* $\varphi \in k(E)$ *when* $Z\varphi = E$.

This calculation shows two things: first, the essence of the *C*-set-*S*-set problem is to examine the zeros of $\varphi \in k(E)$ which are *not* in E; and, second, Tauberian methods, in this case, Proposition 1.1.5b, give *C*-set information from *S*-set data. In fact, for the case $I = \overline{j(E)}$ and $\varphi \in k(E)$, Theorem 2.5.1b and Theorem 2.5.2 tell us that the (generalized) Tauberian techniques at our disposal are those which determine *C*-sets; and that if there are *S*-sets which are not *C*-sets then Tauberian techniques will not find them for us.

Clearly, if $E \subseteq \Gamma$ is an *S*-set then E is a *C*-set if and only if

$$\forall\ \varphi \in k(E), \qquad \exists\ \{\varphi_n : n = 1, \ldots\} \subseteq k(E) \qquad \text{such that}$$
$$\lim_{n \to \infty} \|\varphi - \varphi\varphi_n\|_A = 0.$$

Since the *C*-set property is valid for *S*-sets E and functions ψ with the property that $Z\psi = E$, we have the following fact: *let E be an S-set, and assume that*

$$\forall\ \varphi \in k(E), \qquad \exists\ \psi, \theta \in A(\Gamma), \qquad \text{for which} \qquad \varphi = \psi\theta,$$

where $Z\psi = E$; *then E is a C-set.*

In 1956, [Herz, 1] proved that the Cantor set, C, is an *S*-set; it is not known if C is a *C*-set. We prove what is now a common generalization (viz. Theorem 2.5.3) of Herz's Cantor set result, e.g. [Kahane and Salem, 4, pp. 124–125]. [Herz, 5, pp. 229–230] has formulated the result for any Γ and for the α topology, e.g. Paragraph 2.2. The crux of Herz's proof is contained in the following

Proposition 2.5.3. *Given* $\varphi \in A(\mathbf{T})$. *Define the continuous function* $\varphi_N : [0, 2\pi] \to \mathbf{C}$ *to be* φ *at the points* $2\pi k/N$, $k = 0, \ldots, N$, *and to be extended linearly on the remainder of* $[0, 2\pi]$. *Then* $\varphi_N \in A(\mathbf{T})$ *and* $\lim_{N \to \infty} \|\varphi - \varphi_N\|_A = 0$.

Proof. a) If $\varphi \in C^2(\mathbf{T})$ then φ_N' exists on $[0, 2\pi) \setminus \{2\pi r : r \in \mathbf{Q} \cap (0, 1)\}$ and $\{\varphi_N' : N = 1, \ldots\}$ converges uniformly to φ' on this set.

Integration by parts shows that $C^2(\mathbf{T}) \subseteq A(\mathbf{T})$ (cf. Exercise 2.5.2).

Clearly, $\varphi_N \in A(\mathbf{T})$, and if $\hat{g}_N = \varphi - \varphi_N$ we use Hölder's inequality and the Plancherel theorem to compute

$$\begin{aligned} \|\varphi - \varphi_N\|_A &= |g_N(0)| + \sum_n{}' \left| \frac{1}{n} g_N(n) n \right| \\ &\leqslant |g_N(0)| + \left(\sum_n{}' \frac{1}{n^2} \right)^{1/2} \left(\sum_n{}' |n g_N(n)|^2 \right)^{1/2} \\ &\leqslant \|\varphi - \varphi_N\|_{L^1(\mathbf{T})} + \frac{\pi}{\sqrt{3}} \|\varphi - \varphi_N\|_{L^2(\mathbf{T})}. \end{aligned}$$

Since $\lim \|\varphi - \varphi_N\|_{L^\infty(\mathbf{T})} = 0$ and $\lim \|\varphi' - \varphi_N\|_{L^\infty(\mathbf{T})} = 0$ we conclude now that $\lim \|\varphi - \varphi_N\|_A = 0$.

b) Setting $e_n(\gamma) = e^{in\gamma}$ we now show that $\|(e_n)_N\|_A = 1$.

Define the triangle function $\Delta_{2\pi/N}$; this is the function $\hat{f}_\alpha = \varphi_\alpha$ of Proposition 1.2.2 and is the analogue on $\mathbf{T}$ of the function, $\Delta_\lambda \in A(\mathbf{Я})$, defined in Exercise 1.2.6.

Hence, $\operatorname{supp} \Delta_{2\pi/N} \subseteq [-2\pi/N, 2\pi/N]$ and $\Delta_{2\pi/N}(0) = 1$.

Thus,

$$(e_n)_N(\gamma) = \sum_{k=1}^{N} e_n\left(\frac{2\pi k}{N}\right) \tau_{\frac{2\pi k}{N}} \Delta_{\frac{2\pi}{N}}(\gamma) = \mu_N * \Delta_{\frac{2\pi}{N}}(\gamma)$$

where

$$\mu_N = \sum_{k=1}^{N} e_n\left(\frac{2\pi k}{N}\right) \delta_{\frac{2\pi k}{N}} \in M(\mathbf{T}).$$

Clearly, by the convention (1.3.7), $(f\hat{\mu}_-)^\wedge = \mu * \varphi$ where $\mu \in M(\mathbf{T})$ and $\hat{f} = \varphi \in A(\mathbf{T})$.

Now, $\widehat{\mu_N}_-(m) = N$ for infinitely many $m \in \mathbf{Z}$, and, for the remainder,

$$\widehat{\mu_N}_-(m) = \sum_{k=1}^{N} z^k = z(1 - z^N)/(1 - z) = 0,$$

where $z = e_{(n-m)}(2\pi/N)$.

Consequently, $\widehat{\mu_N}_- \geqslant 0$ and so the Fourier coefficients $g(m)$, of $\hat{g} = \mu_N * \Delta_{2\pi/N}$ are non-negative since "$f_\alpha \geqslant 0$" (as we proved in Proposition 1.2.2).

Thus, $\|(e_n)_N\|_A = \sum g(m) = (e_n)_N(0) = 1$.

c) If $\varphi(\gamma) = \sum f(n) e^{in\gamma} \in A(\mathbf{T})$ then $\varphi_N(\gamma) = \sum f(n)\,(e_n)_N(\gamma)$, and so

$$\|\varphi_N\|_A \leqslant \|\varphi\|_A,$$

by part b).

d) Take $\varphi \in A(\mathbf{T})$ and $\varepsilon > 0$.

Choose $\psi \in C^2(\mathbf{T})$ such that $\|\varphi - \psi\|_A < \varepsilon/3$ and choose M large enough so that

$$\forall\, N \geqslant M, \qquad \|\psi - \psi_N\|_A < \varepsilon/3,$$

by part a).

Then, because $\|\varphi_N - \psi_N\|_A = \|(\varphi - \psi)_N\|_A \leqslant \|\varphi - \psi\|_A$ by part c), we conclude that

$$\forall\, N \geqslant M, \qquad \|\varphi - \varphi_N\|_A < \varepsilon.$$

q.e.d.

The reason to formulate Proposition 2.5.3 becomes clear in the proof of Theorem 2.5.3a.

A closed set $E \subseteq \mathbf{T}$ is an *Herz set* if there is a sequence $\{N_n : n = 1, \ldots\}$ of positive integers tending to infinity such that

(2.5.6) $\qquad \forall\, n = 1, \ldots \quad$ and $\quad \forall\, k = 1, \ldots, N_n,$

$$d\left(E, \frac{2\pi k}{N_n}\right) < \frac{2\pi}{N_n} \quad \Rightarrow \quad \frac{2\pi k}{N_n} \in E,$$

where $d(E, \gamma) = \inf\{d(\lambda, \gamma) : \lambda \in E\}$ and

$$\forall\, \lambda, \gamma \in \mathbf{T}, \qquad d(\lambda, \gamma) = \inf\{|\lambda - \gamma + 2\pi j| : j \in \mathbf{Z}\}.$$

Theorem 2.5.3 a) *Each Herz set E is an S-set.*

b) *If $E \subseteq \mathbf{T}$ is closed then there is $D \subseteq \mathbf{T}$,* card $D \leqslant \aleph_0$, *such that $F = E \cup D$ is closed and is an S-set.*

Proof. a) Given $T \in A'(E)$, $\varphi \in k(E)$, and the sequence $\{N_n : n = 1, \ldots\}$. We'll prove that $\langle T, \varphi \rangle = 0$.

Set

$$T_n = \frac{N_n}{2\pi} T * \Delta_{2\pi/N_n}$$

("Δ" was defined in Proposition 2.5.3) and define

$$\mu_n = \frac{2\pi}{N_n} \sum_{k=1}^{N_n} \delta_{2\pi k/N_n}.$$

The key observation is that

$$\int_{\mathbf{T}} T_n(\gamma)\,\varphi(\gamma)\,\mathrm{d}\mu_n(\gamma) = 0; \tag{2.5.7}$$

in fact, $\operatorname{supp} T_n \subseteq \left\{\gamma : d(E,\gamma) \leqslant \frac{2\pi}{n}\right\}$ so that we obtain (2.5.7) from (2.5.6).

A trivial calculation shows that $\lim T_n = T$ in $\sigma(A'(\mathbf{T}), A(\mathbf{T}))$, and so, since "$\mu_n \to m$", it is reasonable to conjecture that $\langle T, \varphi\rangle = 0$ in light of (2.5.7). This is where we need Proposition 2.5.3.

The left-hand side of (2.5.7) is

$$\sum_{k=1}^{N_n} T * \Delta_{\frac{2\pi}{N_n}}\left(\frac{2\pi k}{N_n}\right)\varphi\left(\frac{2\pi k}{N_n}\right) = \left\langle T_\gamma,\ \sum_{k=1}^{N_n} \Delta_{\frac{2\pi}{N_n}}\left(\frac{2\pi k}{N_n} - \gamma\right)\varphi\left(\frac{2\pi k}{N_n}\right)\right\rangle$$

$$= \langle T, \varphi_{N_n}\rangle;$$

so that since $\lim \|\varphi - \varphi_N\|_A = 0$ we obtain that $\langle T, \varphi\rangle = 0$ from (2.5.7).

b) Take $N_n = n$ and let D be the subset of $X = \left\{\frac{2\pi k}{n} : k = 1, \ldots, n \text{ and } n = 1, \ldots\right\}$ defined by the condition that

$$\frac{2\pi k}{n} \in D \quad \Rightarrow \quad d\left(E, \frac{2\pi k}{n}\right) < \frac{2\pi}{n}.$$

Then $F = E \cup D$ is an Herz set.

q.e.d.

Let $E \subseteq \mathbf{T}$ be an Herz set. If $T \in A'(E)$ then the proof of Theorem 2.5.3 exhibits a sequence $\{\mu_n : n = 1, \ldots\} \subseteq M(E)$ for which $\lim \mu_n = T$ in $\sigma(A'(\mathbf{T}), A(\mathbf{T}))$ and for which

$$\sup_n \|\mu_n\|_{A'} \leqslant \|T\|_{A'}.$$

The fact that we can synthesize with a *sequence* describes the phenomenon of bounded synthesis which we shall discuss in Section 3.2.13. There are generalizations of Herz sets for which bounded synthesis holds and is essentially characterized [Katznelson and Körner, 1].

2.5.5 Pisot numbers and spectral synthesis properties of perfect symmetric sets

Example 2.5.2 a) The Cantor set $C \subseteq \mathbf{T}$ is an S-set. This follows from Theorem 2.5.3 since C is an Herz set for $N_n = 3^n$.

b) If $E \subseteq \mathbf{T}$ is a perfect symmetric set determined by $\{\xi_k : k = 1, \ldots\} \subseteq (0, \frac{1}{2})$ where each $\xi_k = \xi \in \mathbf{Q}$ and $\frac{1}{\xi} \notin \mathbf{Z}$, then $E = E_\xi$ is *not* an Herz set [Rosenthal, 3, Theorem 4].

Remark 1. In light of Example 2.5.2b it is natural to investigate when $E_\xi \subseteq \mathbf{T}$, for an arbitrary $\xi \in (0, \frac{1}{2})$, is an *S*-set. An algebraic integer $\alpha > 1$, each of whose conjugates $\beta \neq \alpha$ satisfies the condition $|\beta| < 1$, is a *Pisot number*. The main interest of these numbers in harmonic analysis stems from Salem's incredible result (1943–1955), e.g. [Salem, 4]: *let* $\xi \in (0, \frac{1}{2})$; E_ξ *is a U-set if and only if* $\alpha = 1/\xi$ *is a Pisot number*. Prior to Salem's work, Bary (1937) proved: *given* $\xi \in \mathbf{Q} \cap (0, \frac{1}{2})$; E_ξ *is a U-set if and only if* $\alpha = \dfrac{1}{\xi} \in \mathbf{Z}$ (cf. Example 2.5.2b). The uniqueness properties of perfect symmetric sets are dealt with in [Bary, 1; 2]. [Meyer, 3; 4; 5; 6] proved that *if* $\alpha > 2$ *is a Pisot number then* E_ξ, $\xi = 1/\alpha$, *is an S-set*. His technique depends on a generalization of the arithmetic progressions, $\{2\pi k/N_n : k = 1, \ldots, N_n\}$, used in Theorem 2.5.3; and he synthesizes in the narrow topology, e.g. Paragraph 2.2. It is not known if E_ξ is a non-*S*-set for some $\xi \in (0, \frac{1}{2})$.

2. [Rosenthal, 3] gives a "structural" characterization of Herz sets.

2.5.6 The finite union of *S*-sets. We know from Theorem 2.5.1c or Exercise 2.5.3b that countable closed sets $E \subseteq \mathbf{T}$ are *C*-sets. With regard to Theorem 2.5.3b it is natural to ask if a *closed* countable set $D \subseteq \mathbf{T}$ can be added to a non-*S*-set $E \subseteq \mathbf{T}$ so that $E \cup D$ is an *S*-set. In fact this situation can *never* arise because of the following result [Warner, 2, Theorem 1.4].

Theorem 2.5.4. *Let* $E_1, E_2 \subseteq \Gamma$ *be closed sets and assume that* $E_1 \cap E_2$ *is a C-set.* $E = E_1 \cup E_2$ *is an S-set if and only if both* E_1 *and* E_2 *are S-sets.*

Warner points out Reiter's influence in the formulation and proof of Theorem 2.5.4. In fact, [Reiter, 6, pp. 557–558] proved the analogue of Theorem 2.5.4 for the case that $E_1 \cap E_2 = \emptyset$. [Calderón, 1, p. 3] and [Herz, 5, p. 228] proved one direction of Theorem 2.5.4, viz. *if* $E_1, E_2 \subseteq \Gamma$ *are S-sets for which* $E_1 \cap E_2$ *is a C-set then* $E_1 \cup E_2$ *is an S-set*. A recent result by [Saeki, 4] (1969) is: *let* $E_1, E_2 \subseteq \Gamma$ *be S-sets and let* $F \subseteq E_1 \cup E_2$ *be a C-set which contains* $\partial E_1 \cap \partial E_2 \cap \partial E$; *then* $E_1 \cup E_2$ *is an S-set* (cf. Proposition 2.5.4). Generally, it is not known if the union of two *S*-sets (even for the case $\Gamma = \mathbf{T}$) is an *S*-set (cf. Exercise 2.5.3b). This union problem is obviously closely related to the *C*-set-*S*-set problem. In this regard, we say that a closed set $E \subseteq \mathbf{T}$ is a *Saeki set* [Saeki, 1, p. 246] if

$$\forall\, \varphi, \psi \in k(E), \qquad \varphi\psi \in \overline{j(E)}.$$

Clearly, *S*-sets are Saeki sets. The following is trivial to check:

Proposition 2.5.4. *Assume that every non-S-set in* $\mathbf{T}$ *is a non-Saeki set.* If $E_1, E_2 \subseteq \mathbf{T}$ are *S-sets then* $E_1 \cup E_2$ *is an S-set.*

2.5.7 The intersection of *S*-sets.

Proposition 2.5.5. *Let* $E \subseteq \mathbb{R}^m$ *be a closed convex set. Then E is an S-set.*

Proof. Suppose $0 \in \mathbb{R}^m$ is the center of E and let $\varphi \in k(E)$.

Let $\psi_n(\gamma) = \left(1 - \frac{1}{n}\right)^m \varphi\left(\left(1 - \frac{1}{n}\right)\gamma\right)$.

Then $\psi_n \in j(E)$ and $\lim_{n\to\infty} \|\psi_n - \varphi\|_A = 0$.

q.e.d.

Thus $B^m = \{\gamma \in \mathbb{R}^m : |\gamma| \leqslant 1\}$ is an S-set and the same argument shows that $(\operatorname{int} B^m)^{\sim}$ is an S-set. Consequently, the intersection of S-sets is not necessarily an S-set because of Exercise 2.5.5 (where we show that $S^2 \subseteq \mathbb{R}^3$ is a non-S-set). [Varopoulos, 5] has shown that this phenomenon about the non-S-set intersection of (perfect) S-sets is valid in the case of any compact Γ. On the other hand, it is not difficult to check that if Γ is compact and $E \subseteq \Gamma$ and $F \subseteq \Gamma$ are C-sets, then $E \cap F$ is a C-set when $(\operatorname{int} E) \cup (\operatorname{int} F) = \Gamma$.

$E \subseteq \mathbb{R}^n$ is a *polygon* if it is a compact connected set obtained from closed half-spaces by a finite number of intersections and unions. Polygons in $\mathbb{R}^n, n > 1$, are *not* strong Ditkin sets; and a polygon $E \subseteq \mathbb{R}^n$, $n > 1$, is convex if and only if $(\operatorname{int} E)^{\sim}$ is a strong Ditkin set [Meyer and Rosenthal, 1].

2.5.8 Sets of spectral resolution and arithmetic conditions. In Theorem 2.4.4 we made use of Kronecker's theorem and we shall give another application in Section 3.2.12. Because of the importance of various independent sets in harmonic analysis (e.g. [Benedetto, 6]), Kronecker's theorem has led to the following definition. A compact set $E \subseteq \Gamma$ is a *Kronecker set* if

$$\forall\ \varepsilon > 0 \text{ and } \forall\ \varphi \in C(E), \text{ for which } |\varphi| = 1, \exists\ x \in G \text{ such that}$$

$$\forall\ \gamma \in E, \quad |\varphi(\gamma) - (\gamma, x)| < \varepsilon.$$

A compact set $E \subseteq \Gamma$ is *independent* if

$$\sum_1^k n_j \gamma_j = 0, \qquad \text{where } n_j \in \mathbf{Z}, \gamma_j \in E, \text{ and } \gamma_i \neq \gamma_j,$$

implies $n_1 = \ldots = n_k = 0$. It is easy to check that Kronecker sets are independent sets; and a simple application of the Radon–Nikodym theorem shows further that if E is a Kronecker set then

$$\forall\ \mu \in M(E), \qquad \|\mu\|_{A'} = \|\mu\|_1$$

(cf. Example 2.5.4). It is interesting to note that *every compact Hausdorff space X is homeomorphic to some Kronecker set E contained in some compact Γ*; in fact, let $\Gamma = \Pi\,\{\mathbf{T}_\varphi : \varphi \in C(X), |\varphi| = 1, \text{ and } \mathbf{T}_\varphi = \mathbf{T}\}$ and let $E = f(X)$ where for each $x \in X, f(x) = \{\varphi(x) : \varphi \in C(X) \text{ and } |\varphi| = 1\}$. We say that Γ is a *torsion module* (over $\mathbf{Z}$) if

$$\forall\ \gamma \in \Gamma \quad \exists\ n \in \mathbf{Z} \setminus \{0\} \qquad \text{such that } n\gamma = 0.$$

Thus D_∞ is a torsion module and so has no independent subsets. If p is a prime number we say that a compact set $E \subseteq \Gamma$ is *p-independent* if

$$\sum_1^k n_j \gamma_j = 0, \qquad \text{where } n_j \in \mathbf{Z},\ \gamma_j \in E, \text{ and } \gamma_i \neq \gamma_j,$$

implies $n_j = 0 \pmod{p}$ for each j. Thus there are points $\gamma \in D_\infty$ for which $\{\gamma\}$ is a p-independent set.

Also we say that a closed set $E \subseteq \Gamma$ is a set of *spectral resolution* if every closed subset $F \subseteq E$ is an S-set; and a closed set $E \subseteq \Gamma$ is a set of *strong spectral resolution* if

$$A'(E) = M(E).$$

There are sets E of spectral resolution for which $A_S'(E) \setminus M(E) \neq \emptyset$ [Varopoulos, 3]. In light of our remarks about the union of S-sets the following result [Saeki, 4] is interesting: *if* $E_1 \subseteq \Gamma$ *is an S-set and* E_2 *is a set of strong spectral resolution then* $E_1 \cup E_2$ *is an S-set*, cf. Exercise 2.5.8.

[Varopoulos, 2] (1965) proved that Kronecker sets in $\mathbf{T}$ are sets of strong spectral resolution, e.g. Example 2.5.3; and the reason we mention these notions now is that it is not known if all sets of strong spectral resolution, or even Kronecker sets, are C-sets. We conjecture that a counterexample for the C-set–S-set problem can be found among the sets of strong spectral resolution, contrary to the devil-may-care opinion we expressed in [Benedetto, 6, p. 159]. As we noted above, it is also not known if the Cantor set, C, is a C-set; C is not a set of spectral resolution [Kahane and Katznelson, 2]. Sets of spectral resolution were introduced by [Malliavin, 4] (1962) for $\Gamma = \mathbf{T}$, and he proved that such sets are U-sets; [Filippi, 1] later extended this result to arbitrary Γ.

2.5.9 Totally disconnected sets and measure theoretic properties of pseudo-measures. Theorem 2.5.1 and Theorem 2.5.2 show the importance of boundary sets to determine spectral synthesis properties; and the potpourri of definitions, unproved results, and open problems listed after these two theorems indicates a need to take a closer look at "irregular" boundary sets. Detailed expositions of the role of "irregular" sets in spectral synthesis are found in [Benedetto, 6; Hewitt and Ross, 1, II; Kahane, 13; Kahane and Salem, 4].

We proceed in the following framework. Let $E \subseteq \Gamma$ be a compact totally disconnected set and define

$$\mathscr{F} = \{F \subseteq E : F \text{ is open in } E \text{ and closed}\}.$$

It is easy to check that since E is compact then the total disconnectedness of E is equivalent to the fact that $\mathscr{F}$ is a topological basis for E. $\mathscr{F}$ is also an algebra of sets. A finite disjoint family $\{F_j : j = 1,\ldots,n\} \subseteq \mathscr{F}$ is a *finite decomposition* of $F \in \mathscr{F}$ if $F = \bigcup_1^n F_j$. For $F \in \mathscr{F}$, we let $\psi_F \in A_c(\Gamma)$ be 1 on a neighborhood of F and 0 on a neighborhood of $E \setminus F$. Then $T \in A'(E)$ is a well-defined finitely additive set function with domain $\mathscr{F}$ and defined by

$$(2.5.8) \qquad \forall\, F \in \mathscr{F}, \qquad T(F) = \langle T, \psi_F \rangle.$$

A *finite decomposition* of $T \in A'(E)$ is a finite set $\{T_j : j = 1, \ldots n\} \subseteq A'(\Gamma)$ and a finite decomposition $\{F_j : j = 1, \ldots, n\} \subseteq \mathscr{F}$ of E such that $T = \sum_1^n T_j$ and $\operatorname{supp} T_j \subseteq F_j$ for each $j = 1, \ldots, n$. If $T \in A'(E)$ and $\{F_j : j = 1, \ldots, n\} \subseteq \mathscr{F}$ is a finite decomposition of E then there is a unique finite decomposition of T given by $T_j = T\psi_{F_j}, j = 1, \ldots, n$. Because E is compact we see that T, as a finitely additive set function on $\mathscr{F}$, is also a countably additive set function on $\mathscr{F}$. It is natural then, especially in light of the notion of strong spectral resolution, to see if we can determine some element $\mu_T \in M(E)$ for which $\mu_T = T$ on $\mathscr{F}$.

Define the semi-norms, $\| \ \|_v$ and $\| \ \|_V$, on $A'(E)$ as:

$$\forall\, T \in A'(E), \qquad \|T\|_v = \sup\{|T(F)| : F \in \mathscr{F}\}$$

and

$$\forall\, T \in A'(E), \qquad \|T\|_V = \sup\{\textstyle\sum |T(F_j)| : \{F_j\} \text{ is a finite decomposition of } E\}.$$

Clearly, $\|T\|_v < \infty$ if and only if $\|T\|_V < \infty$. Define

$$\mathscr{V} = \{T \in A'(E) : \|T\|_v < \infty\},$$

and let $\mathscr{F}_\sigma$ be the smallest σ-algebra containing $\mathscr{F}$. By the Caratheodory–Hopf theorem, e.g. [Hewitt and Stromberg, 1, pp. 141–142; Royden, 1, pp. 253–259], we obtain the following result since $T \in A'(E)$ is countably additive on $\mathscr{F}$:

Proposition 2.5.6. *Let $E \subseteq \Gamma$ be a compact totally disconnected set and let $T \in A'(E)$ be an element of $\mathscr{V}$. Then there is a unique complex valued countably additive set function ν_T defined on $\mathscr{F}_\sigma$ for which $\nu_T = T$ on $\mathscr{F}$.*

2.5.10 Measures associated with pseudo-measures. It is conceivable that there are open sets $U \subseteq E$ which are not in $\mathscr{F}_\sigma$, and thus $\mathscr{F}_\sigma$ is properly contained in the Borel algebra $\mathscr{B}(E)$. Clearly, if E has a countable topological basis then $\mathscr{F}_\sigma = \mathscr{B}(E)$. Problems with axioms of countability are dispensed with by Theorem 2.5.5.

Define

$$(2.5.9) \qquad A_E(\Gamma) = \{\varphi \in A(\Gamma) : \exists\, U_\varphi \supseteq E, \text{ open in } \Gamma, \text{ such that card } \varphi(U_\varphi) < \infty\}.$$

The space $A_E(\Gamma)$ was used by [Katznelson and Rudin, 1] in their study of semi-simple commutative Banach algebras X for which the Stone–Weierstrass theorem is valid; in particular, they proved that *a necessary and sufficient condition in order that every self-adjoint separating subalgebra X of $A(\Gamma)$, with the property*

$$\forall\, \gamma \in \Gamma, \qquad \exists\, \varphi \in X \text{ such that } \varphi(\gamma) \neq 0,$$

be dense in $A(\Gamma)$ is that Γ be totally disconnected.

The following result was used by Varopoulos in his proof that Kronecker sets are sets of strong spectral resolution.

Theorem 2.5.5. *Let $E \subseteq \Gamma$ be a compact totally-disconnected set and let $T \in A'(E)$ be an element of $\mathscr{V}$. Then there is a unique complex measure $\mu_T \in M(\operatorname{supp} T)$ for which $\mu_T = T$ on $\mathscr{F}$ (and $\mu_T = \nu_T$ (defined by Proposition 2.5.6) on $\mathscr{F}_\sigma$).*

Proof. a) $\overline{A_E(\Gamma)} = C_0(\Gamma)$ in the sup norm because of the Stone–Weierstrass theorem and the fact that E is totally disconnected.
b) Define the linear functional

$$(2.5.10) \qquad \begin{aligned} \mu_T : A_E(\Gamma) &\to \mathbf{C}, \\ \varphi &\mapsto \langle T, \varphi \rangle. \end{aligned}$$

For each fixed $\varphi \in A_E(\Gamma)$ with corresponding U_φ (as in (2.5.9)), set

$$(2.5.11) \qquad \forall\, x \in \varphi(E), \qquad U_x = U_\varphi \cap \varphi^{-1}(x).$$

Clearly,

$$(2.5.12) \qquad E \subseteq \cup \{U_x : x \in \varphi(E)\}.$$

It is also easy to see that each U_x is open. In fact, let N be an open neighborhood of x such that

$$N \cap (\varphi(U_\varphi) \setminus \{x\}) = \emptyset.$$

Then $U_\varphi \cap \varphi^{-1}(N) = U_\varphi \cap \varphi^{-1}(x)$ and, obviously, $U_\varphi \cap \varphi^{-1}(N)$ is open.
$\{U_x : x \in \varphi(E)\}$ is a finite disjoint family. If we set $F_x = E \cap U_x$, for $x \in \varphi(E)$, then F_x is closed by (2.5.11) and open in E by the previous observation.
Thus $\{F_x : x \in \varphi(E)\} \subseteq \mathscr{F}$ is a finite decomposition of E. We write $T = \sum_{x \in \varphi(E)} T_x$, where $\{T_x : x \in \varphi(E)\}$ is the corresponding finite decomposition of T.
c) If $\varphi \in A_E(\Gamma)$ we have $\varphi = x$ on U_x and so

$$(2.5.13) \qquad \mu_T(\varphi) = \sum_{x \in \varphi(E)} \langle T\psi_{F_x}, \varphi \rangle = \sum_{x \in \varphi(E)} x \langle T, \psi_{F_x} \rangle = \sum_{x \in \varphi(E)} x\, T(F_x).$$

Thus

$$\forall\, \varphi \in A_E(\Gamma), \qquad |\mu_T(\varphi)| \leqslant \|\varphi\|_{C(E)} \|T\|_V.$$

Therefore, by part a), $\mu_T \in M(\Gamma)$, and by the definition of μ_T we have that $\mu_T = T$ on $\mathscr{F}$. Also by part a), the restriction of $A_E(\Gamma)$ to E is sup norm dense in $C(E)$ and so we have $\operatorname{supp} \mu_T \subseteq E$. Even more, the above estimate allows us to take $\operatorname{supp} \mu_T \subseteq \operatorname{supp} T$.
q.e.d.

The following is easy to check.

Proposition 2.5.7. *Let E be a compact totally disconnected set.*

a) *For all $\nu \in M(E)$, μ_ν, defined by* (2.5.10), *is equal to ν.*

b) *For all $T \in A'(E)$, $\|T\|_V = \|\mu_T\|_1$.*

c) *If* $T \in A'(E)$ *is an element of* $\mathscr{V}$ *then*

$$\|T - \mu_T\|_V = 0.$$

Theorem 2.5.6. *Let* $E \subseteq \Gamma$ *be a compact totally disconnected set and let* $T \in A'(E)$. *If there is* $N_T > 0$ *such that for every finite decomposition* $\{T_j : j = 1, \ldots, n\}$ *of* T,

(2.5.14) $\quad \sum \|T_j\|_{A'} \leqslant N_T,$

then

(2.5.15) $\quad \|T\|_{A'} \leqslant \|T\|_V < \infty$

and, in fact,

$$T = \mu_T.$$

Proof. a) Given $\varepsilon > 0$ and let $\varphi \in A(\Gamma)$ have the properties that $|\varphi| \leqslant 1$ and $|\|T\|_{A'} - \langle T, \varphi\rangle| < \varepsilon/2$.

Take $\lambda \in E$ and a neighborhood U_λ of λ. By Exercise 2.5.1b there is $\theta_\lambda \in A_c(\Gamma)$ such that $\|\theta_\lambda\|_A < \varepsilon/(2N_T)$ and $\theta_\lambda = 0$ on $U_\lambda^\sim$, and there is a neighborhood $V_\lambda \subseteq U_\lambda$ of λ for which

$$\forall\, \gamma \in V_\lambda, \qquad \varphi(\gamma) - \varphi(\lambda) = \theta_\lambda(\gamma).$$

b) Since E is totally disconnected and V_λ is a neighborhood we can choose $F_\lambda \in \mathscr{F}$ such that $\lambda \in F_\lambda$ and $F_\lambda \cap V_\lambda^\sim = \emptyset$. Let $K_\lambda = E \setminus F_\lambda$ so that since F_λ and K_λ are compact subsets of Γ we can find a compact neighborhood $W_\lambda \subseteq V_\lambda$ of F_λ which is disjoint from K_λ. Thus $\partial W_\lambda \cap F_\lambda = \emptyset$ and $\partial W_\lambda \cap K_\lambda = \emptyset$ and, hence, we have found a compact neighborhood $W_\lambda \subseteq V_\lambda$ (in Γ) of λ for which

(2.5.16) $\quad \partial W_\lambda \cap E = \emptyset.$

Since E is compact we choose $\{W_{\lambda_1}, \ldots, W_{\lambda_n}\}$, whose union covers E and such that (2.5.16) is satisfied for each W_{λ_j}.

c) Because of (2.5.16) we can choose $\{W_{\lambda_1}, \ldots, W_{\lambda_n}\}$ to be a disjoint collection (cf. [Benedetto, 3, Lemma 4.1]).

d) $F_{\lambda_j} = W_{\lambda_j} \cap E \in \mathscr{F}, j = 1, \ldots, n$, and we have

$$\langle T, \varphi\rangle = \langle T, \sum \varphi \psi_{F_{\lambda_j}}\rangle = \sum \langle T, \varphi(\lambda_j)\psi_{F_{\lambda_j}}\rangle + \sum \langle T\psi_{F_{\lambda_j}}, \theta_{\lambda_j}\rangle.$$

Consequently,

$$|\langle T, \varphi\rangle| \leqslant \sum |T(F_{\lambda_j})| + \frac{\varepsilon}{2N_T} \sum \|T\psi_{F_{\lambda_j}}\|_{A'}.$$

By hypothesis, then, $\|T\|_{A'} \leqslant \|T\|_V$.

Obviously, for any finite decomposition $\{F_j : j = 1, \ldots, n\} \subseteq \mathscr{F}$ of E,

$$\sum |T(F_j)| \leqslant \sum \|T\psi_{F_j}\|_{A'} \leqslant N_T,$$

and, thus, $\|T\|_V < \infty$.

e) $T = \mu_T$ because of (2.5.15) and Proposition 2.5.7c; in fact,

$$\sum \|T_j - \mu_{T_j}\|_{A'} \leqslant N_T + \|\mu_T\|_1,$$

and so $\|T - \mu_T\|_{A'} \leqslant \|T - \mu_T\|_V = 0$.

q.e.d.

The above result is found in [Benedetto, 3, Theorem 4.1a]. (2.5.14) and (2.5.15) are clear if E is a set of strong spectral resolution, and the content of Theorem 2.5.6 is to obtain the "measure inequality" $\| \ \|_{A'} \leqslant \| \ \|_V$ for pseudo-measures which satisfy (2.5.14).

2.5.11 Varopoulos' lemma. The following result is due to [Varopoulos, 2, Lemme 2; 9, Chapitre IV]; but has its origin in Rajchman's proof that the Cantor set, C, is a U-set; the technique should be compared with Exercise 2.5.1.

Theorem 2.5.7. *Let Γ be compact. For each $\varepsilon > 0$, there is $\delta(\varepsilon) \in (0, \varepsilon)$ such that*

$$\forall\, E \subseteq \Gamma \text{ where } E \text{ is closed}, \ \forall\, x_1, x_2 \in G, \text{ and } \forall\, c \in \mathbf{C} \text{ for which } |c| = 1$$

$$\sup_{\gamma \in E} |(\gamma, x_1) - c(\gamma, x_2)| < \delta(\varepsilon)$$

implies

$$\forall\, T \in A'(E), \quad |\hat{T}(x_1) - c\hat{T}(x_2)| < \varepsilon \|T\|_{A'}.$$

Proof. i) Because of Exercise 2.5.1a there is $\eta(\varepsilon) > 0$ and $\theta \in A_c(\mathbf{T})$ such that

(2.5.17) $\quad \forall\, \gamma \in \mathbf{T}, \quad |1 - e^{i\gamma}| < \eta(\varepsilon) \ \Rightarrow \ \theta(\gamma) = 1 - e^{i\gamma}$

and $\|\theta\|_A < \varepsilon$.

Considering $\mathbf{T}$ as the unit circle in $\mathbf{C}$, (2.5.17) becomes

(2.5.18) $\quad \forall\, z \in \mathbf{C}$ for which $|z| = 1, \quad |1 - z| < \eta(\varepsilon) \ \Rightarrow \ \theta(z) = 1 - z.$

ii) Thus for any fixed $x \in G$ and $d \in \mathbf{C}$ for which $|d| = 1$ we set

$$\forall\, \gamma \in \Gamma, \quad \theta_{x,d}(\gamma) = \theta(d(\gamma, x)).$$

Clearly, if $\theta(z) = \sum a_n z^n$, $|z| = 1$, then

$$\theta_{x,d}(\gamma) = \sum_n a_n d^n (\gamma, nx) = \sum_{\substack{y \in \eta x \\ n \in \mathbf{Z}}} (a_n d^n)(\gamma, y);$$

and thus, not only does $\theta_{x,d} \in A(\Gamma)$, but also

$$\|\theta_{x,d}\|_A = \sum |a_n d^n| = \sum |a_n| < \varepsilon.$$

iii) From (2.5.18) we conclude that for each $\varepsilon > 0$ there is $\eta(\varepsilon) > 0$ such that for all $d \in \mathbf{C}$ for which $|d| = 1$ and for all $x \in G$

(2.5.19) $\quad |1 - d(\gamma, x)| < \eta(\varepsilon) \ \Rightarrow \ 1 - d(\gamma, x) - \theta_{x,d}(\gamma) = 0.$

iv) Now, given c, x_1, and x_2 as in the statement of the theorem, we define

$$\varphi(\gamma) = \bar{c}(\gamma, x_1)\,[1 - c(\gamma, x_2 - x_1) - \theta_{x,c}(\gamma)],$$

where $x = x_2 - x_1$.

Let $\delta(\varepsilon) = \min\left(\varepsilon, \dfrac{\eta(\varepsilon)}{2}\right)$. $\delta(\varepsilon)$ is independent of x_1, x_2, c, and E.

If

$$\forall\, \gamma \in E, \qquad |(\gamma, x_1) - c(\gamma, x_2)| = |1 - c(\gamma, x_2 - x_1)| < \delta(\varepsilon)$$

then $E \subseteq U = \{\gamma \in \Gamma : |1 - c(\gamma, x_2 - x_1)| < \eta(\varepsilon)\}$ and U is open (since $\psi(\gamma) = 1 - c(\gamma, x_2 - x_1)$ is continuous).

By (2.5.19), $\varphi = 0$ on U; and so

$$\forall\, T \in A'(E), \qquad \langle T, \varphi\rangle = 0.$$

v) On the other hand,

$$c\langle T, \varphi\rangle = c\bar{c}\hat{T}(x_1) - c\hat{T}(x_2) - c\bar{c}\langle T_\gamma, (\gamma, x_1)\,\theta_{x,c}(\gamma)\rangle;$$

so that

$$|\hat{T}(x_1) - c\hat{T}(x_2)| \leqslant |\langle T_\gamma, (\gamma, x_1)\,\theta_{x,c}(\gamma)\rangle| < \varepsilon \|T\|_{A'}.$$

q.e.d.

2.5.12 E Kronecker implies $A'(E) = M(E)$.

Example 2.5.3 a) Using Theorem 2.5.7 it is straightforward to prove that if Γ is compact and $E \subseteq \Gamma$ is a Kronecker set then (2.5.14) is valid for each $T \in A'(E)$. In fact, if $\{T_j : j = 1, \ldots, n\}$ is a finite decomposition of $T \in A'(E)$ and $\varepsilon > 0$, there are elements $x_1, \ldots, x_n \in G$ and complex numbers $c_1, \ldots, c_n$ with each $|c_j| = 1$ such that

$$\sum_1^n \|T_j\|_{A'} - \frac{\varepsilon}{2} \leqslant \left|\sum_1^n c_j \hat{T}_j(x_j)\right|.$$

Since E is a Kronecker set we consider the continuous function $\varphi(\gamma) = \sum_{j=1}^n c_j(\gamma, x_j)\chi_{F_j}(\gamma)$ on E, where $F_j = \operatorname{supp} T_j$, and have that

$$\forall\, r > 0\ \exists\, x \in G \qquad \text{such that } \sup_{1 \leqslant j \leqslant n} \{\sup_{\gamma \in F_j} |c_j(\gamma, x_j) - (\gamma, x)|\} < r.$$

Consequently, by Theorem 2.5.7 and for $r = \delta\left(\dfrac{\varepsilon}{2n\|T\|_{A'}}\right)$,

$$\forall\, j = 1, \ldots, n, \qquad |c_j \hat{T}_j(x_j) - \hat{T}_j(x)| < \frac{\varepsilon}{2n}.$$

Therefore,

$$\sum_1^n \|T_j\|_{A'} - \varepsilon \leqslant \|T\|_{A'},$$

and so

$$\sum_1^n \|T_j\|_{A'} = \|T\|_{A'}.$$

b) If E is independent and Γ is non-discrete it is easy to check that $mE = 0$, e.g. [Benedetto, 6, Section 4.4; Rudin, 5, Section 5.3.6] (cf. [Graham, 2] for a generalization). Recall that Kronecker sets are independent.

c) If $T \in A'(E)$ is in $\mathscr{V}$, where $E \supseteq \mathbf{T}$ is closed and $mE = 0$, there is a routine way (e.g. Section 3.2.11) to prove that $T = \mu_T \in M(E)$, where μ_T was defined in Theorem 2.5.5.

d) Combining a) and Theorem 2.5.6, we see that *totally disconnected Kronecker sets* in compact Γ *are sets of strong spectral resolution*. Parts a), b), and c) provide another proof of this result for $\Gamma = \mathbf{T}$. The extension to arbitrary Γ is routine.

2.5.13 $A'(E) = M(E)$.

Remark 1. The idea for Theorem 2.5.7 involves a *composition* of a character and an element of small norm from $A(\mathbf{T})$, so that an operation by a pseudo-measure T yields a difference of two Fourier coefficients of T and preserves the small norm. The statement of Theorem 2.5.7 estimates the differences between Fourier coefficients of pseudo-measures given a sup-norm relation between the corresponding exponentials.

2. Theorem 2.5.7 has had other applications in [Chauve, 1; Drury, 1; Kahane, 12]. We shall mention Kahane's result explicitly. A closed set $E \subseteq \mathbf{T}$ is a *Dirichlet set* if

$$\underline{\lim}_{n \to \infty} \sup_{\gamma \in E} |1 - e^{in\gamma}| = 0,$$

and it is a *strong U-set* if

$$\forall\, T \in A'(E), \qquad \overline{\lim}_{|n| \to \infty} |\hat{T}(n)| = \|T\|_{A'}.$$

It is clear that strong U-sets are U-sets, and trivial to check that Kronecker sets are Dirichlet sets. Kahane proved that Dirichlet sets are strong U-sets; and, for perspective, especially in light of Example 2.5.3, we recall the above-mentioned theorem by Malliavin that sets of spectral resolution are U-sets. Kahane's proof is a direct application of Theorem 2.5.7.

3. In 1969, [Saeki, 4, Theorem 11] was able to prove that any Kronecker set in any Γ is a set of strong spectral resolution. In 1971, [Saeki, 5] generalized this result in the following way. He first showed that *if $E \subseteq \Gamma$ is a Kronecker set then it satisfies the condition*:

$$(K_b)\quad \begin{aligned} &\exists\, b > 0 \text{ such that } \forall\, F_1, F_2 \subseteq E \text{ for which } F_1 \cap F_2 = \emptyset \text{ and each } F_j \text{ is closed,}\\ &\exists\, \varphi \in A(\Gamma) \text{ with the properties that}\\ &\qquad \|\varphi\|_A < b\\ &\text{and}\\ &\qquad \varphi = \begin{cases} 1 \text{ on a neighborhood of } F_1,\\ 0 \text{ on a neighborhood of } F_2. \end{cases} \end{aligned}$$

Then he proved that *$E \subseteq \Gamma$ is a set of strong spectral resolution if and only if there is $b > 0$ for which condition (K_b) is valid.*

Example 2.5.4. In 1953, [Reiter, 2] proved that *if $E \subseteq \Gamma$ is closed, independent, and countable then E is a set of strong spectral resolution*, and, in fact, by the countability, each $T \in A'(E)$ is a discrete measure. The proof, e.g. [Benedetto, 6, Prop. 5.10; Rudin, 5, Section 5.6.7], depends on Kronecker's theorem and shows that $\|\mu\|_{A'} = \|\mu\|_1$ for each $\mu \in M(E)$. It is not too difficult to construct compact countable independent sets satisfying this norm equality but which are not Kronecker sets, e.g. [Benedetto, 6, Example 5.2.] We alluded to Reiter's result at the end of Paragraph 2.2, and it is interesting to compare it now with Theorem 2.2.4 where $\hat{T}(x) = \sum_{\gamma} a_\gamma(\gamma, x)$ converges uniformly instead of absolutely as in this case. We saw that $T \in A'(\Gamma)$ can have countable support and a discontinuous Fourier transform (e.g. Exercise 1.3.2) but that if $\operatorname{supp} T$ is compact and scattered then T is almost periodic.

Exercises 2.5

2.5.1 A norm estimate for $A(\Gamma)$

a) Given $\varepsilon > 0$. Prove that there is $\theta \in A(\mathbf{T})$ and $\delta > 0$ such that $\|\theta\|_A < \varepsilon$ and $\theta(\gamma) = 1 - e^{i\gamma}$ if $|1 - e^{i\gamma}| < \delta$. (*Hint*: Define θ to be $1 - e^{i\gamma}$ if $|\gamma| \leqslant a$, $1 - e^{i(2a-\gamma)}$ if $a < \gamma < 2a$, $1 - e^{-i(2a+\gamma)}$ if $-2a < \gamma < -a$, and 0 if $2a \leqslant |\gamma| \leqslant \pi$. If $\hat{f} = \theta$, then by Hölder's inequality

$$\sum |f(n)| \leqslant \left(\sum \frac{1}{1+n^2}\right)^{1/2} \left(\sum (1+n^2)\,|f(n)|^2\right)^{1/2}$$

$$< b\left(\int_{\mathbf{T}} (|\theta(\gamma)|^2 + |\theta'(\gamma)|^2)\, d\gamma\right)^{1/2}\Big).$$

Compare [Varopoulos, 2, Lemme 1] for a different proof of this result.

b) We now give a generalization of part a). Given $\varphi \in A(\Gamma)$, $\lambda \in \Gamma$, $\varepsilon > 0$, and U a neighborhood of λ. Prove that there is $\theta \in A_c(\Gamma)$ and a neighborhood $V \subseteq U$ of λ such that $\theta = 0$ on $U^{\sim}$, $\|\theta\|_A < \varepsilon$, and

$$\forall\, \gamma \in V, \qquad \varphi(\gamma) - \varphi(\lambda) = \theta(\gamma).$$

(*Hint*: Choose $\varphi_\lambda \in A(\Gamma)$ for which $\varphi_\lambda = \varphi(\lambda)$ on a neighborhood V_1 of λ. By Theorem 1.2.2a we can choose $\psi \in A_c(\Gamma)$ such that $\psi = 1$ in a neighborhood $V_2 \subseteq U$ of λ, $\psi = 0$ on $U^{\sim}$, and $\|(\varphi - \varphi_\lambda)\,\psi\|_A < \varepsilon$. Set $\theta = (\varphi - \varphi_\lambda)\psi$ and $V = V_1 \cap V_2$). Unfortunately, we have no control over the size of V and it may be too small to obtain good norm estimates.

c) Given $\varepsilon > 0$ and a compact set $K \subseteq \Gamma$. Prove that there is $\theta \in A_c(\Gamma)$ such that $0 \leqslant \theta \leqslant 1$, $\theta = 1$ on K, and

$$\|\theta\|_A < 1 + \varepsilon.$$

(*Hint*: By Proposition 1.1.4, choose $\varphi \in A_c(\Gamma)$ such that $\varphi = 1$ on K and $0 \leqslant \varphi \leqslant 1$. By Theorem 1.2.1 take $\psi \in A_c(\Gamma)$, such that $\|\psi\|_A = 1$ and $\|\varphi - \varphi\psi\|_A < \varepsilon$. Set $\theta = \psi + \varphi - \varphi\psi$. It is trivial to check that θ has the desired properties).

2.5.2 $\mathrm{Lip}_\alpha(\mathbf{T}) \subseteq A(\mathbf{T})$, $\alpha > 1/2$

Using integration by parts we see that $C^\infty(\mathbf{T}) \subseteq A(\mathbf{T})$ and $C_c^\infty(\mathbf{R}^n) \subseteq A(\mathbf{R}^n)$. Prove that $\overline{C^\infty(\mathbf{T})} = A(\mathbf{T})$ and $\overline{C_c^\infty(\mathbf{R}^n)} = A(\mathbf{R}^n)$ in the $\|\ \|_A$-topology (cf. Proposition 1.1.3). (*Hint*: Take $\{\varphi_m : m = 1, \ldots\} \subseteq C^\infty(\mathbf{T})$ such that $\int_{\mathbf{T}} \varphi_m(\gamma)\,d\gamma = 1$, $\varphi_m \geqslant 0$, and $\operatorname{supp} \varphi_m \subseteq \left[-\frac{\pi}{m}, \frac{\pi}{m}\right]$; then a direct calculation, using the dominated convergence theorem, shows that

$$\forall\, \varphi \in A(\mathbf{T}), \qquad \lim_{m \to \infty} \|\varphi - \varphi * \varphi_m\|_A = 0.$$

For the $\mathbf{R}^n$ case, let $\{K_m : m = 1, \ldots\} \subseteq \mathbf{R}^n$ be an increasing sequence of compact neighborhoods such that $\mathbf{R}^n = \bigcup_m K_m$. Choose $\varphi_m \in C_c^\infty(\mathbf{R}^n)$ such that $\varphi_m > 0$ on $\operatorname{int} K_m$ and $\operatorname{supp} \varphi_m \subseteq K_m$; and let $I \subseteq A(\mathbf{R}^n)$ be the closed ideal generated by $\{\varphi_m : m = 1, \ldots\}$. Then $I = A(\mathbf{R}^n)$ by Theorem 1.2.4).

The fact, $C^\infty(\mathbf{T}) \subseteq A(\mathbf{T})$, can be extensively refined as we indicated in our remark about $\mathrm{Lip}_\alpha(\mathbf{T})$ in Exercise 1.3.5. We'll now be more specific. Define

$$\forall\, \varphi \in C(\mathbf{T}), \qquad \omega_\varphi(\delta) = \sup\{|\varphi(\lambda) - \varphi(\gamma)| : \lambda, \gamma \in \mathbf{T} \text{ and } d(\lambda, \gamma) \leqslant \delta\}$$

("d" was defined before Theorem 2.5.3). $\varphi \in C(\mathbf{T})$ is an element of $\mathrm{Lip}_\alpha(\mathbf{T}), \alpha > 0$, if

$$\omega_\varphi(\delta) = O(\delta^\alpha), \qquad \delta \to 0 \text{ and } \delta > 0.$$

S. Bernstein proved that *if* $\sum \omega_\varphi\left(\frac{1}{n}\right)/\sqrt{n} < \infty$ *then* $\varphi \in A(\mathbf{T})$ and, thus, if $\varphi \in \mathrm{Lip}_\alpha(\mathbf{T})$, $\alpha > 1/2$, then $\varphi \in A(\mathbf{T})$ (cf. Exercise 2.1.2e and [Kahane, 13, pp. 13–20; Kahane and Salem, 4, Chapitre X]).

The Hardy–Littlewood function,

$$\varphi(\gamma) \sim \sum_{n=1}^{\infty} \frac{e^{in \log n}}{n} e^{in\gamma},$$

e.g. [Zygmund, 2, I, pp. 197–199], and the Rudin–Shapiro function, e.g. [Kahane and Salem, 4, pp. 129–130; Katznelson, 5, pp. 33–34], are examples of elements in $(\mathrm{Lip}_{1/2}(\mathbf{T})) \setminus A(\mathbf{T})$. As is well known,

$$\psi(\gamma) = \sum_{1}^{\infty} \frac{\varepsilon_n}{n} \sin n\gamma \in C(\mathbf{T})$$

if $\{\varepsilon_n : n = 1, \ldots\}$ decreases to 0 (the series converges uniformly) and so $\psi \notin A(\mathbf{T})$ if $\sum_1 \varepsilon_n/n$ diverges, cf. Exercise 1.3.2.

2.5.3 Properties of C-sets

The properties of C-sets have been developed mainly by [Calderón, 1; Herz, 5, Section 6; Rudin, 5, Section 7.5; Warner, 2]; Theorem 2.5.1c is found in [Reiter, 8, Chapter 2, Section 5.3; Warner, 1].

a) Prove Proposition 2.5.2. (*Hint*: Take $\varphi \in k(E)$. To show $\varphi \in \overline{\varphi j(E)}$. Since $\varphi \in k(\partial E)$ and ∂E is a C-set we have that

$$\exists \{\varphi_n : n = 1, \ldots\} \subseteq j(\partial E) \cap A_c(\Gamma) \qquad \text{such that } \lim_{n \to \infty} \|\varphi - \varphi\varphi_n\|_A = 0.$$

Note that $U_n = ((\Gamma \setminus E) \cap \operatorname{supp} \varphi_n)^\sim \supseteq E$ is open. Choose $\psi_n \in A(\Gamma)$ such that $\psi_n = 0$ on $U_n^\sim$ and $\psi_n = 1$ on $E \cap \operatorname{supp} \varphi_n$. It is easy to check that $\varphi_n - \varphi_n \psi_n = 0$ on U_n and $\varphi\varphi_n\psi_n = 0$. Thus, $\varphi\varphi_n = \varphi(\varphi_n - \varphi_n\psi_n)$ and $\varphi_n - \varphi_n\psi_n \in j(E) \cap A_c(\Gamma)$).

b) Let $\{E_n : n = 1, \ldots\} \subseteq \Gamma$ be a sequence of C-sets and assume that $E = \bigcup_1^\infty E_n$ is closed. Prove that E is a C-set. (*Hint*: Take $\varphi \in k(E)$ and $\varepsilon > 0$. By hypothesis and an inductive procedure we have

$$\forall\, n \geqslant 1,\ \exists\, \varphi_m \in j(E_m) \cap A_c(\Gamma),\ m = 1, \ldots, n, \qquad \text{such that}$$
$$\|\varphi\varphi_1 \ldots \varphi_{n-1} - \varphi\varphi_1 \ldots \varphi_n\|_A < \varepsilon/2^n.$$

Let $U_m \supseteq E_m$ be an open set for which φ_m vanishes on U_m. Since $E \cap \operatorname{supp} \varphi_1$ is compact and

$$E \cap \operatorname{supp} \varphi_1 \subseteq \bigcup_1^\infty U_m$$

we see that

$$E \cap \operatorname{supp} \varphi_1 \subseteq \bigcup_1^N U_m$$

for some N. Set $\psi = \varphi_1 \ldots \varphi_N$. Note that $\psi \in j(E)$ and $\|\varphi - \varphi\psi\|_A < \varepsilon$).

Another important C-set criterion is the fact that closed subgroups of Γ are C-sets, e.g. [Rudin, 5, Section 7.5.] We also refer to the treatment of C-sets in [Reiter, 8], where C-sets are called Wiener–Ditkin sets.

2.5.4 A classical generalization of Wiener's Tauberian theorem

We give a generalization due to Beurling and [Moh, 1] of the classical Wiener Tauberian theorem. The basic technique goes back to [Beurling, 2]. Let h be a non-negative Lebesgue measurable function on $\mathbf{R}$ such that

(E2.5.1) $$\lim_{x\to\infty} h(x)/x = 0$$

and

(E2.5.2) $$\forall\, y \in X \subseteq \mathbf{R}, \quad \text{for which } m(\mathbf{R}\setminus X) = 0, \qquad \lim_{x\to\infty} \frac{h(x+yh(x))}{h(x)} = 1.$$

Prove the following result. Given h which satisfies (E2.5.1) and (E2.5.2), $f \in L^1(\mathbf{R})$ for which $\hat{f}$ never vanishes, and $\Phi \in L^\infty(\mathbf{R})$; if

$$\lim_{x\to\infty} \int_{\mathbf{R}} \Phi(y) f\left(\frac{x-y}{h(x)}\right)\frac{dy}{h(x)} = r\int_{\mathbf{R}} f(y)\,dy$$

then

$$\forall\, g \in L^1(\mathbf{R}), \qquad \lim_{x\to\infty} \int_{\mathbf{R}} \Phi(y) g\left(\frac{x-y}{h(x)}\right)\frac{dy}{h(x)} = r\int_{\mathbf{R}} g(y)\,dy.$$

(*Hint*: Take $r = 0$ and define

$$F(x) = \int_{\mathbf{R}} \Phi(y) f\left(\frac{x-y}{h(x)}\right)\frac{dy}{h(x)} \quad \text{and} \quad G(x) = \int_{\mathbf{R}} \Phi(y) g\left(\frac{x-y}{h(x)}\right)\frac{dy}{h(x)}.$$

Assuming that $\lim_{x\to\infty} F(x) = 0$ and that there is a sequence $\{x_n : n = 1, \ldots\} \subseteq \mathbf{R}$ increasing to infinity such that $\underline{\lim}\, |G(x_n)| = \eta > 0$, we shall obtain a contradiction. To this end we define

$$F_n(x) = F(x_n + h(x_n)\,x) \quad \text{and} \quad G_n(x) = G(x_n + h(x_n)\,x).$$

Using the dominated convergence theorem and (E2.5.2), calculate that

$$\forall\, x \in \mathbf{R}, \qquad \lim_{n\to\infty} (f * G_n(x) - g * F_n(x)) = 0.$$

By (E2.5.1) and the hypothesis, $\lim_{y\to\infty} F(y) = 0$, we see that $\lim_{n\to\infty} g * F_n(x) = 0$ and so

$$\forall\, x \in \mathbf{R}, \qquad \lim_{n\to\infty} f * G_n(x) = 0.$$

The definition of G and properties of the weak $*$ topology yield the fact that $\lim G_{n_m} = \Psi \in L^\infty(\mathbf{R})$ in the $\sigma(L^\infty(\mathbf{R}), L^1(\mathbf{R}))$ topology. Thus $f * \Psi = 0$. Note that $\Psi(0)$ exists and $|\Psi(0)| \geqslant \eta$. Consequently, once Ψ is shown to be continuous in a neighborhood of the origin we obtain the desired contradiction because of Theorem 1.3.1c. Actually Moh effects this part of the proof, as Beurling did in 1945, by calculating that Ψ is uniformly continuous and appealing to Theorem 2.2.1, cf. part c) of the proof of Theorem 2.2.1).

2.5.5 Schwartz's example

a) Let $n > 1$ and set $B^n = \{\gamma \in \hat{\mathbf{R}}^n : |\gamma| \leqslant 1\}$. It is well-known that $m_n B^n = \pi^{n/2}/\Gamma\left(1 + \frac{n}{2}\right)$, where m_n is Haar measure on $\hat{\mathbf{R}}^n$; we don't make use of the specific value of this constant. Let $\varphi : S^{n-1} \to \mathbf{C}$ be continuous, where S^{n-1} is considered as a subspace of $\hat{\mathbf{R}}^n$. Define $\psi : \hat{\mathbf{R}}^n \to \mathbf{C}$ as follows: $\psi(0) = 0$; $\psi(r\gamma) = \varphi(\gamma)$ if $0 < r \leqslant 1$ and $\gamma \in S^{n-1}$; and $\psi(r\gamma) = 0$ if $r > 1$ and $\gamma \in S^{n-1}$. Clearly, $\psi \in L^\infty(\hat{\mathbf{R}}^n)$ and $\psi = 0$ on $B^{n\sim}$; in particular, the integral,

$$\text{(E2.5.3)} \qquad \langle \mu, \varphi \rangle = \frac{1}{m_n B^n} \int_{\hat{\mathbf{R}}^n} \psi(\lambda)\, \mathrm{d}m_n(\lambda),$$

exists for each $\varphi \in C(S^{n-1})$. Because of (E2.5.3) we readily see that $\mu \in M(S^{n-1})$, $\mu \geqslant 0$, and $\langle \mu, 1 \rangle = 1$. If $B_\varepsilon \subseteq \hat{\mathbf{R}}^n$ is an open ball of radius $\varepsilon \in (0,1)$ and $B_\varepsilon \cap S^{n-1} \neq \emptyset$, prove that $\mu(B_\varepsilon \cap S^{n-1}) > 0$.

b) Let $\varphi \in k(S^{n-1}) \cap C_c^\infty(\hat{\mathbf{R}}^n)$, $n > 1$, have the property that $\partial\varphi/\partial\gamma_1 \geqslant 0$ on S^{n-1} and $\partial\varphi/\partial\gamma_1 > 0$ for some point of S^{n-1}.

Prove that $\partial\mu/\partial\gamma_1$, defined by

$$\forall\, \theta \in C_c^\infty(\hat{\mathbf{R}}^n), \left\langle \frac{\partial\mu}{\partial\gamma_1}, \theta \right\rangle = -\left\langle \mu, \frac{\partial\theta}{\partial\gamma_1} \right\rangle,$$

is in $D(\hat{\mathbf{R}}^n)$, and that

$$\left\langle \frac{\partial\mu}{\partial\gamma_1}, \varphi \right\rangle \neq 0.$$

c) Recall that $C_c^\infty(\hat{\mathbf{R}}^n) \subseteq A(\hat{\mathbf{R}}^n)$. Prove that $T = \partial\mu/\partial\gamma_1 \in A'(\hat{\mathbf{R}}^n)$, if $n \geqslant 3$. Once we prove this it is clear from part b), the fact that $\operatorname{supp} \dfrac{\partial\mu}{\partial\gamma_1} \subseteq S^{n-1}$, and the definition of φ, that $T \in A'(S^{n-1}) \setminus A_S'(S^{n-1})$. (*Hint*: It is sufficient to prove that $\sup\{|x\hat{\mu}(x)| : x \in \mathbf{R}^n\} < \infty$. If $R : \hat{\mathbf{R}}^n \to \hat{\mathbf{R}}^n$ is an orthogonal transformation, verify the change of variable formula

$$\text{(E2.5.4)} \qquad \forall\, \varphi \in L^1_\mu(S^{n-1}), \qquad \int_{S^{n-1}} \varphi \circ R\, \mathrm{d}\mu = \int_{S^{n-1}} \varphi\, \mathrm{d}\mu.$$

Using (E2.5.3), (E2.5.4), and the orthogonal transformation

$$R:\mathbb{R}^3 \to \mathbb{R}^3$$
$$(\gamma_1,\gamma_2,\gamma_3) \mapsto (0,0,|\gamma|),$$

a direct calculation shows that

$$\text{(E2.5.5)} \qquad \hat{\mu}(x) = \frac{1}{m_3 B^3} \int_{B^3} \exp i\, \frac{\gamma_3 |x|}{|\gamma|}\, dm_3(\gamma).$$

By considering the spherical coordinates, $\gamma_1 = c\sin a\cos b$, $\gamma_2 = c\sin a\sin b$, $\gamma_3 = c\cos a$, compute that

$$\forall\, x \in \mathbb{R}^3, \qquad |x\hat{\mu}(x)| \leqslant \frac{4\pi}{3m_3\, B^3}\bigg).$$

This exercise completes the proof of the original counterexample to synthesis due to [Schwartz, 2] (1948) that we referred to in Section 1.4.8. Analogous examples are found in [Dixmier, 2].

d) Let $f \in L^1(\mathbf{R}^n)$ be a radial function and write $\Phi(\rho) = \hat{f}(\gamma)$ as in Exercise 1.4.4a. Prove that $\Phi \in C^1(0,\infty)$, and that for each $\rho > 0$ there is $K_n(\rho) > 0$ *independent* of f, such that $|\Phi'(\rho)| \leqslant K_n(\rho)\|f\|_1$. (*Hint*: Compare (E1.4.5)).

e) Let $f \in L^1(\mathbf{R}^n)$, $\hat{f} = \varphi$ and $n \geqslant 3$, be a radial function for which $xf(x) \in L^1(\mathbf{R}^n)$, and assume that $\Phi(1) = 0$ and $\Phi'(1) > 0$. Some relatively simple examples of such functions are given in [Reiter, 5, p. 469]. Note that $\varphi \in k(S^{n-1})$. Prove that $I_{\varphi^2} \subsetneq I_\varphi$ [Reiter, 5, Theorem 1]. Cf. the condition "$I_{\varphi^2} \subsetneq I_\varphi$" with the results of Paragraph 3.1.

f) Prove that $S^{n-1} \times \mathbb{R}^m \subseteq \mathbb{R}^{n+m}$, $n \geqslant 3$, is a non-S-set.

g) The *order of* $T \in D_c(\mathbb{R}^n)$, $\operatorname{ord} T$, is the smallest value of m for which T is a continuous linear functional on the space $C^m(\mathbb{R}^n)$, taken with its natural topology, e.g. [Schwartz, 5]. Prove that for each $n \geqslant 1$ there is $T_n \in D_c(\mathbb{R}^n) \cap A'(\mathbb{R}^n)$ such that $\lim \operatorname{ord} T_n = \infty$. (*Hint*: The technique of part c) can be used).

$S^{n-1} \subseteq \mathbb{R}^n$, $n \geqslant 3$, is a non-S-set in the following strong way: *if* $E \subseteq \mathbb{R}^n$ *is closed and* $S^{n-1} \setminus E \neq \emptyset$ *then* $E \cup S^{n-1}$ *is a non-S-set* [Reiter, 5, Theorem 2] (cf. Theorem 2.5.3). The fact that $S^{n-1} \subseteq \mathbb{R}^n$, $n \geqslant 3$, is a non-S-set is equivalent to the statement that primary ideals are not necessarily maximal ideals in the Banach algebra of radial functions on $\mathbb{R}^n$ [Reiter, 5, Section 3; Varopoulos, 7; Varopoulos, 9, Chapter V]. References for examples of non-S-sets which essentially generalize the $S^{n-1} \subseteq \mathbb{R}^n$, $n \geqslant 3$, example are given in [Reiter, 8, p. 37]. [Herz, 4] proved that $S^1 \subseteq \mathbb{R}^2$ is an S-set, and it is not known if $S^1 \subseteq \mathbb{R}^2$ is a C-set. Even though $S^{n-1} \subseteq \mathbb{R}^n$, $n \geqslant 3$, is not an S-set, [Varopoulos, 7] was able to use the method in [Herz, 4] to prove:

$$\forall\, \varepsilon > 0 \quad \text{and} \quad \forall\, \varphi \in k(S^{n-1})\ \exists\, \psi \in C_c^\infty(\mathbb{R}^n) \cap k(S^{n-1}) \text{ such that}$$

$$\|\varphi - \psi\|_A < \varepsilon.$$

2.5.6 Maximal primary ideals in Banach algebras

Let X be a regular semi-simple commutative Banach algebra. Prove that each closed primary ideal in X is maximal if and only if for each $x \in X$ and for each maximal ideal γ for which the Gelfand transform of x at γ, $\hat{x}(\gamma)$, vanishes, there is a sequence $\{x_n : n = 1, \ldots\} \subseteq X$ such that each $\hat{x}_n$ equals $\hat{x}$ on a neighborhood V_n of γ and

$$\lim_{n \to \infty} \|x_n\| = 0$$

(cf. Theorem 1.2.2 and Exercise 1.2.4). $A(\Gamma)$ is such an algebra.

2.5.7 A classical equivalent form of Wiener's Tauberian theorem

Let $f \in C_b(\mathbf{R})$ have the property that

$$\|f\| = \sum_{n=-\infty}^{\infty} \sup_{x \in [n, n+1]} |f(x)| < \infty,$$

and let W(**Я**) be the class of Fourier transforms of these functions. W(**Я**) is a Banach algebra where the norm of $\hat{f} = \varphi$ is defined as $\|\varphi\| = \|f\|$. Prove that Wiener's Tauberian theorem (in the form of Theorem 1.2.4) is equivalent to the statement: let $I \subseteq W$(**Я**) be a closed ideal and assume that for each $\gamma \in$ **Я** there is $\varphi \in I$ such that $\varphi(\gamma) \neq 0$; then $I = W$(**Я**), e.g. [Wiener, 7]. The complete proof is found in [Edwards, 2] and one direction was given in [Hardy, 1, Section 12.7]. There is a generalization in [Bochner and Chandrasekharan, 1, Chapter VI].

2.5.8 The union of S-sets

a) Given $T \in A'(\Gamma)$ and $\varphi \in A(\Gamma)$ and assume that $\varphi \in k(\operatorname{supp} T)$ and that $T\varphi$ is almost periodic. Prove that $T\varphi = 0$. (*Hint*: By Theorem 2.2.3a it is sufficient to prove that $\operatorname{supp}_{AP} T\varphi = \emptyset$. Assume not and let $\gamma \in \operatorname{supp}_{AP} T\varphi$. Then the mean value has absolute value $|M(\widehat{T\varphi}(x)\overline{(\gamma, x)})| = r > 0$. Take $\varepsilon \in (0, r)$. Since $\varphi = 0$ on $\operatorname{supp} T$, $\operatorname{supp}_{AP} T\varphi \subseteq \operatorname{supp} T$, and points are C-sets, we can choose $\psi_\varepsilon \in j(\{\gamma\})$ such that $\|\varphi - \varphi\psi_\varepsilon\|_A < \varepsilon / \|T\|_{A'}$. Then, observing that $\gamma \notin \operatorname{supp}_{AP} T\varphi\psi_\varepsilon$, compute that $r < \varepsilon$).

b) Let $E_1 \subseteq \Gamma$ be an S-set and assume that $E_2 \subseteq \Gamma$ contains no non-empty perfect subsets. Let $E_1 \cup E_2$ be closed. If $T \in A'(E_1 \cup E_2)$ and $\varphi \in k(E_1 \cup E_2)$ prove that $T\varphi$ is almost periodic. (*Hint*: Let $\{\psi_n : n = 1, \ldots\} \subseteq j(E_1)$ have the property that $\lim_{n\to\infty} \|\varphi - \psi_n\|_A = 0$, and, hence, $\operatorname{supp} T\psi_n \subseteq E_2$; then $\operatorname{supp} T\psi_n$ is scattered. By Theorem 2.2.4 and the fact that $\widehat{T\psi_n}$ is uniformly continuous we conclude that $T\psi_n$ is almost periodic. Therefore $T\varphi$ is almost periodic since $\lim_{n \to \infty} \|T\varphi - T\psi_n\|_{A'} = 0$).

c) Parts a) and b) combine to yield: *if $E_1 \subseteq \Gamma$ is an S-set, if $E_2 \subseteq \Gamma$ contains no non-empty perfect subsets, and if $E_1 \cup E_2$ is closed, then $E_1 \cup E_2$ is an S-set*. Cf. Theorem 2.5.1c, Theorem 2.5.4, and the discussion after Theorem 2.5.4. This result has also been proved by Rosenthal (for E_2 countable) and [Veech, 2, p. 421]; Veech's proof is interesting but complicated in that it uses properties of minimal and almost automorphic functions.

3 Results in spectral synthesis

3.1 Non-synthesizable phenomena

3.1.1 Introductory remarks on non-synthesis. The fact that there are non-synthesizable pseudo-measures, discovered for $\mathbb{R}^3$ by [Schwartz, 2] in 1948 and for any non-discrete Γ by [Malliavin, 1; 2; 3] in 1959, has warranted a finer study of the structure of pseudo-measures, first as to their synthesizable properties and ultimately to the general problem of relating various aspects of the support of a pseudo-measure to the behavior of its Fourier transform. Thus, questions, as to whether various sets with arithmetic properties, such as Kronecker sets, are synthesizable or not, become meaningful. We refer to the fact that there are non-S-sets in any non-discrete Γ as *Malliavin's theorem.*

A large literature has evolved giving many diverse examples of non-synthesizable phenomena. In Theorem 3.1.1 we shall give the general method developed by Malliavin to determine such non-synthesis. Generally, it is necessary to construct specific elements of $A(\Gamma)$ in order to employ Theorem 3.1.1 and we shall refer to—and not provide the details to—an assortment of these constructions. We gave Schwartz's example, that $S^{n-1} \subseteq \mathbb{R}^n$, $n \geqslant 3$, is a non-S-set, in Exercise 2.5.5; and we shall give Varopoulos' proof of Malliavin's theorem starting from this example and employing tensor algebra techniques.

3.1.2 Malliavin's idea. Malliavin's original idea [Malliavin, 1, Section 5; Katznelson, 5, Section 8.8] is rooted in previously quoted (Paragraph 2.4) work on the symbolic calculus by [Kahane, 3] and [Katznelson, 1]. Basically, Malliavin's procedure yields a result in the "individual symbolic calculus" which finally comes down to extending a linear functional *and* preserving its support; a different twist on this latter issue is found in [Benedetto, 8]. An interesting analysis of the role of these support preserving extensions and of differentiation (e.g. Exercise 2.5.5c and (3.1.1)–(3.1.2) below) in the construction of non-S-sets has been given by [Atzmon, 1], cf. [Katznelson and Rudin, 1; Rudin, 5, Section 7.7.1]. Without going into detail about the symbolic calculus we now outline Malliavin's idea in the following way.

The problem is to find $T \in A'(\Gamma)$ and $\varphi \in A(\Gamma)$ such that $\varphi \in k\,(\operatorname{supp} T)$ and $\langle T, \varphi \rangle \neq 0$. For $\delta' \in D(\mathbb{R})$ we have

$$(3.1.1) \qquad \langle \delta'(\gamma), \gamma \rangle = -1,$$

and, by a formal Fourier inversion formula,

$$\delta'(\gamma) = \int_{\mathbf{R}} e^{-ix\gamma}(ix)\,dx. \tag{3.1.2}$$

We can generalize the formulas (3.1.1) and (3.1.2) by considering functions

$$\varphi : \Gamma \;\rightarrow\; \mathbf{R}$$

instead of the identity function

$$id : \mathbf{R} \;\rightarrow\; \mathbf{R}.$$

Then (3.1.1) and (3.1.2) become

$$\langle \delta'(\varphi), \varphi \rangle \neq 0 \tag{3.1.1)$'$}$$

and

$$\delta'(\varphi) = \frac{1}{2\pi} \int_{-\infty}^{\infty} e^{-ix\varphi}(ix)\,dx, \tag{3.1.2)$'$}$$

respectively. We know, of course, that $\operatorname{supp} \delta' = \{0\} \subseteq \mathbf{R}$; considered in terms of the identity function, *id*, we could say that $\delta'(\gamma)$ has its support contained in the zero-set of the function, $id(\gamma) = \gamma$. With this latter interpretation, we say that

(3.1.3) $\delta'(\varphi)$ has its support contained in $Z\varphi$.

Thus, if $\varphi \in A(\Gamma)$ can be chosen in such a way that the formal expression, $\delta'(\varphi)$, given in (3.1.2)$'$ is a pseudo-measure then (3.1.1)$'$ and (3.1.3) yield a formal solution to the problem. In fact, we now prove that all of these formulas can be justified.

3.1.3 Malliavin's operational calculus. First of all note that if Γ is compact then $A(\Gamma) \subseteq A'(\Gamma)$ and the map

$$\begin{aligned} \mathbf{R} &\rightarrow A'(\Gamma) \\ u &\mapsto (iu)^k e^{iu\varphi(\gamma)} \end{aligned}$$

is continuous. Thus, if

$$\int_{-\infty}^{\infty} |u|^k \, \|e^{iu\varphi}\|_{A'}\, du < \infty, \tag{3.1.4}$$

then the elementary results of vector-valued Riemann integration theory yield the fact that the integrals

$$T_{\varphi,j} = \frac{1}{2\pi} \int_{-\infty}^{\infty} (iu)^j e^{iu\varphi}\, du, \qquad j = 0, \ldots k \tag{3.1.5}$$

are elements of $A'(\Gamma)$. We now verify that $T_{\varphi,j} \in A'(\Gamma)$ without using vector-valued integration theory; and then are able to give rigorous proofs of (3.1.1)′ and (3.1.3).

Theorem 3.1.1. *Let Γ be compact, and assume that $\varphi \in A(\Gamma)$ is real-valued and satisfies (3.1.4) for a fixed non-negative integer k. Then*

a) *For each $\psi \in A(\Gamma)$ and for each $j = 0,\ldots,k$,*

$$\langle T_{\varphi,j}, \psi\rangle = \frac{1}{2\pi}\int_{\Gamma}\int_{-\infty}^{\infty} (iu)^j e^{iu\varphi(\gamma)}\psi(\gamma)\,du\,d\gamma$$

exists and $T_{\varphi,j}\colon A(\Gamma) \to \mathbf{C}$ is in $A'(\Gamma)$.

b) *For $j = 0,\ldots,k$, $\operatorname{supp} T_{\varphi,j} \subseteq Z\varphi$.*

c) *Assume*

(3.1.6) $\qquad \hat{T}_{\varphi,0}(0) \neq 0.$

Then, for each $j = 1,\ldots,k$,

(3.1.7) $\qquad \langle T_{\varphi,j}, \varphi^j\rangle \neq 0 \quad$ *and* $\quad T_{\varphi,j}\varphi^{j+1} = 0.$

Proof. a) Given $e^{iu\varphi(\gamma)} = \sum_{x\in G} a(x,u)(\gamma,x) \in A(\Gamma)$ and $\psi(\gamma) = \sum_{x\in G} b(x)(\gamma,x) \in A(\Gamma)$, where all but a countable number of terms in each sum is zero.

Then

$$\frac{1}{2\pi}\int_{\Gamma}\int_{-\infty}^{\infty} |(iu)^j e^{iu\varphi(\gamma)}\psi(\gamma)|\,du\,d\gamma$$

$$= \frac{1}{2\pi}\int_{\Gamma}\int_{-\infty}^{\infty} \left|u^j\left(\sum_{x,y\in G} a(x,u)\,b(y)\,(\gamma,y)\right)\right| du\,d\gamma$$

(3.1.8)
$$\leqslant \frac{1}{2\pi}\|\psi\|_A \int_{-\infty}^{\infty} |u|^j \|e^{iu\varphi}\|_{A'}\,du < \infty.$$

a) follows from (3.1.8)

b) Write

$$T_{\varphi,j,\varepsilon}(\gamma) = \frac{1}{2\pi}\int_{-\infty}^{\infty} (iu)^j e^{iu\varphi(\gamma)} e^{-\varepsilon^2 u^2}\,du, \qquad \varepsilon > 0,$$

so that if

$$\psi_\varepsilon(r) = \frac{1}{2\pi}\int_{-\infty}^{\infty} e^{iur} e^{-\varepsilon^2 u^2}\,du$$

then

$$\psi_\varepsilon^{(j)}(\varphi(\gamma)) = T_{\varphi,j,\varepsilon}(\gamma).$$

The fact that $T_{\varphi,j,\varepsilon} \in A(\Gamma)$ follows from the Wiener–Lévy theorem.

By Exercise 1.1.1a,

$$\psi_\varepsilon(r) = \frac{\sqrt{\pi}}{\varepsilon}\, e^{-r^2/4\varepsilon^2},$$

and so if $K \subseteq \mathbf{R}$ is compact and $0 \notin K$ then

$$\lim_{\varepsilon \to 0} \psi_\varepsilon^{(j)} = 0, \qquad \text{uniformly on } K.$$

Consequently,

$$\lim_{\varepsilon \to 0} T_{\varphi,j,\varepsilon} = 0, \tag{3.1.9}$$

uniformly on every compact subset of Γ disjoint from $Z\varphi$.

Using the Lebesgue dominated convergence theorem, (3.1.4), and computing as in (3:1.8), we see that

$$\lim_{\varepsilon \to 0} \|T_{\varphi,j} - T_{\varphi,j,\varepsilon}\|_{A'} = 0, \tag{3.1.10}$$

since

$$\forall\, \psi \in A(\Gamma), \qquad |\langle T_{\varphi,j} - T_{\varphi,j,\varepsilon}, \psi\rangle|$$
$$\leqslant \frac{1}{2\pi}\|\psi\|_A \int_{-\infty}^{\infty} |u^j(1 - e^{-\varepsilon^2 u^2})|\, \|e^{iu\varphi}\|_{A'}\, du.$$

(3.1.9) and (3.1.10) combine to yield b).

c) Using (3.1.6) we verify that $\langle T_{\varphi,j}, \varphi^j\rangle \neq 0$, for $j = 1,\ldots,k$.

$$\langle T_{\varphi,j}, \varphi^j\rangle = \lim_{\varepsilon \to 0} \frac{1}{2\pi} \int_\Gamma \int_{-\infty}^{\infty} (iu)^j e^{iu\varphi(\gamma)} \varphi(\gamma)^j e^{-\varepsilon^2 u^2}\, du\, d\gamma. \tag{3.1.11}$$

Since $\frac{d}{du} e^{iu\varphi(\gamma)} = i\varphi(\gamma) e^{iu\varphi(\gamma)}$, integration by parts shows that the right-hand side of (3.1.11) is

$$-\lim_{\varepsilon \to 0} \frac{1}{2\pi} \int_\Gamma \int_{-\infty}^{\infty} e^{iu\varphi(\gamma)} (iu\varphi(\gamma))^{j-1} e^{-\varepsilon^2 u^2}[j - 2\varepsilon^2 u^2]\, du\, d\gamma$$

$$= -j\langle T_{\varphi,j-1}, \varphi^{j-1}\rangle = (-1)^j j!\, \hat{T}_{\varphi,0}(0).$$

Finally, we prove that

$$\langle T_{\varphi,j}, \varphi^{j+1}\rangle = 0, \qquad \text{for } j=1,\ldots,k.$$

In fact,

$$\langle T_{\varphi,1}, \varphi^2\rangle = \lim_{\varepsilon\to 0}\frac{1}{2\pi}\int_{\Gamma}\int_{-\infty}^{\infty} u\varphi(\gamma)\,\mathrm{e}^{-\varepsilon^2u^2}\frac{\mathrm{d}}{\mathrm{d}u}\,\mathrm{e}^{\mathrm{i}u\varphi(\gamma)}\,\mathrm{d}u\,\mathrm{d}\gamma$$

$$= -\lim_{\varepsilon\to 0}\frac{1}{2\pi}\int_{\Gamma}\int_{-\infty}^{\infty}\varphi(\gamma)\,\mathrm{e}^{\mathrm{i}u\varphi(\gamma)}\,\mathrm{e}^{-\varepsilon^2u^2}\,\mathrm{d}u\,\mathrm{d}\gamma$$

$$= \lim_{\varepsilon\to 0}\frac{\mathrm{i}}{2\pi}\int_{\Gamma}\int_{-\infty}^{\infty}\left(\frac{\mathrm{d}}{\mathrm{d}u}\,\mathrm{e}^{\mathrm{i}u\varphi(\gamma)}\right)\mathrm{e}^{-\varepsilon^2u^2}\,\mathrm{d}u\,\mathrm{d}\gamma$$

$$= -\lim_{\varepsilon\to 0}\frac{2\mathrm{i}\varepsilon^2}{2\pi}\int_{\Gamma}\int_{-\infty}^{\infty} u\mathrm{e}^{-\varepsilon^2u^2}\mathrm{e}^{\mathrm{i}u\varphi(\gamma)}\,\mathrm{d}u\,\mathrm{d}\gamma = 0.$$

q.e.d.

3.1.4 Non-spectral functions and Lipschitz conditions.

Remark 1. Assume that $\varphi \in A(\Gamma)$ is a real function that satisfies (3.1.4). We shall now prove that the condition (3.1.6) is no restriction in the statement of Theorem 3.1.1c. In fact, if we set

$$f(u) = \langle \mathrm{e}^{\mathrm{i}u\varphi(\gamma)}, 1\rangle$$

then $f \in L^1(\mathbf{R})$ is continuous and $f(0) \neq 0$, and so $\hat{f}(r) \neq 0$ for some real r. Thus

$$\hat{T}_{\varphi+r,0}(0) = \frac{1}{2\pi}\int_{\Gamma}\int_{-\infty}^{\infty}\mathrm{e}^{\mathrm{i}ur}\mathrm{e}^{\mathrm{i}u\varphi(\gamma)}\,\mathrm{d}u\,\mathrm{d}\gamma \neq 0,$$

and, clearly, the analogue of (3.1.4) holds for $\varphi + r$ if it holds for φ.

2. In light of Theorem 3.1.1c and the above remark, the construction of functions $\varphi \in A(\Gamma)$ which satisfy (3.1.4) emerges as an important part of the problem to determine non-S-sets; we shall refer to such functions φ as *non-spectral* functions. Malliavin's original non-spectral function φ did not satisfy any Lipschitz condition of positive order. Using probabilistic methods, [Kahane, 4; 6] (1959–60) was able to ameliorate the situation by providing non-spectral $\varphi \in \mathrm{Lip}_\alpha(\mathbf{T})$, for each $\alpha < 1/4$ (cf. [Kaufman, 1]). If $E \subseteq \mathbf{T}$ is closed we write

(3.1.12) $$\forall\, \lambda > 0, \qquad E_\lambda = \{\gamma \in \mathbf{T} : d(\gamma, E) \leqslant \lambda\}$$

("d" was defined before Theorem 2.5.3); and note that if $\varphi \in \mathrm{Lip}_\alpha(\mathbf{T})$, $\alpha > 0$, then

$$\sup_\gamma \{|\varphi(\gamma)| : \gamma \in E_\lambda\} = O(\lambda^\alpha), \lambda \to 0 \tag{3.1.13}$$

(the converse is not true!). In 1967, by a refinement of his earlier method [Kahane, 6], [Kahane, 10; 13, pp. 64–68] proved: *for each $\alpha < 1/2$ there is a non-spectral φ which satisfies* (3.1.13); this result is important in light of the Beurling–Pollard type result: *each $\varphi \in A(\mathbf{T})$ which satisfies* (3.1.13) *for $\alpha = 1/2$ is synthesizable* (e.g. Section 3.2.5). The first explicit construction of a non-spectral element $\varphi \in \mathrm{Lip}_\alpha(\mathbf{T})$, $\alpha > 0$, is due to [Kahane and Katznelson, 2] (1963). Another, more or less combinatoric construction, is due to [Richards, 1] (1967). The simplest example of a non-spectral function is due to [Katznelson, 5, Chapter VIII, Section 7.6]. Katznelson works on $\Gamma = D_\infty$ and, as such, is able to give Varopoulos' proof of Malliavin's theorem without using Schwartz's example. His method extends easily to $\Gamma = \mathbf{T}$.

3.1.5 Principal ideals and non-synthesis. Malliavin verified (3.1.4) for compact Γ in the following strong way:

$$\text{(M)} \quad \begin{array}{l} \forall\, s > 1, \quad \exists\, \varphi \in A(\Gamma), \text{ real-valued, and } \exists\, \delta > 0 \text{ such that} \\ \|e^{iu\varphi}\|_{A'} = O(e^{-\delta|u|^{1/s}}), \ |u| \to \infty \end{array}$$

(cf. [Benedetto, 9, Theorem 2.1]). Condition (M) obviously implies (3.1.4) for each k. Condition (M), Remark 1 and Theorem 3.1.1 combine to yield the following strengthening of Malliavin's theorem for compact Γ (e.g. [Rudin, 4]).

Theorem 3.1.2. *Let Γ be compact and let $\varphi \in A(\Gamma)$ satisfy condition* (M). *There is an interval (a,b), possibly of infinite length, such that for each $r \in (a,b)$ $\{I_{(\varphi+r)^n} : n = 1, \ldots\}$ is a family consisting of distinct closed principal ideals; and, in particular, for each $r \in (a,b)$, $Z(\varphi + r)$ is a non-S-set.*

A standard argument can be used to obtain the same conclusion for any non-discrete Γ (e.g. Theorem 3.1.7). As indicated at the beginning of Paragraph 3.1 we shall not prove the likes of (M).

Remark. We use the notation of Theorem 3.1.2. Note that $Z(\varphi + r_1) \cap Z(\varphi + r_2) = \emptyset$ for $r_1 \neq r_2$ and so $mZ(\varphi + r) = 0$ for some $r \in (a,b)$. For such an r let $\psi = (\varphi + r)^2$ so that if $\hat{g} = \psi$ then $g \in L^1(G)$; consequently, $g \in L^p(G)$ for all $p \geqslant 1$. It is not difficult to adapt the above techniques to be able to choose $T \in A'(Z\psi)$ for which $\langle T, \psi \rangle \neq 0$ and such that $\hat{T} \in L^q(G)$ for a given $q > 2$. Thus, by the Hahn–Banach theorem, we have a proof of Segal's theorem mentioned in Remark 1 after Exercise 2.1.6.

3.1.6 Malliavin's theorem and a property of $A(\Gamma)$. Condition (M) can essentially not be strengthened:

Proposition 3.1.1. *Let Γ be compact. There is no real-valued element $\varphi \in A(\Gamma)$ for which $\delta > 0$ can be found such that*

$$\|e^{iu\varphi}\|_{A'} = O(e^{-\delta|u|}), \ |u| \to \infty. \tag{3.1.14}$$

Proof. Let $e^{iu\varphi(\gamma)} = \sum_{x \in G} a(x,u)(\gamma,x) \in A(\Gamma)$ and for each $q \in [1,\infty]$ let u_q be the $L^q(G)$ norm of the sequence $\{a(x,u): x \in G\}$.

Since $\|e^{iu\varphi}\|_{L^2(\Gamma)} = 1$ we have $u_2 = 1$ by the Plancherel theorem.

By the spectral radius formula, $\lim_{n\to\infty} \|(e^{i\varphi})^n\|_A^{1/n} = \|e^{i\varphi}\|_\infty = 1$, and so $\lim_{u\to\infty} u_1^{1/u} = 1$.

Let $\delta > 0$. Then $\lim_{u\to\infty}(u_1/e^{\delta u})^{1/u} = e^{-\delta}$ and hence

$$\lim_{u\to\infty} \frac{1}{u} \log\left(\frac{u_1}{e^{\delta u}}\right) = -\delta.$$

Since $\lim_{u\to\infty}(1/u) = 0$ we see that $\lim_{u\to\infty} \log\left(\frac{u_1}{e^{\delta u}}\right) = -\infty$, and, consequently,

$$\text{(3.1.15)} \qquad \lim_{u\to\infty} \frac{u_1}{e^{\delta u}} = 0.$$

Clearly,

$$u_2^2 \leqslant u_1 u_\infty,$$

so that, because $u_2 = 1$, we have

$$\frac{1}{u_1} \leqslant u_\infty.$$

Therefore, $e^{\delta u}/u_1 \leqslant u_\infty e^{\delta u}$ which, in light of (3.1.15), tells us that $\lim_{u\to\infty}(\|e^{iu\varphi}\|_{A'}/e^{-\delta u}) = \infty$.

q.e.d.

3.1.7 Helson's theorem on non-S-sets and the structure of ideals. In 1952, [Helson, 2] proved that if $J, K \subseteq A(\Gamma)$ are closed ideals for which

$$\text{(3.1.16)} \qquad E = ZJ = ZK \quad \text{and} \quad J \subsetneq K,$$

then there is a closed ideal $I \subseteq A(\Gamma)$ for which

$$\text{(3.1.17)} \qquad J \subsetneq I \subsetneq K.$$

Consequently, once Malliavin's theorem was proved a weak form of Theorem 3.1.2 was immediately available; the strength of Theorem 3.1.2 is that the ideals can be chosen as closed principal ideals. Helson's proof depended on a result by Godement on unitary representations of abelian groups. Katznelson (1966) (unpublished) strengthened (3.1.17) and we shall prove Katznelson's result in Theorem 3.1.3a; his basic tool is Ditkin's $D(\varphi, I)$ lemma. [Saeki, 1] (1968) also proved Helson's theorem by using Ditkin's lemma (cf. Reiter, 7; Stegeman, 1]).

3.1.8 Katznelson's proof of Helson's theorem.

Theorem 3.1.3 a) *Let $J, K \subseteq A(\Gamma)$ be closed ideals which satisfy* (3.1.16) (*and therefore Γ is non-discrete*). *There is a continuum, $\{I_\alpha\}$, of distinct closed ideals of $A(\Gamma)$ such that*

$$\forall \alpha, \qquad J \subsetneq I_\alpha \subsetneq K,$$

and there are pairs $\{I_1, I_2\} \subseteq \{I_\alpha\}$ such that

$$I_1 \cap I_2 = J.$$

b) *Denumerable decreasing and increasing sequences can be chosen from $\{I_\alpha\}$.*

c) *If each point of ZJ has a countable open basis in the relative topology of E then there is a decreasing sequence $\{I_n : n = 1, \ldots\} \subseteq \{I_\alpha\}$ such that*

$$\cap I_n = J.$$

Proof. a.i) Define $D(K,J)$ to be the set of points $\gamma \in \Gamma$ such that for each neighborhood N_γ of γ there is $\varphi \in K$ such that

$$\forall \psi \in J, \quad \exists \lambda \in N_\gamma \text{ for which } \psi(\lambda) \neq \varphi(\lambda).$$

Note that if $K = k(E)$ and $J = \overline{j(E)}$ then $D(k,j) \subseteq D(K,J)$, where $D(k,j)$ was defined after Proposition 2.5.1.

The proof of Proposition 2.5.1b shows that $D(K,J)$ is perfect.

Since $J \subsetneq K$ we see that $D(K,J) \neq \emptyset$ because of Theorem 2.4.1.

Let $P \subsetneq D(K,J)$ be a compact perfect non-empty set which is the closure of its interior taken relative to $D(K,J)$.

We shall associate a closed ideal I_P, with the required properties, to each such perfect set, P.

ii) Take P as in part i) and set $Q = D(K,J) \setminus P$.

Define

$$I_P = \{\varphi \in K : \forall \gamma \in Q, \quad \varphi \in J_{\mathrm{loc}}(\gamma)\}.$$

Clearly $I_P \subseteq A(\Gamma)$ is an ideal and $J \subseteq I_P \subseteq K$.

We shall now prove that I_P is closed.

Take $\psi \in \bar{I}_P$ and choose any $\gamma \in Q$. Let $\theta_\gamma \in A_c(\Gamma)$ be 1 on a neighborhood of γ and assume that it vanishes on a neighborhood of P.

Let $\lim \|\psi_n - \psi\|_A = 0$ for $\{\psi_n : n = 1, \ldots\} \subseteq I_P$.

If $\lambda \in Q$ then $\psi_n \theta_\gamma \in J_{\mathrm{loc}}(\lambda)$ since $\psi_n \in I_P$.

If $\lambda \notin D(K,J)$ then, since $\psi_n \theta_\gamma \in K$, we have $\psi_n \theta_\gamma \in J_{\mathrm{loc}}(\lambda)$.

Consequently, by the way we have defined θ_γ, we see that $\psi_n\theta_\gamma \in J$ because $\psi_n\theta_\gamma \in J_{\mathrm{loc}}(\lambda)$ for each $\lambda \in \Gamma$.

Thus, $\psi\theta_\gamma \in J$ and so $\psi \in J_{\mathrm{loc}}(\gamma)$. Therefore $\psi \in I_P$ since γ was an arbitrary element of Q.

iii) We show that $I_P \setminus J \neq \emptyset$.

Choose $\lambda \in \Gamma$ in the interior of P relative to the topology on $D(K,J)$, and let $V_\lambda \subseteq D(K,J)$ be an open neighborhood of λ (in the relative topology of $D(K,J)$) such that $V_\lambda \cap Q = \emptyset$.

Take an open set $N_\lambda \subseteq \Gamma$ for which $N_\lambda \cap D(K,J) = V_\lambda$; in particular, $N_\lambda \cap Q = \emptyset$ since $V_\lambda \cap Q = \emptyset$.

Let $B_\lambda \subseteq N_\lambda$ be a compact neighborhood of λ, and choose $\theta \in A_c(\Gamma)$ such that $\theta(B_\lambda) = 1$ and $\operatorname{supp}\theta \subseteq N_\lambda$. Thus $\theta(Q) = 0$.

For λ and B_λ choose $\varphi \in K$ as in the definition of $D(K,J)$.

Then $\varphi\theta$ is not equal to any element of J on B_λ.

On the other hand, $\varphi\theta \in J_{\mathrm{loc}}(\gamma)$ for each $\gamma \in Q$; and so $\varphi\theta \in I_P$.

iv) Note that $D(I_P,J) \subseteq D(K,J)$.

We shall prove

$$P = D(I_P,J). \tag{3.1.18}$$

from which it follows immediately that $I_P \subsetneq K$.

The proof of part iii) tells us that $P \subseteq D(I_P,J)$. To see this first let λ be in the interior of P, relative to the topology on $D(K,J)$, and let N_λ be a neighborhood of λ. As in part iii) we can take N_λ so that $N_\lambda \cap Q = \emptyset$.

Then choose B_λ, θ, and φ as in part iii). Thus $\varphi\theta \in I_P$, and, by applying the definition of $D(I_P,J)$, we see that $P \subseteq D(I_P,J)$.

To prove that $D(I_P,J) \subseteq P$ we take $\lambda \in Q \cap D(I_P,J)$ and obtain a contradiction.

Choose N_λ, a compact neighborhood of λ, disjoint from P. By the definition of $D(I_P,J)$ there is $\varphi \in I_P$ so that φ is not equal to any element of J on N_λ.

Let $\theta \in A_c(\Gamma)$ be 1 on N_λ and assume $(\operatorname{supp}\theta) \cap P = \emptyset$.

Then $\theta\varphi \in I_P$ and $\theta\varphi \in J_{\mathrm{loc}}(\gamma)$ for each $\gamma \in P$.

Thus, by Theorem 2.4.1, $\theta\varphi \in J$; and this contradicts the fact that φ is not equal to any element of J on N_λ.

v) If P_1, P_2 are disjoint compact perfect non-empty subsets of $D(K,J)$ then $I_{P_1} \cap I_{P_2} = J$.

b) is now obvious, and c) is routine.

q.e.d.

A constructive procedure has been given in [Osipov, 1] to determine a continuum, $\{I_\alpha\}$, of distinct closed ideals where $ZI_\alpha = S^2 \subseteq \mathbb{R}^3$.

3.1.9 The projective tensor product. In 1965, in an important series of papers (e.g. [Varopoulos, 4]), Varopoulos proved Malliavin's theorem using Schwartz's example and projective tensor products of spaces of continuous functions.

Expositions of Varopoulos' work are due to [Herz, 7; 8] and [Kahane, 8]; and the basic reference for the connection between harmonic analysis and tensor algebras is [Varopoulos, 8]. The relation of tensor products to Hilbert spaces was discovered by Grothendieck in what he called the "fundamental theorem on the metric theory of tensor products", e.g. [Lindenstrauss and Pelczynski, 1]. Analogously, Varopoulos' results show the relation of tensor products to harmonic analysis.

Let E_1 and E_2 be compact Hausdorff spaces. The *projective tensor product of* $C(E_1)$ *and* $C(E_2)$, denoted by

$$V(E_1, E_2) \quad \text{or} \quad C(E_1) \hat{\otimes} C(E_2)$$

(we write $V(E)$ instead of $V(E, E)$), is the set of elements $\Phi \in C(E_1 \times E_2)$ which have the form

$$\Phi = \sum \varphi_n \psi_n, \qquad \varphi_n \in C(E_1) \text{ and } \psi_n \in C(E_2), \tag{3.1.19}$$

where

$$\sum \|\varphi_n\|_\infty \|\psi_n\|_\infty < \infty. \tag{3.1.20}$$

Because of (3.1.20), $V(E_1, E_2)$ can be normed by

$$\forall\, \Phi \in V(E_1, E_2), \qquad \|\Phi\|_\otimes = \inf\{\sum \|\varphi_n\|_\infty \|\psi_n\|_\infty\},$$

where the infimum is taken over all representations (3.1.19) of Φ; as such $V(E_1, E_2)$ is a commutative Banach algebra (under pointwise multiplication) with unit.

3.1.10 Grothendieck's characterization of the projective tensor product. We shall now sketch some properties of tensor products. These will not be necessary to read Varopoulos' applications of tensor algebra methods to harmonic analysis in Sections 3.1.12–3.1.15; but are only presented for some added perspective.

Let X_1, X_2 be complex Hausdorff locally convex topological vector spaces with tensor product, $X_1 \otimes X_2$, as defined at the end of Paragraph 2.4. Also, let $\{p_i : i \in I\}$ and $\{q_j : j \in J\}$, where I and J are index sets, be families of semi-norms on X_1 and X_2, respectively, which define their respective topologies. Preserving the notation, $\alpha : X_1 \times X_2 \to X_1 \otimes X_2$, of Paragraph 2.4 it is standard to prove (e.g. [Benedetto, 6, pp. 231–233]) that: *there is a finest locally convex Hausdorff topology*, T_p, *on* $X_1 \otimes X_2$ *such that* α *is continuous; and the semi-norms which characterize* T_p *are of the form*

$$\forall\, u \in X_1 \otimes X_2,\ p_i \otimes q_j(u) = \inf\left\{\sum_n p_i(s_n) q_j(t_n) : u = \sum_n s_n \otimes t_n\right\}, \tag{3.1.21}$$

$i \in I, j \in J$. $(X_1 \otimes X_2, \mathrm{T}_p)$ is the *projective topological tensor product*. If X_1 and X_2 are normed spaces then $(X_1 \otimes X_2, \mathrm{T}_p)$ is normed by (3.1.21) (cf. (3.1.20)). Since $(X_1 \otimes X_2, \mathrm{T}_p)$ is a Hausdorff locally convex space it is also a uniform space, and so we can discuss its completion, $X_1 \hat{\otimes} X_2$. We give Pietsch's proof of the following theorem due to Grothendieck (cf. the definition of $V(E_1, E_2)$).

Theorem 3.1.4. *Let X_1 and X_2 be metrizable locally convex vector spaces. Then each $u \in X_1 \hat{\otimes} X_2$ has the representation*

$$u = \sum_{n=1}^{\infty} c_n s_n \otimes t_n, \qquad c_n \in \mathbf{C},\ s_n \in X_1,\ t_n \in X_2, \tag{3.1.22}$$

where $\lim s_n = 0$ *in* X_1, $\lim t_n = 0$ *in* X_2, $\sum |c_n| < \infty$, *and the convergence of the sum in* (3.1.22) *is in the topology of* $X_1 \hat{\otimes} X_2$.

Proof. Assume without loss of generality that the sequences, $\{p_i : i = 1, \ldots\}$ and $\{q_j : j = 1, \ldots\}$, of semi-norms describing the topologies of X_1 and X_2 are increasing; and also assume that $\{p_k \otimes q_k : k = 1, \ldots\}$ is increasing.

Let r_k be the continuous extension of $p_k \otimes q_k$ to $X_1 \hat{\otimes} X_2$. Take $u \in X_1 \hat{\otimes} X_2$ and choose a sequence $\{w_m : m = 1, \ldots\}$ contained in $X_1 \otimes X_2$ for which $\lim w_m = u$ in $X_1 \hat{\otimes} X_2$. Clearly, we can choose a sub-sequence $\{u_k : k = 1, \ldots\} \subseteq \{w_m : m = 1, \ldots\} \subseteq X_1 \otimes X_2$ such that

$$\forall\, k, \qquad r_k(u - u_k) < k^{-2} 2^{-(k+1)}.$$

Set

$$u_1 = \sum_{n=1}^{n_1} c_n s_n \otimes t_n \tag{3.1.23}$$

to be any representation of u_1, and define

$$\forall\, k, \qquad v_k = u_{k+1} - u_k.$$

Thus, for each k,

$$\begin{aligned} r_k(v_k) &\leqslant r_k(u - u_k) + r_k(u - u_{k+1}) \leqslant r_k(u - u_k) + r_{k+1}(u - u_{k+1}) \\ &< k^{-2} 2^{-k}\left(2^{-1} + \left(\frac{k}{k+1}\right)^2 2^{-2}\right) < k^{-2} 2^{-k}, \end{aligned} \tag{3.1.24}$$

since $r_{k+1} \geqslant r_k$.

Because of (3.1.23) and the definition of v_k we have

$$u_2 = \sum_{n=1}^{n_2} c_n s_n \otimes t_n$$

where

$$v_1 = (u_2 - u_1) = \sum_{n=n_1+1}^{n_2} c_n s_n \otimes t_n.$$

Generally, then

$$v_k = \sum_{n=n_k+1}^{n_{k+1}} c_n s_n \otimes t_n.$$

Now, because of (3.1.21) and (3.1.24), the above representation of v_k can be chosen so that

$$\sum_{n=n_k+1}^{n_{k+1}} |c_n| \leqslant 2^{-k}$$

and

$$\forall\, n \in (n_k, n_{k+1}], \quad p_k(s_n) \leqslant 1/k \quad \text{and} \quad q_k(t_n) \leqslant 1/k.$$

This completes the proof since $u = u_1 + \sum_{k=1}^{\infty} v_k$.

q.e.d.

3.1.11 Further properties of topological tensor products. Without getting involved in too much detail we now define the other most popular topology on $X_1 \otimes X_2$ for purposes of comparison.

Let $\mathscr{U}_j$ be a neighborhood basis at $0 \in X_j$ of the Hausdorff locally convex space X_j and define

$$\forall\, U_j \in \mathscr{U}_j, \quad U_j^0 = \{R \in X_j' : \forall\, r \in U_j, |\langle R, r\rangle| \leqslant 1\}.$$

Consider the tensor product $X_1' \otimes X_2'$ with corresponding map, $\alpha' : X_1' \times X_2' \to X_1' \otimes X_2'$; and let T_ε be the topology in $X_1 \otimes X_2$ of uniform convergence on all sets having the form $\alpha'(U_1^0, U_2^0) \subseteq X_1' \otimes X_2'$ where $U_j \in \mathscr{U}_j$. T_ε is aptly called the topology of *equicontinuous convergence* on $X_1 \otimes X_2$. $(X_1 \otimes X_2, T_\varepsilon)$ is a Hausdorff locally convex space and we designate its completion by $X_1 \check{\otimes} X_2$.

We have $T_\varepsilon \subseteq T_p$ in $X_1 \otimes X_2$ so that, generally, $X_1 \hat{\otimes} X_2 \subseteq X_1 \check{\otimes} X_2$. An Hausdorff locally convex space X_1 is *nuclear* if $X_1 \hat{\otimes} X_2 = X_1 \check{\otimes} X_2$ for each Hausdorff locally convex space X_2. An infinite dimensional Banach space is never nuclear.

Proposition 3.1.2. *Let X_1 be a complex vector space and let X_2 be a complex vector space of* **C***-valued functions defined on a set E. Then there is a vector space isomorphism of $X_1 \otimes X_2$ onto a complex vector space, $F(E, X_1)$, of functions,*

$$E \to X_1,$$

which take their values in finite dimensional subspaces of X_1.

Proof. The correspondence for the required isomorphism is given by

$$\begin{aligned} X_1 \otimes X_2 &\to F(K, X_1) \\ \sum s_n \otimes t_n &\mapsto \sum s_n t_n \end{aligned}$$

where

$$\forall\, \gamma \in E, \qquad (\sum s_n t_n)(\gamma) = \sum s_n t_n(\gamma).$$

It is straightforward to check the necessary details.

q.e.d.

As a corollary we have: *let X be an Hausdorff locally convex space and let E be a compact space; then $X \otimes C(E)$ is a space of continuous functions on E taking values in X.* Now, let $\{p_i : i \in I\}$ be a family of semi-norms characterizing the topology on an Hausdorff locally convex space X, and let $C(E, X)$ be the space of continuous functions on the compact Hausdorff space E taking values in X. As such it is natural to topologize $C(E, X)$ with the family $\{\tilde{p}_i : i \in I\}$ of semi-norms defined by

$$(3.1.25) \qquad \forall\, \Phi \in C(E, X) \text{ and } \forall\, i \in I, \qquad \tilde{p}_i(\Phi) = \sup_{\gamma \in E} p_i(\Phi(\gamma)).$$

We know that $X \otimes C(E) \subseteq C(E, X)$ and it turns out that *the induced topology of* (3.1.25) *on $X \otimes C(E)$ is T_ε.* Even more is true:

$$\overline{X \otimes C(E)} = C(E, X)$$

and if X is complete then

$$(3.1.26) \qquad X \check{\otimes} C(E) = C(E, X),$$

algebraically and topologically.

We now go one step further.

Proposition 3.1.3. *Let E_1 and E_2 be compact Hausdorff spaces and take $C(E_i)$, $i = 1, 2$, and $C(E_1 \times E_2)$ with the sup norm topology. Then*

$$C(E_1 \times E_2) = C(E_1) \check{\otimes} C(E_2),$$

algebraically and topologically.

Proof. In light of (3.1.26), which we haven't proved, it is sufficient to show that

$$C(E_1 \times E_2) = C(E_2, C(E_1)).$$

We define the map

$$\begin{aligned} C(E_2, C(E_1)) &\to C(E_1 \times E_2) \\ \Phi &\mapsto \tilde{\Phi} \end{aligned}$$

where

$$\forall\, \gamma_i \in E_i, \qquad \tilde{\Phi}(\gamma_1, \gamma_2) = [\Phi(\gamma_2)](\gamma_1).$$

It is straightforward to check the necessary details.

q.e.d.

In light of these remarks, we see that *if* E_1 *and* E_2 *are compact Hausdorff spaces then*

$$V(E_1, E_2) = C(E_1) \hat{\otimes} C(E_2) \subseteq C(E_1 \times E_2),$$

where the injection is a norm-decreasing linear map. Of course, this brings us back to (3.1.19), (3.1.20), and the norm $\| \quad \|_{\otimes}$.

As we mentioned in Section 3.1.9, $V(E_1, E_2)$ is a commutative Banach algebra with unit, and it is straightforward to check that it is also semi-simple, regular, self-adjoint, and with maximal ideal space,

$$V(E_1, E_2)^m = E_1 \times E_2.$$

3.1.12 Varopoulos' idea. Let Γ be infinite and compact. The initial relations between harmonic analysis and tensor products that were discovered by Varopoulos can be described by the two statements:

$(V.1)$ $\qquad A(\Gamma)$ is a closed subalgebra of $V(\Gamma)$

and

$(V.2)$ $\qquad A(E_1 + E_2) = V(E_1, E_2)$

for certain sets $E_1, E_2 \subseteq \Gamma$. By means of $(V.1)$ information about $A(\Gamma)$ can be translated into information about $V(\Gamma)$. In particular, the fact that $S^2 \subseteq \mathbf{Я}^3$ is a non-S-set for $A(\mathbf{Я}^3)$ (Exercise 2.5.5) or, equivalently, for $A(\mathbf{T}^3)$ will tell us that there are non-S-sets for $V(\mathbf{T}^3)$ (where the notion of an "S-set for $V(\mathbf{T}^3)$" is defined in a natural way). By means of $(V.2)$ information about $V(E_1, E_2)$ can be translated into information about $A(E_1 + E_2)$. In particular, the fact, which we expand on in a moment, that there are non-S-sets for $V(D_\infty)$, will tell us that there are non-S-sets for the Banach algebra $A(E_1 + E_2)$, where, among other conditions, we demand that E_j be homeomorphic to D_∞; it is then easy to see that $E_1 + E_2$ contains non-S-sets for $A(\Gamma)$. As we mentioned earlier, it is then not too difficult to prove that there are non-S-sets in any non-discrete LCAG, given the fact that there are non-S-sets in every infinite compact Γ.

There is one interesting hitch in this scheme of tensor-talking Schwartz's example into Malliavin's theorem. Basically the above plan has to include a method of transferring the non-S-set for $V(\mathbf{T}^3)$ into the non-S-set for $V(D_\infty)$. The question is whether the solution to this latter problem is any easier than transferring Schwartz's non-S-set $E \subseteq \mathbf{T}^3$ directly into a non-S-set for $A(\Gamma)$, without first climbing down into $V(\mathbf{T}^3)$ and then climbing up from $V(D_\infty)$. It turns out, by a general theorem of Reiter, Herz, and De Leeuw, which we discuss below, that if there is a continuous homomorphic injection

$$\alpha: \mathbf{T}^3 \rightarrow \Gamma,$$

then $\alpha(E) \subseteq \Gamma$ is a non-*S*-set. Of course, if we use the above-mentioned (in Section 3.1.5) example of a non-*S*-set for $A(D_\infty)$, due to Katznelson, then we only need $(V.1)$ and $(V.2)$ to obtain Malliavin's theorem. The issue of transferring non-*S*-sets from one group to another still maintains its importance.

3.1.13 Imbeddings of group algebras into tensor algebras. Let Γ be infinite and compact. Clearly, the following linear maps are well-defined [Herz, 7; 8]:

$$\begin{aligned} M: C(\Gamma) &\to C(\Gamma \times \Gamma) \\ \varphi &\mapsto M\varphi = \Phi, \end{aligned}$$

where

$$\Phi(\lambda, \gamma) = \varphi(\lambda + \gamma),$$

and

$$\begin{aligned} P: C(\Gamma \times \Gamma) &\to C(\Gamma) \\ \Phi &\mapsto P\Phi = \varphi, \end{aligned}$$

where

$$\varphi(\gamma) = \int_\Gamma \Phi(\gamma - \lambda, \lambda)\, \mathrm{d}\lambda.$$

Theorem 3.1.5. *Let Γ be infinite and compact.*

a) $MA(\Gamma) \subseteq V(\Gamma)$.

b) *The map*

(3.1.27) $\quad M:(A(\Gamma), \| \ \|_A) \to (V(\Gamma), \| \ \|_\otimes)$

is an (injective) isometry, and so $MA(\Gamma)$ is a closed subalgebra of $(V(\Gamma), \| \ \|_\otimes)$.

c) $A(\Gamma) = \{\varphi \in C(\Gamma): M\varphi \in V(\Gamma)\}$ *(where the inclusion, $MA(\Gamma) \subseteq V(\Gamma)$, is part* a)).

Proof. a) Given $\varphi(\gamma) = \sum_{x \in G} a_x(\gamma, x) \in A(\Gamma)$.

Then

$$\Phi(\lambda, \gamma) = M\varphi(\lambda, \gamma) = \sum_{x \in G} a_x(\lambda, x)(\gamma, x).$$

By the definition of $\| \ \|_\otimes$, $\|\Phi\|_\otimes \leqslant \sum_{x \in G} |a_x|$; and so $MA(\Gamma) \subseteq V(\Gamma)$.

b.i) First note that $P \circ M: C(\Gamma) \to C(\Gamma)$ is the identity since

$$\forall\, \varphi \in C(\Gamma), \qquad P(M\varphi)(\gamma) = \int_\Gamma (M\varphi)(\gamma - \lambda, \lambda)\, \mathrm{d}\lambda = \int_\Gamma \varphi(\gamma)\, \mathrm{d}\lambda = \varphi(\gamma).$$

Consequently, M is injective.

ii) We now prove that $P(V(\Gamma)) \subseteq A(\Gamma)$, *and that* $\|P\| \leqslant 1$ where $\|P\|$ is the canonical norm for the map

$$P:(V(\Gamma), \|\ \ \|_{\otimes}) \rightarrow (A(\Gamma), \|\ \ \|_A).$$

Let $\Phi = \varphi \otimes \psi$, where $\varphi, \psi \in C(\Gamma)$. Thus

$$P\Phi(\gamma) = \int_{\Gamma} \varphi(\gamma - \lambda)\psi(\lambda)\,\mathrm{d}\gamma \in A(\Gamma)$$

since $L^2(\Gamma) * L^2(\Gamma) = A(\Gamma)$ (a bona fide application of Cor. 1.1.2!).

Also

(3.1.28) $\quad \|P\Phi\|_A = \|\varphi * \psi\|_A \leqslant \|\varphi\|_2 \|\psi\|_2 \leqslant \|\varphi\|_\infty \|\psi\|_\infty$

and so

(3.1.29) $\quad \|P\Phi\|_A \leqslant \|\Phi\|_{\otimes}.$

Because of (3.1.28), (3.1.29), the definition of $\|\ \ \|_{\otimes}$, and the fact that P is linear we have $PV(\Gamma) \subseteq A(\Gamma)$ and $\|P\| \leqslant 1$.

iii) We now prove that $\|M\| \leqslant 1$ where $\|M\|$ is the canonical norm for (3.1.27).

Take $x \in G$ and compute

$$\|M(\cdot, x)\|_{\otimes} \leqslant 1 = \|(\cdot, x)\|_A$$

since $(M(\ \cdot, x))(\lambda, \gamma) = (\lambda, x)(\gamma, x)$.

Consequently,

$$\forall\, \varphi \in A(\Gamma), \qquad \|M\varphi\|_{\otimes} \leqslant \|\varphi\|_A,$$

and so $\|M\| \leqslant 1$.

iv) The isometry,

$$\forall\, \varphi \in A(\Gamma), \qquad \|\varphi\|_A = \|M\varphi\|_{\otimes},$$

is now established by parts b.i), b.ii), and b.iii) since

$$\forall\, \varphi \in A(\Gamma), \qquad \|\varphi\|_A = \|P \circ M(\varphi)\|_A \leqslant \|M\varphi\|_{\otimes} \leqslant \|\varphi\|_A.$$

c) Take $\varphi \in C(\Gamma)$ for which $M\varphi \in V(\Gamma)$.

By part b.ii), $P \circ M(\varphi) \in A(\Gamma)$, and by part b.i), $P \circ M(\varphi) = \varphi$. Thus,

$$\{\varphi \in C(\Gamma) : M\varphi \in V(\Gamma)\} \subseteq A(\Gamma)$$

and we have c).

q.e.d.

A closed set $F \subseteq \Gamma \times \Gamma$ is an *S-set for* $V(\Gamma)$ if for each $\Phi \in V(\Gamma)$ which vanishes on F there is a sequence $\{\Phi_n : n = 1, \ldots\} \subseteq V(\Gamma)$ each of whose elements vanishes on a neighborhood of F such that

$$\text{(3.1.30)} \qquad \lim_n \|\Phi - \Phi_n\|_\otimes = 0.$$

Proposition 3.1.4. *Let Γ be infinite and compact, and let $E \subseteq \Gamma$ be a non-S-set (for $A(\Gamma)$). Then*

$$F = \{(\lambda, \gamma) \in \Gamma \times \Gamma : \lambda + \gamma \in E\} \subseteq \Gamma \times \Gamma$$

is a non-S-set for $V(\Gamma)$.

Proof. Choose $\varphi \in k(E) \setminus \overline{j(E)}$ and set $\Phi = M\varphi$.
Suppose $\Phi_n \in V(\Gamma)$ vanishes on a neighborhood of F and (3.1.30) is valid.
Then, by Theorem 3.1.5, $P\Phi_n \in j(E)$ and

$$\|\varphi - P\Phi_n\|_A = \|P(M\varphi - \Phi_n)\|_A \leqslant \|\Phi - \Phi_n\|_\otimes.$$

This yields the desired contradiction.

q.e.d.

Herz proved the converse of Proposition 3.1.4, e.g. [Varopoulos, 8, pp. 98–99]: *if $E \subseteq \Gamma$ is an S-set for $A(\Gamma)$ then F is an S-set for $V(\Gamma)$.*

3.1.14 Imbeddings of tensor algebras into restrictions of group algebras. Let Γ be infinite and compact. An initial, crucial, and simple-to-prove observation by Varopoulos is the fact that *if $E_1, E_2 \subseteq \Gamma$ are Helson sets then*

$$A(E_1 \times E_2) = V(E_1, E_2);$$

this led to conditions of the form (V.2).
If $E_1, E_2 \subseteq \Gamma$ are closed we generalize the map M of Section 3.1.13 by defining

$$\text{(3.1.31)} \qquad \begin{aligned} M : C(E_1 + E_2) &\to C(E_1 \times E_2) \\ \varphi &\mapsto M\varphi = \Phi, \end{aligned}$$

where

$$\Phi(\lambda, \gamma) = \varphi(\lambda + \gamma);$$

thus, M is the canonical homomorphism induced by the continuous map

$$\begin{aligned} E_1 \times E_2 &\to E_1 + E_2 \\ (\lambda, \gamma) &\mapsto \lambda + \gamma. \end{aligned}$$

As in parts a) and b.iii) of Theorem 3.1.5 it is easy to check that $MA(E_1 + E_2) \subseteq V(E_1, E_2)$ and

(3.1.32) $$M: A(E_1 + E_2) \to V(E_1, E_2)$$

is continuous with $\|M\| \leqslant 1$. We now strengthen (3.1.32) to obtain the likes of (V.2).

Theorem 3.1.6. *Let Γ be infinite and compact. Assume that Γ contains a Kronecker set which is homeomorphic to D_∞. Then there are closed sets $E_1, E_2 \subseteq \Gamma$ such that*

(3.1.33) $$V(D_\infty) = A(E_1 + E_2),$$

isometrically and algebraically.

Proof. i) If E is the given Kronecker set we let E_1 and E_2 be disjoint non-empty relatively open subsets of E whose union is E. Since E is independent the canonical maps

$$D_\infty \times D_\infty \to E_1 \times E_2 \to E_1 + E_2$$

are bijections; and so the canonical homomorphism, M, given by (3.1.31), actually identifies $C(E_1 + E_2)$ and $C(D_\infty \times D_\infty)$. Thus, the map (3.1.32) is an injective continuous linear map, norm bounded by 1; and we shall now prove that it is surjective (the proof of surjectivity is completed in part v)).

ii) Take $\Phi \in V(D_\infty)$ and let $\varphi \in C(E_1 + E_2)$ have the property that $M\varphi = \Phi$. We write

(3.1.34) $$\Phi = \sum \varphi_n \otimes \psi_n, \qquad \varphi_n \in C(E_1) \text{ and } \psi_n \in C(E_2),$$

so that

$$\forall\, \lambda \in E_1, \quad \gamma \in E_2, \qquad \varphi(\lambda + \gamma) = \sum \varphi_n(\lambda)\, \psi_n(\gamma).$$

For each $\varepsilon \in (0, 1)$ choose a representation (3.1.34) so that

$$\sum \|\varphi_n\|_{C(E_1)} \|\psi_n\|_{C(E_2)} \leqslant \|\Phi\|_\otimes + \frac{\varepsilon}{3}.$$

Since E is a Kronecker set we can use an argument involving the Tietze extension theorem to prove that each E_j, $j = 1, 2$, is a Kronecker set, and then we are able to prove that

$$\forall\, n, \qquad \|\varphi_n\|_{C(E_1)} = \|\varphi_n\|_{A(E_1)}$$

and

$$\|\psi_n\|_{C(E_2)} = \|\psi_n\|_{A(E_2)}.$$

(Actually, if we only use the more evident fact that each E_j is an Helson set (e.g. [Benedetto, 6, pp. 120–121] or the appropriate remark in Section 2.5.8), the proof of the entire theorem goes through except that instead of proving norm equality in (3.1.33) we prove norm equivalence.)

Thus,

(3.1.35) $$\sum_n \|\varphi_n\|_{A(E_1)} \|\psi_n\|_{A(E_2)} \leqslant \|\Phi\|_{\otimes} + \frac{\varepsilon}{3}.$$

Given φ_n and ψ_n we choose $\tilde{\varphi}_n$ and $\tilde{\psi}_n$ in $A(\Gamma)$ such that

$$\varphi_n = \tilde{\varphi}_n \text{ on } E_1 \qquad \text{and} \qquad \psi_n = \tilde{\psi}_n \text{ on } E_2.$$

Next, take representations

$$\forall\, \lambda \in E_1, \qquad \varphi_n(\lambda) = \sum_{x \in G} a_x(n)\,(\lambda, x)$$

and

$$\forall\, \gamma \in E_2, \qquad \psi_n(\gamma) = \sum_{y \in G} b_y(n)\,(\gamma, y)$$

for which

$$\sum_{x \in G} |a_x(n)| \leqslant \|\tilde{\varphi}_n\|_A, \qquad \sum_{y \in G} |b_y(n)| \leqslant \|\tilde{\psi}_n\|_A$$

and

$$\sum_{x \in G} |a_x(n)| \leqslant \|\varphi_n\|_{A(E_1)} + \frac{\varepsilon}{3 \cdot 2^n \|\tilde{\psi}_n\|_A},$$

$$\sum_{y \in G} |b_y(n)| \leqslant \|\psi_n\|_{A(E_2)} + \frac{\varepsilon}{3 \cdot 2^n \|\tilde{\varphi}_n\|_A}.$$

Using (3.1.35) we compute that

$$\forall\, \lambda \in E_1 \text{ and } \forall\, \gamma \in E_2, \qquad \varphi(\lambda + \gamma) = \sum_{x, y \in G} c_{x,y}(\lambda, x)\,(\gamma, y),$$

where $c_{x,y} = \sum_n a_x(n)\, b_y(n)$ and

(3.1.36) $$\sum |c_{x,y}| \leqslant \|\Phi\|_{\otimes} + \varepsilon.$$

iii) For each $x, y \in G$ define $\varphi_{x,y} \in C(E)$ as

$$\varphi_{x,y}(\gamma) = \begin{cases} (\gamma, x) & \text{for} \quad \gamma \in E_1, \\ (\gamma, y) & \text{for} \quad \gamma \in E_2. \end{cases}$$

Since E is a Kronecker set and $|\varphi_{x,y}| = 1$ we see that

$$\forall\, x, y \in G \text{ and } \forall\, \eta > 0 \;\; \exists\, u(\eta, x, y) = u \in G \text{ such that}$$

$$\sup_{\gamma \in E} |\varphi_{x,y}(\gamma) - (\gamma, u)| < \eta.$$

Thus, for these values of $x, y \in G$, $\eta > 0$, and $u \in G$, we have

$$\begin{aligned}\|(\lambda, x)(\gamma, y) - (\lambda + \gamma, u)\|_{\otimes} &\leqslant \|(\gamma, y)[(\lambda, x) - (\lambda, u)] + (\lambda, u)[(\gamma, y) - (\gamma, u)]\|_{\otimes} \\ &\leqslant \|(\gamma, y)\|_{C(E_2)}\|(\lambda, x) - (\lambda, u)\|_{C(E_1)} \\ &\quad + \|(\lambda, u)\|_{C(E_1)}\|(\gamma, y) - (\gamma, u)\|_{C(E_2)} < 2\eta.\end{aligned}$$

iv) Take $\eta = \varepsilon/(2 \sum_{x,y \in G} |c_{x,y}|)$ in part iii) and define

$$\psi_{\varepsilon}(\lambda) = \sum_{x,y \in G} c_{x,y}(\lambda, u(\eta, x, y)) \in A(\Gamma),$$

so that $\psi_{\varepsilon}(\lambda + \gamma) = \Psi_{\varepsilon}(\lambda, \gamma) \in M(A(E_1 + E_2))$.
Note that

(3.1.37) $\quad \|\Phi - \Psi_{\varepsilon}\|_{\otimes} \leqslant \varepsilon,$

because of (3.1.36) and part iii), and that

(3.1.38) $\quad \|\psi_{\varepsilon}\|_A \leqslant \|\Phi\|_{\otimes} + \varepsilon$

because of (3.1.36).
v) We now use (3.1.37) and (3.1.38) in the following way.
Given $\varepsilon > 0$. Choose $\theta_1 \in A(E_1 + E_2)$ such that for $\Theta_1 = M\theta_1$

$$\|\Phi - \Theta_1\|_{\otimes} \leqslant \frac{\varepsilon}{2}$$

and

$$\|\theta_1\|_{A(E_1 + E_2)} \leqslant \|\Phi\|_{\otimes} + \frac{\varepsilon}{2}.$$

Then, setting $\Phi_1 = \Phi - \Theta_1 \in V(D_{\infty})$, choose $\theta_2 \in A(E_1 + E_2)$ such that for $\Theta_2 = M\theta_2$,

$$\|\Phi_1 - \Theta_2\|_{\otimes} \leqslant \frac{\varepsilon}{2^2}$$

and

$$\|\theta_2\|_{A(E_1 + E_2)} \leqslant \|\Phi_1\|_{\otimes} + \frac{\varepsilon}{2^2}.$$

Proceeding in this way, we set $\varphi = \sum_1^{\infty} \theta_n$.

Note that

$$\forall\, n \geqslant 2, \qquad \|\theta_n\|_{A(E_1 + E_2)} \leqslant \|\Phi_{n-1}\|_{\otimes} + \frac{\varepsilon}{2^n} \leqslant \varepsilon\left(\frac{1}{2^{n-1}} + \frac{1}{2^n}\right),$$

and so $\varphi \in A(E_1 + E_2)$ since

$$\|\varphi\|_{A(E_1+E_2)} \leqslant \|\theta_1\|_{A(E_1+E_2)} + \sum_2^\infty \|\theta_n\|_{A(E_1+E_2)} \leqslant \left(\frac{\varepsilon}{2} + \|\Phi\|_\otimes\right) + 3\varepsilon. \tag{3.1.39}$$

It is also clear that $M\varphi = \Phi$ since

$$\|\Phi - (\Theta_1 + \ldots + \Theta_n)\|_\otimes = \|\Phi_{n-1} - \Theta_n\|_\otimes \leqslant \frac{\varepsilon}{2^n};$$

and thus we obtain the surjectivity that we promised in part i).

The norm equality follows from (3.1.39).

q.e.d.

3.1.15 Remarks on Varopoulos' method. By means of some technical modifications, Theorem 3.1.6 can be strengthened to the form—

Theorem 3.1.6′. *Let Γ be infinite and compact. Then there are closed sets $E_1, E_2 \subseteq \Gamma$ such that*

$$V(D_\infty) = A(E_1 + E_2),$$

topologically and algebraically.

Instead of providing the details required to prove Theorem 3.1.6′ we shall make some relevant remarks and move on.

Remark 1. In the mid-1950's, using a construction due to Carleson, Rudin was the first to construct perfect non-empty Kronecker sets in **T**. It then became apparent that some groups Γ did not contain any Kronecker sets homeomorphic to D_∞, and a related notion, that of a *K_p-set*, was devised. Just as Kronecker sets are independent, so K_p-sets are p-independent. Also, *for any non-discrete Γ there is a Kronecker or K_p-set, p prime, contained in Γ and homeomorphic to D_∞* [Rudin, 5, Chapter 5] (cf. the remark in Section 2.5.8 on Kronecker set homeomorphs to compact spaces). This fact is precisely what is needed to go from Theorem 3.1.6 to Theorem 3.1.6′. The construction of Kronecker sets as well as Kaufman's existential proof of perfect Kronecker sets are dealt with in several places, e.g. [Benedetto, 6; Hewitt and Ross, 1, II; Kahane, 13; Katznelson, 5].

2. Because of Theorem 3.1.6, a non-S-set for $V(D_\infty)$ is transferred into a non-S-set for $A(E_1 + E_2)$, and, consequently, $E_1 + E_2$ is not a set of spectral resolution (for $A(\Gamma)$). Thus in order to complete the proof of Malliavin's theorem for compact Γ, using Theorem 3.1.5 and Theorem 3.1.6, we need only prove that a non-S-set for $V(\mathbf{T}^3)$ is transferred into a non-S-set for $V(D_\infty)$; we do this in Exercise 3.1.5. The extension of Malliavin's theorem to the non-compact non-discrete case is proved by means of Theorem 3.1.7.

3. In light of the fact that we can find non-S-sets in $E_1 + E_2$ when $E = E_1 \cup E_2$ is a Kronecker set, etc., it is interesting to generalize Theorem 3.1.6 to a result with the following form: $A(E_1 + E_2) = V(D_\infty)$ if E_1 and E_2 are disjoint sets homeomorphic to D_∞ and if $E_1 \cup E_2$ is an Helson set. Theorems such as this have been proved recently by [Kaijser, 1] (cf. [Saeki, 6]). Actually, Varopoulos has proved that *if Γ is compact and if E_1 and E_2 are closed uncountable metrizable subsets of Γ then $E_1 + E_2$ is not a set of spectral resolution.*

4. Varopoulos' proof of Malliavin's theorem leaves the interesting question of determining which non-S-sets for $A(\mathbf{T})$, say, are "tensor images" of Schwartz's example.

5. Using these methods it can also be shown (for E_1 countable) that $E_1 + E_2$ is a perfect set of spectral resolution as well as being a set of analyticity.

3.1.16 Mappings of restriction algebras. Let Γ be a non-discrete LCAG. If G with the discrete topology is denoted by G_d then $\widehat{G_d} = \beta\Gamma$ is the *Bohr compactification* of Γ. As such (e.g. [Rudin, 5, Chapter 1]), $\beta\Gamma$ is a compact group and there is a continuous homomorphic injection, $b: \Gamma \to \beta\Gamma$, such that $\overline{b\Gamma} = \beta\Gamma$; the natural transpose map, $G_d \to G$, is, of course, a continuous bijection.

If $E_1 \subseteq \Gamma$ and $E_2 \subseteq \Gamma$ are disjoint compact sets homeomorphic to D_∞ and if $E_1 \cup E_2$ is a Kronecker or K_p-set, then it is clear that bE_1 and bE_2 are disjoint compact subsets of $\beta\Gamma$ such that each is homeomorphic to D_∞ and for which $b(E_1 \cup E_2) = bE_1 \cup bE_2$ is a Kronecker or K_p-set. To avoid notational confusion we write $A(E, X) = A(X)/k(E)$ where X is a LCAG, $E \subseteq X$ is closed, and $k(E) = \{\varphi \in A(X): \varphi = 0 \text{ on } E\}$. Because of Theorem 3.1.6 (or Theorem 3.1.6′) we can prove that there are non-S-sets for $A(b(E_1 + E_2), b\Gamma)$; thus, once we establish Theorem 3.1.7b, we shall have proved that there are non-S-sets for $A(E_1 + E_2, \Gamma)$, and, hence, that $E_1 + E_2 \subseteq \Gamma$ is not a set of spectral resolution (for $A(\Gamma)$). This completes the proof of Malliavin's theorem for any non-discrete Γ.

A form of Theorem 3.1.7a was first used by Wiener [Wiener, 7, p. 80] for the case of $\mathbf{Я}$ and $\mathbf{T}$.

Theorem 3.1.7. *Let Γ be a* LCAG *and let $E \subseteq \Gamma$ be compact.*

a) *If $\varphi \in A(E)$ and $r > \|\varphi\|_{A(E)}$ then there are sequences $\{x_n : n = 1, \ldots\} \subseteq G$ and $\{a_n : n = 1, \ldots\} \subseteq \mathbf{C}$ such that*

$$\sum |a_n| < r$$

and

$$\forall\, \gamma \in E, \qquad \varphi(\gamma) = \sum a_n(\gamma, x_n).$$

b) *The map*

$$\hbar : A(bE, \beta\Gamma) \rightarrow A(E, \Gamma)$$
$$\varphi \mapsto \varphi \circ b$$

is a bijective isometry and algebraic homomorphism.

Proof. a.i) Given $\{b_n : n = 1, \ldots\} \subseteq \mathbf{C}$ and $\{y_n : n = 1, \ldots\} \subseteq G$ for which $\sum |b_n| < \infty$. We prove that

$$(3.1.40) \qquad \|\sum b_n y_n\|_{A(E)} \leqslant \sum |b_n|.$$

Take $\varepsilon > 0$ and let $\mu = \sum b_n \delta_{y_n}$. Choose $f \in L^1(G)$ such that $\hat{f} = 1$ on E and $\|f\|_1 < 1 + \varepsilon$. Then $\widehat{f * \mu} \in A(\Gamma)$ and $\widehat{f * \mu} = \sum b_n y_n$ on E.

The norm inequality, (3.1.40), follows since

$$\|\hat{f}\hat{\mu}\|_{A(\Gamma)} < (1 + \varepsilon) \sum |b_n|$$

implies that

$$\| \sum b_n y_n \|_{A(E)} \leqslant (1 + \varepsilon) \sum |b_n|$$

for every $\varepsilon > 0$.

ii) By the Hahn–Banach theorem,

$$\{\sum b_n y_n : \sum |b_n| \leqslant 1 \text{ and } y_n \in G \text{ for } n = 1, \ldots\}$$

is $A(E)$-norm dense in the unit ball of $A(E)$.

iii) Choose $\varepsilon > 0$ such that $\|\varphi\|_{A(E)} + \varepsilon < r$. Take $\psi_1 = \sum_n b_{1,n} y_{1,n}$ for which $\|\varphi - \psi_1\|_{A(E)} < \frac{\varepsilon}{2}$ and $\sum_n |b_{1,n}| < r - \varepsilon$.

Let $\varphi_1 = \varphi - \psi_1$ and choose $\psi_2 = \sum_n b_{2,n} y_{2,n}$ according to the general rule that if $j \geqslant 2$ and φ_{j-1} is given then

$$\psi_j = \sum_n b_{j,n} \, y_{j,n}$$

is chosen so that $\|\varphi_{j-1} - \psi_j\|_{A(E)} < \frac{\varepsilon}{2^j}$ and $\sum_n |b_{j,n}| < \frac{\varepsilon}{2^{j-1}}$.

Consequently, $\varphi(\gamma) = \sum_{j,n} b_{j,n}(\gamma, y_{j,n})$ on E and $\sum_{j,n} |b_{j,n}| < r$.

(There are more complicated proofs of this fact.)

b.i) Let $\psi \in A(E, \Gamma)$. By part a) we have

$$\forall\, \gamma \in E, \qquad \psi(\gamma) = \sum a_n(\gamma, x_n),\, x_n \in G \text{ and } \sum |a_n| < \infty.$$

If $\lambda \in bE$ we define

$$\varphi(\lambda) = \sum a_n(\lambda, x_n),$$

and so $\hbar$ is surjective.

Once we show that $\hbar$ is well-defined it follows that it is an algebraic homomorphism and an injection.

ii) We prove

$$(3.1.41) \qquad \forall\, \varphi \in A(bE, \beta\Gamma), \qquad \|\hbar\,\varphi\|_{A(E,\Gamma)} \leqslant \|\varphi\|_{A(bE,\beta\Gamma)};$$

this yields the continuity of $\hbar$, and, consequently, the fact that it is well-defined.

Given $\varphi \in A(bE, \beta\Gamma)$ and $\varepsilon > 0$. In order to verify (3.1.41) it is sufficient to prove that

$$(3.1.42) \qquad \|\hbar\,\varphi\|_{A(E,\Gamma)} \leqslant \|\varphi\|_{A(bE,\beta\Gamma)} + \varepsilon.$$

Write

$$\forall\, \lambda \in bE, \qquad \varphi(\lambda) = \sum a_n(\lambda, x_n),$$

where $x_n \in G_d$ and $\sum |a_n| \leqslant \|\varphi\|_{A(bE,\beta\Gamma)} + \varepsilon$.

Because of (3.1.40) we have

$$\forall\, \gamma \in E, \qquad (\hbar\,\varphi)(\gamma) = \sum a_n(\gamma, x_n),$$

where

$$\|\hbar\,\varphi\|_{A(E)} \leqslant \sum |a_n|.$$

This yields (3.1.42).

iii) Take $\varphi \in A(bE, \beta\Gamma)$ and let $r > \|\varphi \circ b\|_{A(E,\Gamma)}$. We'll prove that $r > \|\varphi\|_{A(bE,\beta\Gamma)}$. This fact combined with (3.1.41) yields the isometry.

From part a) we have

$$\forall\, \gamma \in E, \qquad \varphi \circ b(\gamma) = \sum a_n(\gamma, x_n)$$

where $x_n \in G$ and $\sum |a_n| < r$.

Consequently,

$$\forall\, \lambda \in bE, \qquad \varphi(\lambda) = \sum a_n(\lambda, x_n)$$

where $x_n \in G_d$.

Therefore

$$\|\varphi\|_{A(bE,\beta\Gamma)} \leqslant \sum |a_n| < r.$$

q.e.d.

Remark. In Theorem 3.1.7a we proved that if $\sum |b_n| < \infty$, $E \subseteq \mathfrak{R}$ is compact, and $\{y_n : n = 1, \ldots\} \subseteq \mathbf{R}$, then $\sum b_n y_n \in A(E)$. [Katznelson and McGehee, 2] have constructed a denumerable compact set $E \subseteq \mathfrak{R}$ with one limit point γ_0 such that $\sum b_n y_n \in C(E) \setminus A(E)$ for some choice of $\{b_n \in \mathbf{C} : \sum |b_n| < \infty\}$ and $\{y_n : n = 1, \ldots\} \subseteq \beta\mathbf{R}$; their construction uses the fact that there are non-synthesis sets in $\mathfrak{R}$ as well as Herz's procedure to prove that the Cantor set, $C \subseteq \mathbf{T}$, is an S-set (e.g. Paragraph 2.5). Using this set, E, [Graham, 3] has proved that

$$\overline{(E \setminus \{\gamma_0\})}^{\beta} \setminus (E \setminus \{\gamma_0\}) \subseteq \beta(\mathfrak{R}_d)$$

is a non-S-set in $\beta(\mathfrak{R}_d)$, where "$-\beta$" indicates closure in $\beta(\mathfrak{R}_d)$ (cf. [Drury, 1]). Note that $\mathfrak{R} \subseteq \beta(\mathfrak{R}_d) \subseteq \beta\mathfrak{R}$.

[De Leeuw and Herz, 1, Theorem 1] (1963) have strengthened Theorem 3.1.7b in the following way.

Theorem 3.1.8. *Let Γ_1 and Γ_2 be* LCAG *and let $\alpha : \Gamma_1 \to \Gamma_2$ be an injective continuous homomorphism.*

If E_1 is compact and $\alpha E_1 = E_2$ then the canonical transpose,

(3.1.43) $\qquad h : A(E_2, \Gamma_2) \;\to\; A(E_1 \Gamma_1),$

is a well-defined bijective isometric isomorphism.

Proof. We give a fairly thorough sketch of the proof. The details are to be provided in Exercise 3.1.6.

i) Since α is a continuous homomorphism it is not difficult to prove that $h(B(\Gamma_2)) \subseteq B(\Gamma_1)$, e.g. [Rudin, 5, p. 79].

ii) If $E \subseteq \Gamma$ is compact then $B(E) = A(E)$; this is proved using the fact that for each $\varepsilon > 0$ there is $\varphi \in A(\Gamma)$ for which $\varphi = 1$ on E and $\|\varphi\|_A < 1 + \varepsilon$.

iii) Since α is a continuous homomorphism, (3.1.43) is a well-defined norm decreasing map.

To prove iii), take $\varphi \in A(E_2, \Gamma_2)$ and use i) and ii) in succession to obtain the inclusion, $h(A(E_2, \Gamma_2)) \subseteq A(E_1, \Gamma_1)$.

The norm inequality follows by the estimates used to obtain the inclusion.

iv) It is easy to check that h is injective.

v) Starting from the hypothesis that α is a continuous homomorphism we can define the canonical transpose (of h)

$$h' : A_S'(E_1) \;\to\; A_S'(E_2).$$

We now prove in a straightforward way that h' is an isometry since α is injective.

vi) The fact that h is a surjective isometry is a consequence of the following general theorem about Banach spaces (e.g. [Rudin, 5, pp. 259–260]): if X_1 and X_2 are complex Banach spaces and $h \in L(X_1, X_2)$ then h is a surjective isometry if h is injective and if the transpose, h', is an isometry.

q.e.d.

The fact that $\hbar$ is surjective is the major issue in Theorem 3.1.8.

3.1.17 A characterization of synthesis preserving mappings. In light of Theorem 3.1.8 and the problem of transferring S or non-S-sets between different groups, it is natural to ask when a map $\alpha: \Gamma_1 \to \Gamma_2$ determines a bijective isometric isomorphism $\hbar$: $A_j(E_2) \to A_j(E_1)$, where $E_j \subseteq \Gamma_j$ is compact. The proof of the following result establishes criteria of this kind and uses Theorem 3.1.8; the proof is difficult.

Theorem 3.1.9. ([De Leeuw and Herz, 1]). *Let Γ_1 and Γ_2 be* LCAG *and let $\alpha: \Gamma_1 \to \Gamma_2$ be an injective continuous homomorphism. Assume that $E_1 \subseteq \Gamma_1$ is compact and set $E_2 = \alpha E_1$. Then E_1 is an S-set* (*for $A(\Gamma_1)$*) *if and only if E_2 is an S-set* (*for $A(\Gamma_2)$*).

Related results were proved independently by [Spector, 1, e.g. Proposition 8] (cf. [Spector, 2]). [Reiter, 3; 4] initiated research on the problem solved by Theorem 3.1.9; (in the notation of Theorem 3.1.9) he proved that if α is also a relatively open map of Γ_1 onto a closed subgroup of Γ_2 then $E_1 \subseteq \Gamma_1$ is an S-set if and only if $E_2 \subseteq \Gamma_2$ is an S-set. Glicksberg proved one direction of Theorem 3.1.9, viz. if E_2 is an S-set then E_1 is an S-set, independently of De Leeuw and Herz.

Actually, the hypothesis in Theorem 3.1.9 that α be injective can be weakened to the assumption that α be injective on a neighborhood of E_1. On the other hand, the following observation [De Leeuw and Herz, 1, p. 224] shows that it is not sufficient to assume that α is a continuous homomorphism, injective on E_1.

Example 3.1.1. Define $\alpha: \mathbb{R}^3 \to \mathbb{R}^2$ by the map $\alpha(\gamma_1, \gamma_2, \gamma_3) = (\gamma_1, \gamma_2)$. Let $E_1 = \{(\gamma_1, \gamma_2 \gamma_3) \in \mathbb{R}^3 : \gamma_1^2 + \gamma_2^2 + \gamma_3^2 = 1 \text{ and } \gamma_3 \geqslant 0\}$. Thus, $E_2 = \{(\gamma_1, \gamma_2) \in \mathbb{R}^2 : \gamma_1^2 + \gamma_2^2 \leqslant 1\} = B^2$. α is injective on E_1, E_1 is a non-S-set in $\mathbb{R}^3$ because of Schwartz's example, and E_2 is an S-set in $\mathbb{R}^2$ by Proposition 2.5.5.

Γ_1 and Γ_2 are *locally isomorphic* if there are neighborhoods $V_i \subseteq \Gamma_i$ of $0 \in \Gamma_i$, $i = 1,2$, and a homeomorphism α of V_1 onto V_2 such that

$$\forall\, \lambda, \gamma \in V_1, \qquad \lambda + \gamma \in V_1 \quad \Rightarrow \quad \alpha(\lambda) + \alpha(\gamma) = \alpha(\lambda + \gamma).$$

With regard to the homomorphism problem discussed in Section 2.4.9, Spector [Spector, 1, Proposition 9] proved the following result using the same methods that he employed in his contribution to Theorem 3.1.9: *assume that Γ_1 and Γ_2 are locally isomorphic; then $A(\Gamma_1)/j(\{0_1\})$ and $A(\Gamma_2)/j(\{0_2\})$ are algebraically isomorphic* (where 0_j is the 0-element of Γ_j). It is natural to ask about the converse situation: if $A(\Gamma_1)/j(\{0_1\})$ and $A(\Gamma_2)/j(\{0_2\})$ are algebraically isomorphic is it true that Γ_1 and Γ_2 are locally isomorphic? Jerison has provided an affirmative answer when Γ_1 and Γ_2 are locally connected metrizable groups; a fancy topological contribution to the problem is found in [Jerison, Siegel, and Weingram, 1].

3.1.18 The structure of $A(\Gamma)$ and the union of Helson sets. We now use Theorem 3.1.8 as the point of departure for a brief exposition of an important recent theorem giving information on the structure of $A(\mathbf{T})$; basically we shall just introduce the reader to some of the literature.

We rephrase Theorem 3.1.8 in the following way

Theorem 3.1.8′. *Let Γ_1 and Γ_2 be* LCAG, *let $\alpha: \Gamma_1 \to \Gamma_2$ be an injective continuous homomorphism, and let E_1 be compact so that $E_2 = \alpha E_1$ is compact. For each $\varphi \in A(\Gamma_1)$ and for each $r > 1$ there is $\psi \in A(\Gamma_2)$ such that*

$$\|\psi\|_{A(\Gamma_2)} \leqslant r\|\varphi\|_{A(\Gamma_1)}$$

and

$$\forall \gamma \in E_2, \qquad \psi(\gamma) = \varphi \circ \alpha^{-1}(\gamma).$$

Proof. Let $r = 1 + \delta$ and take $0 < \varepsilon < \delta\|\varphi\|_{A(\Gamma_1)}$.
Let $\tilde{\varphi} \in A(E_1, \Gamma_1)$ have the property that $\varphi = \tilde{\varphi}$ on E_1.
By Theorem 3.1.8, $\hbar^{-1}\,\tilde{\varphi} \in A(E_2, \Gamma_2)$ and

$$\|\hbar^{-1}\tilde{\varphi}\|_{A(E_2,\Gamma_2)} = \|\tilde{\varphi}\|_{A(E_1,\Gamma_1)}.$$

Take $\psi \in A(\Gamma_2)$ with the property that $\psi = \hbar^{-1}\,\tilde{\varphi}$ on E_2 and

$$\|\psi\|_{A(\Gamma_2)} < \|\hbar^{-1}\,\tilde{\varphi}\|_{A(E_2,\Gamma_2)} + \varepsilon.$$

q.e.d.

If $E \subseteq \Gamma$ then

$$H(E) = \sup\{\|\varphi\|_{A(E)}/\|\varphi\|_{C_0(E)} : \varphi \in A(E) \setminus \{0\}\}$$

is the *Helson constant of E* (where $\|\varphi\|_{C_0(E)} = \sup_{\gamma \in E} |\varphi(\gamma)|$). $H(E)$ is finite if and only if E is an Helson set.

The following generalization of Theorem 3.1.8′ for the case that $H(E_2) < \infty$ is important because of the information obtained about the behavior of the functions outside of E_2.

Theorem 3.1.10. *Let Γ_1 and Γ_2 be* LCAG, *let $\alpha: \Gamma_1 \to \Gamma_2$ be a continuous homomorphism, and let $E_2 \subseteq \Gamma_2$ be an Helson set. Assume that*

(3.1.44) $\qquad \beta: E_2 \;\to\; \Gamma_1$

is a continuous function with the property that

(3.1.45) $\qquad \alpha \circ \beta: E_2 \;\to\; E_2$

is the identity function. Then for any $\varphi \in A(\Gamma_1)$, $r > H(E_2)$, $\varepsilon > 0$, and U a neighborhood of $0 \in \Gamma_2$, there is $\psi \in A(\Gamma_2)$ such that $\|\psi\|_{A(\Gamma_2)} \leqslant r^2\|\varphi\|_{A(\Gamma_1)}$,

$$\forall \gamma \in E_2, \qquad |\varphi \circ \beta(\gamma) - \psi(\gamma)| < \varepsilon,$$

and

$$\forall\, \gamma \in \Gamma_2, \quad |\psi(\gamma)| \leqslant r^2 \sup\{|\varphi(\lambda)| : \alpha(\lambda) \in \gamma + U\}.$$

Conditions (3.1.44) and (3.1.45) are obviously satisfied in Theorem 3.1.8′. In the above form, Theorem 3.1.10 is due to [Herz, 11]. The major application of Theorem 3.1.10 is to prove

Theorem 3.1.11. *There is a continuous function ω: $(0,1] \to [1,\infty)$ with the following property. Given Γ, an Helson set $E \subseteq \Gamma$, $r > H(E)$, $\varphi \in C_0(E)$, and a closed subset $F \subseteq \Gamma$ disjoint from E. Then for each $\varepsilon \in (0,1]$ there is $\psi \in A(\Gamma)$ such that*

$$\|\psi\|_{A(\Gamma)} \leqslant r^2 \|\varphi\|_\infty \omega(\varepsilon),$$

$$\forall\, \gamma \in E, \quad \psi(\gamma) = \varphi(\gamma),$$

and

$$\forall\, \gamma \in F, \quad |\psi(\gamma)| \leqslant r^2\, \varepsilon \|\varphi\|_\infty$$

(cf. *Remark* 3 *after Example* 2.5.3).

The main motivation for the development of Theorem 3.1.11 was to prove that the finite union of Helson sets in Γ is an Helson set. For the case that Γ is discrete this union problem was solved by [Drury, 2] by using essentially a special case of Theorem 3.1.10. Varopoulos then extended the theorem to metrizable Γ and Lust made the final step to every Γ. Stegeman and Herz are responsible for sharp estimates on $\omega(\varepsilon)$.

In any case, Drury's idea has been the major influence on this subject. [McGehee, 3] has an historical note and discussion on these matters; and [McGehee, 2] contains a detailed exposition of [Herz, 11]. In light of the proof of Theorem 3.1.6 it is interesting to note that the union of two Kronecker sets (in $\mathbf{T}$) is not necessarily a Kronecker set, e.g. [Bernard and Varopoulos, 1].

Exercises 3.1

3.1.1 Non-synthesis of discrete spaces

Malliavin's theorem and Exercise 1.4.3 combine to yield: *Γ is discrete if and only if E is an S-set for each closed set $E \subseteq \Gamma$.*

A natural question is whether this result is valid for every commutative semi-simple Banach algebra X (instead of $A(\Gamma)$) with maximal ideal space X^m (instead of Γ).

a) Let X be the Banach algebra defined in Exercise 1.1.2. Prove that $X^m = \mathbf{Z}$ contains non-S-sets (e.g. [Mirkil, 1]).

b) Let X be a commutative semi-simple Banach algebra whose maximal ideal space X^m is a discrete abelian group. Assume

i) $\{\varphi \in X : \operatorname{card} \operatorname{supp} \varphi < \infty\}$ is dense in X;

ii) For each $\varphi \in X$ and each $x \in \widehat{X^m}$,

$$(\cdot, x)\varphi(\cdot) \in X.$$

Prove that each $E \subseteq X^m$ is an S-set. Note that the example of part a) satisfies condition i). With regard to this example we also mention [Graham, 5; Wermer, 1].

3.1.2 $A(\Gamma)$ and finitely generated closed ideals
Prove that $A(\mathbb{R})$ doesn't have any non-0 countably generated closed ideals (we distinguish, for example, between "principal closed ideals" and "closed principal ideals"). Compare this with Douady's result: *let X be any Banach algebra and let $I \subseteq X$ be an ideal; if $\bar{I}$ is finitely generated then $I = \bar{I}$.*

3.1.3 Multipliers and spectral synthesis

$\hbar \in L(L^p(\Gamma), L^p(\Gamma))$, $p \in [1, \infty]$, is a *convolutor on* $L^p(\Gamma)$ if

$$\forall\ \varphi \in L^p(\Gamma) \text{ and } \forall\ x \in G, \qquad \hbar(\tau_x \varphi) = \tau_x \hbar(\varphi).$$

Consequently, the "L^1 multiplier theorem" quoted in the suggested proof of Exercise 2.4.8c asserts that *$M(\Gamma)$ is the set of all convolutors on $L^1(\Gamma)$*. In 1923, Fekete essentially established the *multiplier problem*, viz. to determine the convolutors of $L^p(\Gamma)$. For the case of $p = 1$, the multiplier problem is related in the obvious way to the homomorphism problem discussed in Section 2.4.9 (cf. Exercise 2.4.8). Regarding this nexus, a fundamental result which can be used to prove the above L^1 multiplier theorem, and which can be derived using Bochner's theorem and Exercise 2.5.1c, is the following: *if $X \supseteq B(G)$ is a commutative ring of functions (on G) under pointwise multiplication and if $A(G)$ is an ideal in X then $X = B(G)$* (the result remains valid, if instead of the hypothesis that $A(G)$ is an ideal in X, we assume that $X \cdot A(G) \subseteq B(G)$); and some fundamental papers are [Edwards, 1; Helson, 3; Wendel, 1; 2]. R. E. Edwards established the "modern" approach to the multiplier problem in the late 1950's and [Hörmander, 1] appeared in 1960 (in this regard, see [Figà-Talamanca and Gaudry, 1]); expositions of the field are found in [Brainerd and Edwards, 1: Edwards, 5, II, Chapter 16; Larsen, 1].

a) For each $p \in [1, \infty]$ let $A'_p(\Gamma)$ be the set of elements $T \in A'(\Gamma)$ for which

(E3.1.1) $\quad \exists\ K_{T,p} > 0$ such that $\forall\ \varphi \in C_c(\Gamma) \subseteq A'(\Gamma)$

$$\|T * \varphi\|_p \leqslant K_{T,p} \|\varphi\|_p.$$

Prove that if $T \in A'_p(\Gamma)$ and $\varphi \in L^p(\Gamma)$, where $p \in [1, \infty]$, then T is a convolutor on $L^p(\Gamma)$ and $T * \varphi \in L^p(\Gamma)$ (for $p = \infty$, $T * \varphi \in C_0(\Gamma)$).

b) Let $\hbar \in L(L^p(\Gamma), L^p(\Gamma))$, $p \in [1, \infty)$, be a convolutor on $L^p(\Gamma)$. Prove that there is $T \in A'_p(\Gamma)$ such that

$$\forall\ \varphi \in L^p(\Gamma), \qquad \hbar(\varphi) = T * \varphi.$$

(*Hint*: First do the case $p = 2$; then use the Riesz–Thorin theorem). $\hat{T}$ is the *multiplier* corresponding to the convolutor T.

Let $A_p(\Gamma)$, $1 < p < \infty$, be the space of continuous functions

$$(E3.1.2) \qquad \Phi = \sum_{k=1}^{\infty} \varphi_k * \psi_k, \qquad \varphi_k \in L^p(\Gamma) \text{ and } \psi_k \in L^q(\Gamma),$$

where $\frac{1}{p} + \frac{1}{q} = 1$ and $\sum_k \|\varphi_k\|_p \|\psi_k\|_q < \infty$.

$$A_p(\Gamma) = L^p(\Gamma) \hat{\otimes} L^q(\Gamma)$$

is a Banach space normed by

$$\forall\, \Phi \in A_p(\Gamma), \qquad \|\Phi\| = \inf \left\{ \sum_k \|\varphi_k\|_p \|\psi_k\|_q \right\},$$

where the infimum is taken over all representations of Φ having the form of (E3.1.2). $A_1(\Gamma) = C_0(\Gamma)$ (resp. $A_2(\Gamma) = A(\Gamma)$) and the corresponding $\| \;\; \|$-norm is the supnorm (resp. $\| \;\; \|_A$ norm). [Figà-Talamanca, 1] proved that *if* $A_p'(\Gamma)$, $1 < p < \infty$, *defined in part* a), *is given the canonical norm topology induced from* $L(L^p(\Gamma), L^p(\Gamma))$ *then* $A_p'(\Gamma)$ *is the dual of* $A_p(\Gamma)$. Later, by an adjustment of a technique in [Herz, 7], e.g. [Eymard, 2, Théorème 1], Herz proved that $A_p(\Gamma)$, $1 < p < \infty$, is a Banach algebra (cf. [Herz, 10] for a virtuoso performance, as well as other work by Herz on $A_p(\Gamma)$).

c) Prove that

$$A_1'(\Gamma) = A_\infty'(\Gamma) = M(\Gamma) \subseteq A_p'(\Gamma) = A_q'(\Gamma) \subseteq A_2'(\Gamma) = A'(\Gamma),$$

where $p \in (1, \infty)$ and $\frac{1}{p} + \frac{1}{q} = 1$.

The spectral synthesis problem for the $(A_p(\Gamma), A_p'(\Gamma))$ duality was posed by Eymard in 1968 (e.g. [Eymard, 1; 2]). He determined values of p and n for which $S^{n-1} \subseteq \mathbb{R}^n$ is and is not a "p S-set", e.g. [Eymard, 2, Théorème 5]. Later, [Lohoué, 1] showed: *there is* $\varphi \in A(\Gamma)$ *such that for each* $p \in (1, \infty)$, I_φ, *with the closure taken in the* $A_p(\Gamma)$ *topology, is not self-adjoint; in particular, there is a closed set* $E \subseteq \Gamma$ *such that, for each* $p \in (1, \infty)$, E *is not a "p S-set"* (the remark and references about differentiation at the beginning of Section 3.1.2 are relevant here). In this regard, and related to Meyer's work discussed in Remark 1 after Example 2.5.2, see [Lohoué, 2].

3.1.4 A topology for which synthesis always holds

[Schwartz, 3], [Ditkin, 1], Beurling, [Pollard, 2], and [Herz, 5, Theorem 4.5, pp. 210–212 and pp. 225–226] have proved the following: *let* Γ *be* $\mathbb{R}$ *or* $\mathbb{R}^2$; *then there is a topology on* $C_b(G)$, *finer than the* κ *topology, such that spectral analysis and spectral synthesis always hold* (and, yet, spectral synthesis fails in $\mathbb{R}$ and $\mathbb{R}^2$ for the weak $*$ topology!). This result is strengthened in [Herz, 6]; in this regard and with regard to the technique introduced in Section 3.2.2, we refer to [Herz, 9]. Prove that if $E \subseteq \mathbf{R}^3$ is the surface defined by

$$\sin^2\frac{\gamma_1}{2} + \sin^2\frac{\gamma_2}{2} + \sin^2\frac{\gamma_3}{2} = r, \qquad r \in (0,\tfrac{1}{2}),$$

then E is not an S-set for the topology of pointwise convergence, e.g. [Herz, 5].

3.1.5 Malliavin's theorem–the completion of Varopoulos' proof

a) Define the map

$$\alpha: D_\infty \quad \to \quad \mathbf{T}^3$$

which takes an element $\gamma \in (\gamma_1, \ldots) \in D_\infty$ into the element

$$(2\pi(.r_1 r_4 r_7 \ldots (2)),\ 2\pi(.r_2 r_5 r_8 \ldots (2)),\ 2\pi(.r_3 r_6 r_9 \ldots (2))) \in \mathbf{T}^3$$

where $r_j = e^{\gamma_j} = e^{\pi k_j}$, for $k_j = 0$ or $k_j = 1$, .abc...(2) indicates the binary expansion of an element $\lambda \in [0,1]$, and $2\pi(.\mathrm{abc}\ldots(2)) = 2\pi\lambda \in [0, 2\pi]$. Prove that α is a continuous bijection, where the injection follows once we eliminate a set of measure 0 from both D_∞ and $\mathbf{T}^3$.

b) Because of part a) we induce the homomorphisms

$$C(\mathbf{T}^3) \quad \to \quad C(D_\infty) \quad \to \quad L^\infty(\mathbf{T}^3),$$

from α and α^{-1}; and consequently we have the well-defined homomorphisms

(E3.1.3) $\quad V(\mathbf{T}^3) \quad \to \quad V(D_\infty) \quad \to \quad L^\infty(\mathbf{T}^3) \hat{\otimes} L^\infty(\mathbf{T}^3).$

Using (E3.1.3), prove that the first homomorphism of (E3.1.3), viz. $V(\mathbf{T}^3) \to V(D_\infty)$, is an isometric homomorphism; and that, as such, we obtain a continuous map

(E3.1.4) $\quad D_\infty \times D_\infty \quad \to \quad \mathbf{T}^3 \times \mathbf{T}^3$

between the corresponding maximal ideal spaces.

c) Prove that if $E \subseteq \mathbf{T}^3 \times \mathbf{T}^3$ is a non-S-set for $V(\mathbf{T}^3)$, then its inverse image by (E3.1.4) is a non-S-set for $V(D_\infty)$. (*Hint*: Suppose the result is false and use (E3.1.3) to obtain synthesis in $L^\infty(\mathbf{T}^3) \hat{\otimes} L^\infty(\mathbf{T}^3)$; then regularize the situation in $L^\infty(\mathbf{T}^3) \hat{\otimes} L^\infty(\mathbf{T}^3)$ to provide synthesis in $V(\mathbf{T}^3)$, the desired contradiction). As indicated in Section 3.1.15, this completes the proof of Malliavin's theorem.

In light of our discussion of the Reiter, De Leeuw, Herz theorem, it is instructive to state the general result which [Varopoulos, 8] gave in order to prove Exercise 3.1.5c; the proof is easy. *Let X_1 and X_2 be regular Banach algebras with units, φ_1 and φ_2, let $h: X_1 \to X_2$ be an isometric homomorphism onto a closed subalgebra for which $h(\varphi_1) = \varphi_2$, and let $\alpha: X_2^m \to X_1^m$ be the induced continuous transpose of h* (cf. Proposition 2.4.1); *then*

i) $\qquad$ *α is injective* (by the regularity)

and

ii) *if* $E_1 \subseteq X_1^m$ *is closed and* $E_2 = \alpha^{-1} E_1$ *then*

$$k(E_1) = \hbar^{-1}(k(E_2)).$$

If $\hbar$ satisfies two further conditions, which it does in the case of Exercise 3.1.5c, then we can conclude that

iii) $$\overline{j(E_1)} = \hbar^{-1}(\overline{j(E_2)})$$

("k" and "j" are defined in the expected way).

3.1.6 Conditions for which $A(E_1, \Gamma_1) = A(E_2, \Gamma_2)$

Fill in the details for the suggested proof of Theorem 3.1.8.

3.1.7 A relation between $A(\mathbf{Я})$ *and* $A(\mathbf{T})$, *and a result in synthesis*

a) We've indicated a connection between Theorem 3.1.7a and a theorem by Wiener. Actually, Wiener proved the following theorem (for the case of $\theta = 1$):

Let $\varphi : [-\pi, \pi] \to \mathbf{C}$ be bounded and assume that there is $\delta > 0$ for which $\varphi(\gamma) = 0$ when $\pi - \delta \leqslant |\gamma| \leqslant \pi$. Define $\psi : \mathbf{Я} \to \mathbf{C}$ as

$$\psi(\gamma) = \begin{cases} \varphi(\gamma) & \text{for } |\gamma| \leqslant \pi \\ 0 & \text{for } |\gamma| > \pi, \end{cases}$$

and let $h : \mathbf{R} \to \mathbf{C}$ satisfy the following conditions: $h \geqslant 0$ is increasing on $\{x : x > 0\}$, $h(-x) = h(x)$ for each $x \in \mathbf{R}$, and there is a constant k such that

$$\forall\, x \in \mathbf{R}, \qquad h(2x) \leqslant k h(x).$$

Then

$$\varphi(\gamma) = \sum a_n e^{in\gamma} \qquad \text{where} \qquad \sum |a_n| h(n) < \infty$$

if and only if

$$\psi(\gamma) = \int_{-\infty}^{\infty} f(x) e^{ix\gamma} dx \qquad \text{where} \qquad \int_{-\infty}^{\infty} |f(x)| h(x) dx < \infty.$$

Prove this result.

b) Let $\hat{f} = \varphi \in A(\mathbf{T})$ satisfy the condition that

(E3.1.5) $$\sum{}' |f(n)| \log |n| < \infty.$$

Prove that if $\varphi\left(\pm \frac{\pi}{2}\right) = 0$ then $\varphi \chi_{(-\pi/2, \pi/2)} \in A(\mathbf{T})$; and using part a) prove that $\varphi \chi_{(\lambda, \gamma)} \in A(\mathbf{T})$ whenever $\varphi(\lambda) = \varphi(\gamma) = 0$ for $-\pi \leqslant \lambda < \gamma < \pi$, e.g. [Kahane, 2; Wik, 2].

3.2 Synthesizable phenomena

3.2.1 Integral representation and spectral synthesis. In this section we shall prove results aimed at determining whether or not a given pseudo-measure is synthesizable or not. The techniques that we use are related to ideas developed by Beurling.

It turns out that there is a fundamental relationship between the existence of an integral representation for the operation $\langle T, \varphi \rangle$, where $T \in A'(\Gamma)$ and $\varphi \in A(\Gamma)$, and the synthesizability of T or φ.

We shall be more specific on this point, shortly, but, first, we establish the setting and some notation. For technical convenience we shall generally work on $\Gamma = \mathbf{T}$ and with the space

$$A' = \{T \in A'(\mathbf{T}) : \hat{T}(0) = 0\}.$$

By working in A' there is no restriction in the generality of any of our results since one can always add in a δ and/or make a translation without affecting a synthesizable phenomenon. If $T \in A'$ has a Fourier series representation

$$T \sim \sum{}' c_n e^{in\gamma}$$

then

$$F \sim \sum{}' \frac{c_n}{in} e^{in\gamma} \in L^2(\mathbf{T})$$

since $\{c_n : n \in \mathbf{Z}\}$ is bounded; and

$$F' = T,$$

distributionally. In this situation, we shall write

$$F \sim T \quad \text{or} \quad F \sim T \in A'$$

for the correspondence between F and T; obviously, then, if $F \sim T$ and k is a constant then $(F - k) \sim T$.

We know that each $\mu \in M(\mathbf{T})$ is synthesizable and so

$$\forall\, \varphi \in k(\operatorname{supp} \mu), \qquad \langle \mu, \varphi \rangle = 0.$$

If $F \sim \mu \in A' \cap M(\mathbf{T})$, then F is a function of bounded variation and

$$\forall\, \varphi \in k(\operatorname{supp} \mu), \qquad \int_{\mathbf{T}} \varphi \, dF = \langle \mu, \varphi \rangle = 0;$$

$BV(\mathbf{T})$ will denote the class of functions F having bounded variation for which $F \sim \mu \in A' \cap M(\mathbf{T})$. In this case, it is easy to verify that $\int \varphi \, dF = 0$ from first definitions. In fact, F is constant on any open interval contiguous to $\operatorname{supp}\mu$ and so when we write out the approximating sums for the Stieltjes integral, $\int_{\mathbf{T}} \varphi\, dF$, we see that

$$\sum \varphi(\xi_j)(F(\gamma_j) - F(\gamma_{j-1})) = 0$$

when γ_j and γ_{j-1} are in the same contiguous interval. In the limit, the fact that $\varphi = 0$ on $\operatorname{supp}\mu$ assures that the other terms in such sums are 0.

Consequently, we see that if the inner product $\langle T, \varphi\rangle$, $T \in A'(\mathbf{T})$ and $\varphi \in A(\mathbf{T})$, has a "Stieltjes integral" representation and $\varphi = 0$ on $\operatorname{supp} T$ then there is a good chance that synthesis holds, i.e., $\langle T, \varphi\rangle = 0$.

One approach to rigorize this idea on integral representation is given in Section 3.2.2–Section 3.2.4.

3.2.2 The Beurling integral and a criterion for synthesis. The basic and essentially elementary calculation that we use is the following: for $F \sim T \in A'$ and $\hat{f} = \varphi \in A(\mathbf{T})$,

$$\text{(3.2.1)} \qquad \langle T, \varphi\rangle = \sum \hat{T}(n) f(n) = \mathrm{i} \sum n\hat{F}(n) f(n) = \mathrm{i} \sum{}' \hat{F}(n) f(n)\, n \left(\frac{4}{\pi|n|} \int_0^\infty \frac{\sin^2(nr/2)}{r^2}\, \mathrm{d}r \right)$$

$$= \frac{1}{\pi} \int_0^\infty \frac{1}{r^2} \left(\sum \hat{F}(n)\, \hat{H}(n) f(n) (\mathrm{e}^{\mathrm{i}nr} - 1)(\mathrm{e}^{-\mathrm{i}nr} - 1) \right) \mathrm{d}r,$$

where $H \in A'$ is the "conjugate distribution"

$$H \sim \mathrm{i} \sum{}' \operatorname{sgn} n\, \mathrm{e}^{\mathrm{i}n\gamma}, \qquad \operatorname{sgn} n = n/|n|;$$

note that by Parseval's formula the expression (3.2.1) becomes

$$\text{(3.2.2)} \qquad B(\varphi, F) = \frac{1}{\pi} \int_0^\infty \frac{1}{r^2} \left(\int_{\mathbf{T}} [H * (\tau_r F - F)(\gamma)]\, [(\tau_{-r}\varphi - \varphi)(\gamma)]\, \mathrm{d}\gamma \right) \mathrm{d}r.$$

$B(\varphi, F)$ is the *Beurling integral of φ with respect to F.*

Since

$$\sum{}' |f(n)\, \hat{F}(n)| \int_0^\infty \frac{\sin^2(nr/2)}{r^2}\, \mathrm{d}r < \infty,$$

the above calculation is valid, and we have

Proposition 3.2.1. *Given $F \sim T \in A'$ and $\varphi \in A(\mathbf{T})$. Then*

$$\text{(3.2.3.)} \qquad B(\varphi, F) = \langle T, \varphi\rangle.$$

The integral $B(\varphi, F)$ exists for $F \sim T \in A'$ and $\varphi \in C(\mathbf{T})$ if

$$|B|(\varphi, F) = \frac{1}{\pi}\int_0^\infty \frac{1}{r^2}\left(\int_{\mathbf{T}} |H * (\tau_r F - F)(\gamma)(\tau_{-r}\varphi - \varphi)(\gamma)|\, d\gamma\right) dr < \infty.$$

In order to prove Theorem 3.2.1, below, we need to introduce the following quantities:

$$\forall\, g \in L^2(\mathbf{T}), \qquad \|g, r\| = \left(\int_{\mathbf{T}} |g(\gamma + r) - g(\gamma)|^2\, d\gamma\right)^{1/2}$$

and

$$\forall\, \varphi \in C(\mathbf{T}) \qquad \text{and} \qquad \forall\, F \in L^2(\mathbf{T}),$$

$$\|B\|(\varphi, F) = \frac{1}{\pi}\int_0^\infty \frac{1}{r^2}\|\varphi, r\|\,\|F, r\|\, dr.$$

Clearly, $\|B\|(\varphi, F)$ may be infinite, and $\|B\|(\varphi, F) < \infty$ if and only if

$$\int_0^{2\pi} \frac{1}{r^2}\|\varphi, r\|\,\|F, r\|\, dr < \infty.$$

Also

$$\forall\, \varphi \in C(\mathbf{T}) \text{ and } \forall\, F \in L^2(\mathbf{T}), \qquad |B(\varphi, F)| \leqslant |B|(\varphi, F) \leqslant \|B\|(\varphi, F).$$

For any function $\varphi \in L^2(\mathbf{T})$ we define (as [Beurling, 6; 11; 12] has done) the *circular contraction*

$$\varphi_\rho(\gamma) = \begin{cases} \varphi(\gamma) & \text{for } |\varphi(\gamma)| \leqslant \rho \\ \rho\dfrac{\varphi(\gamma)}{|\varphi(\gamma)|} & \text{for } |\varphi(\gamma)| \geqslant \rho, \end{cases}$$

where $\rho > 0$. A straightforward calculation shows that

(3.2.4) $\qquad \forall\, \rho > 0$ and $\forall\, \lambda, \gamma, \qquad |\varphi_\rho(\gamma + \lambda) - \varphi_\rho(\gamma)| \leqslant |\varphi(\gamma + \lambda) - \varphi(\lambda)|.$

Proposition 3.2.2. *Given $F \sim T \in A'$ and $\varphi \in C(\mathbf{T})$ for which $|B|(\varphi, F) < \infty$. Then $|B|(\varphi_\rho, F) < \infty$ for each $\rho > 0$, and*

(3.2.5) $\qquad \lim_{\rho \to 0} |B|(\varphi_\rho, F) = 0.$

Proof. $|B|(\varphi_\rho, F) \leqslant |B|(\varphi, F)$ because of (3.2.4).

The result follows by the definition of φ_ρ and the Lebesgue dominated convergence theorem.

q.e.d.

We are now in a position to relate these calculations to the problem of synthesis. Take $F \sim T \in A'$ and $\varphi \in A(\mathbf{T})$ for which $\operatorname{supp} T \subseteq Z\varphi$, and assume that $|B|(\varphi, F) < \infty$. Because of Proposition 3.2.1 and Proposition 3.2.2 we have (3.2.3) and (3.2.5). On the other hand, $\varphi_\rho = \varphi$ on a neighborhood of $\operatorname{supp} T$. In light of (3.2.3) it is not unreasonable to expect that with a strengthening of our hypothesis, $|B|(\varphi, F) < \infty$, we could conclude that

$$(3.2.6) \qquad B(\varphi, F) = B(\varphi_\rho, F);$$

and consequently we would obtain the annihilation, $\langle T, \varphi \rangle = 0$. Note that $\varphi_\rho \in C(\mathbf{T})$ when $\varphi \in A(\mathbf{T})$; and that if we knew that $\varphi_\rho \in A(\mathbf{T})$ then the condition, $|B|(\varphi, F) < \infty$, would be sufficient to ensure that $\langle T, \varphi \rangle = 0$ (cf. Exercise 3.1.7).

Theorem 3.2.1. *Given $F \sim T \in A'$ and $\varphi \in A(\mathbf{T})$ for which* $\operatorname{supp} T \subseteq Z\varphi$. *If $\|B\|(\varphi, F) < \infty$ then $\langle T, \varphi \rangle = 0$.*

Proof. In light of what we've done it is sufficient to prove that $\|B\|(\varphi, F) < \infty$ implies (3.2.6).

Set $\psi = \varphi_p - \varphi$ so that $\psi \in j(Z\varphi)$.

Let $\theta_\lambda \in A(\mathbf{T})$ be an approximate identity for which $\psi_\lambda = \psi * \theta_\lambda \in j(Z\varphi)$. For example, let

$$(3.2.7) \qquad \theta_\lambda = \frac{4\pi}{\lambda} \Delta_\lambda \qquad \text{for } |\lambda| < \pi,$$

where Δ_λ was defined in Exercise 1.2.6.

Since $\|\theta_\lambda\|_{A'} \leqslant 1$, we have

$$\|\psi_\lambda, r\|^2 = 4 \sum_n |\hat{\psi}(n)\, \hat{\theta}_\lambda(n)|^2 \sin^2 \frac{nr}{2} \leqslant \|\psi, r\|^2.$$

Because we are assuming $\|B\|(\varphi, F) < \infty$ we can use the Lebesgue dominated convergence theorem, and compute that

$$\lim_{\lambda \to 0} B(\psi_\lambda - \psi, F) = 0.$$

$\psi_\lambda \in A(\mathbf{T})$ and so $B(\psi_\lambda, F) = \langle T, \psi_\lambda \rangle$ from Proposition 3.2.1; thus $B(\psi_\lambda, F) = 0$ since $\psi_\lambda \in j(Z\varphi)$ and $\operatorname{supp} T \subseteq Z\varphi$.

Hence $B(\psi, F) = 0$, and this is (3.2.6).

q.e.d.

3.2.3 Technical lemmas. In order to apply Theorem 3.2.1 we consider some elementary conditions in order that

$$(3.2.8) \qquad \frac{1}{2\pi} \int_0^{2\pi} \frac{1}{\tau^2} \|F, \tau\|\, \|\varphi, \tau\|\, d\tau < \infty.$$

Proposition 3.2.3. *Given $F \in L^1(\mathbf{T})$ for which $\hat{F}(0) = 0$. If $F \sim T \in A'$ then*

$$\|F, r\| = O(r^{1/2}), \qquad r \to 0$$

(cf. Exercise 2.1.2a).

Proof. Since $F \sim T \in A'$ and

$$\|F, r\|^2 = 4 \sum{}' |\hat{F}(n)|^2 \sin^2 \frac{nr}{2}$$

for $F \in L^2(\mathbf{T})$ (by Parseval's formula) we have

$$\text{(3.2.9)} \qquad \|F, r\|^2 \leqslant K \sum_1^\infty \frac{1}{n^2}(1 - \cos nr).$$

The Fourier series on the right-hand side of (3.2.9) represents the even function $\varphi(r) = \pi r - \frac{r^2}{2}$ on $[0, \pi)$.

q.e.d.

The technique of Proposition 3.2.3 is strengthened in Exercise 3.2.1 to characterize pseudo-measures in terms of quantities such as $\|F, r\|$.

Notationally, we set

$$\omega_p(\varphi, \lambda) = \sup_{0 \leqslant r \leqslant \lambda} \{\|\tau_r \varphi - \varphi\|_p\}$$

and so $\|\varphi, \lambda\| \leqslant \omega_2(\varphi, \lambda)$. Also, $\varphi \in \mathrm{Lip}_\alpha(\mathbf{T})$ if and only if $\varphi \in C(\mathbf{T})$ and

$$\text{(3.2.10)} \qquad \omega_\varphi(\lambda) = \omega_\infty(\varphi, \lambda) = O(\lambda^\alpha), \qquad \lambda \to 0,$$

where $\lambda > 0$. $\omega_p(\varphi, \lambda)$ is an increasing function as either p or λ increase, and

$$\forall\, \varphi \in L^p(\mathbf{T}), \qquad \lim_{\lambda \to 0} \omega_p(\varphi, \lambda) = 0.$$

Further, if $\varphi \in C(\mathbf{T})$ then

$$\lim_{p \to \infty} \omega_p(\varphi, \lambda) = \omega_\infty(\varphi, \lambda)$$

Proposition 3.2.4. a) *Given $\hat{f} = \varphi \in A(\mathbf{T})$ for which*

$$\sum |f(n)|^2 |n| < \infty.$$

Then

$$\forall\, \alpha < \frac{3}{2}, \qquad \frac{1}{2\pi} \int_0^{2\pi} \frac{1}{\lambda^\alpha} \|\varphi, \lambda\| \, d\lambda < \infty.$$

b) *Given* $\varphi \in \mathrm{Lip}_\alpha(\mathbf{T})$, $\alpha > 1/2$. *Then*

$$\frac{1}{2\pi}\int_0^{2\pi} \frac{1}{\lambda^{3/2}}\|\varphi,\lambda\|\,d\lambda \leqslant \frac{1}{2\pi}\int_0^{2\pi} \frac{1}{\lambda^{3/2}}\omega_\infty(\varphi,\lambda)\,d\lambda < \infty.$$

Proof. a) Take $\alpha < 3/2$.

$$\frac{1}{2\pi}\int_0^{2\pi} \lambda^{-\alpha}\|\varphi,\lambda\|\,d\lambda = \frac{1}{2\pi}\int_0^{2\pi} \lambda^{1-\alpha}\left(\lambda^{-2}\sum|f(n)|^2\sin^2\frac{n\lambda}{2}\right)^{1/2} d\lambda$$

$$\leqslant \left(\int_{\mathbf{T}} \lambda^{2-2\alpha}\,d\lambda\right)^{1/2}\left[\sum|f(n)|^2\int_{\mathbf{T}}\lambda^{-2}\sin^2\frac{n\lambda}{2}\,d\lambda\right]^{1/2}$$

$$\leqslant K\sum|f(n)|^2|n|.$$

b) The result is immediate from (3.2.10).

q.e.d.

3.2.4 Applications of the Beurling integral. Define, as we did in Section 2.4.10,

$$(3.2.11)\qquad A(\varphi) = \frac{1}{2\pi}\int_0^{2\pi} \lambda^{-3/2}\|\varphi,\lambda\|\,d\lambda$$

for $\varphi \in L^2(\mathbf{T})$. Theorem 3.2.1 and Proposition 3.2.3 yield—

Theorem 3.2.2. *Given* $\varphi \in A(\mathbf{T})$. *If* $A(\varphi) < \infty$ *then* φ *is synthesizable.*

As we noted in Section 2.4.10, $A(\varphi) < \infty$ for $\varphi \in L^2(\mathbf{T})$ actually implies that $\varphi \in A(\mathbf{T})$.

There are two classes of functions which come to mind immediately as possible applications of Theorem 3.2.2: first, we'd like to test various $A(\mathbf{T}) \cap \mathrm{Lip}_\alpha(\mathbf{T})$, $\alpha > 0$; and second we'd like to test $A(\mathbf{T}) \cap BV(\mathbf{T})$ since, as is well known, $\varphi \in BV(\mathbf{T})$ if and only if

$$(3.2.12)\qquad \|\tau_r\varphi - \varphi\|_1 = O(|r|), \qquad r \to 0$$

(e.g. [Benedetto, 12]).

For the first case we use Proposition 3.2.4b and obtain–

Corollary 3.2.2.1. *If*

$$(3.2.13)\qquad \varphi \in \mathrm{Lip}_\alpha(\mathbf{T}), \qquad \alpha > 1/2,$$

then φ *is synthesizable* (*of course, if* $\alpha > 1/2$ *then* $\mathrm{Lip}_\alpha(\mathbf{T}) \subseteq A(\mathbf{T})$ *by Bernstein's theorem, as we noted in Exercise* 2.5.2).

For the second case, (3.2.10) and (3.2.11) combine to yield–

Corollary 3.2.2.2. *If $\varphi \in A(\mathbf{T}) \cap \mathrm{Lip}_\alpha(\mathbf{T}) \cap BV(\mathbf{T})$, where $\alpha > 0$, then φ is synthesizable.*

If $\varphi \in A(\mathbf{T})$ is an element of $\mathrm{Lip}_{1/2}(\mathbf{T})$ or $BV(\mathbf{T})$ then the integral, (3.2.11), has the form $\int_0^{2\pi} \frac{d\lambda}{\lambda}$. In both of these cases the L^2-quantity, $\|\varphi, \lambda\|$, is inappropriate for the functions we are considering, and there is no reason to expect the integral representation, $B(\varphi, F)$, to provide information about synthesis. Before further discussion on the method of integral representation we shall settle the specific problems concerning $\mathrm{Lip}_{1/2}(\mathbf{T})$ (Paragraph 3.2.5) and $BV(\mathbf{T})$ (Paragraph 3.2.7).

Remark. The contraction method in spectral synthesis is due to Beurling, but important contributions are found in [Kinukawa, 1; 2] (cf. [Kinukawa, 3, Theorem 3]). Beurling's theorems provide synthesis for weaker than weak * topologies, and parallel results without the theory of contraction are found in [Benedetto, 11]. In this regard, see the remark at the end of [Beurling, 6, Section 2] and [Domar, 3].

3.2.5 The Beurling–Pollard theorem. Given a closed set $E \subseteq \mathbf{T}$ and recall the definition of E_λ in (3.1.12) and the related remark in (3.1.13). The following result is essentially due to Beurling and [Pollard, 2] (cf. [Kahane, 10, pp. 77–78]).

Theorem 3.2.3. *Given $\varphi \in A(\mathbf{T})$ and set $E = Z\varphi$. Each of the following conditions implies that φ is synthesizable:*

a) $\lim_{\lambda \to 0} \left(\sqrt{\frac{m(E_\lambda \setminus E)}{\lambda}} \sup_{\gamma \in E_\lambda} |\varphi(\gamma)| \right) = 0, \; \lambda > 0.$

b) $\sup_{\gamma \in E_\lambda} |\varphi(\gamma)| = O(\lambda^{1/2}), \; \lambda \to 0, \; \lambda > 0.$

c) $\varphi \in \mathrm{Lip}_\alpha(\mathbf{T}), \; \alpha \geqslant 1/2.$

Proof. It is obviously sufficient to prove that φ is synthesizable when condition a) is satisfied.

Given $T \in A'(E)$ we use (3.2.7) and set

$$T_\lambda = T * \theta_\lambda.$$

Note that $T_\lambda \in C(\mathbf{T})$ and $\mathrm{supp}\, T_\lambda \subseteq E_\lambda$.

Clearly, $\|T_\lambda\|_{A'} \leqslant \|T\|_{A'}$ and so by the Lebesgue dominated convergence theorem, $T_\lambda \to T$ in $\sigma(A'(\mathbf{T}), A(\mathbf{T}))$.

Thus,

$$\langle T, \varphi \rangle = \lim_{\lambda \to 0} \langle T_\lambda, \varphi \rangle = \lim_{\lambda \to 0} \int_{E_\lambda \setminus E} T_\lambda(\gamma)\, \varphi(\gamma)\, d\gamma. \tag{3.2.14}$$

Observe that

$$\|T_\lambda\|_2 \leqslant \|T\|_{A'} \|\theta_\lambda\|_2 = O(\lambda^{-1/2}), \quad \lambda \to 0, \lambda > 0. \tag{3.2.15}$$

By (3.2.14),

$$|\langle T_\lambda, \varphi\rangle| \leqslant \| T_\lambda \|_2 \,(\sup_{\gamma\in E_\lambda} |\varphi(\gamma)|)(m(E_\lambda \setminus E))^{1/2}. \tag{3.2.16}$$

(3.2.14), (3.2.15), (3.2.16), and *a* combine to give the result.

q.e.d.

The theorem can be made slightly finer by using only the L^2-estimate in (3.2.16).

3.2.6 Constant values of the primitive of a pseudo-measure. Proposition 3.2.5b is a technical fact which we need in what follows. The statement itself is intuitively reasonable but some proof is required.

Proposition 3.2.5. a) *Let $F \in L^1[0,1]$ be real-valued and assume that*

$$\int_0^1 F(\gamma)\,\varphi(\gamma)\,\mathrm{d}\gamma = 0$$

for all continuous functions φ which vanish at 0 *at* 1. *Then $F = 0$* a.e.
b) *Given $F \sim T \in A'$ and assume that $T = 0$ on the open interval $(\lambda, \gamma) \subseteq$ Я, considered as an open subset of* **T**. *Then there is a constant k such that $F = k$* a.e. *on (λ, γ).*

Proof. a) Assume the result is not true so that $|F| > 0$ on some set $X \subseteq (0,1)$ having positive measure. Let

$$\int_0^1 |F(\gamma)|\,\mathrm{d}\gamma > \varepsilon > 0.$$

There is an open set $U_\varepsilon \subseteq (0,1)$ such that

$$\int_{(0,1)\setminus U_\varepsilon} |F(\gamma)|\,\mathrm{d}\gamma < \frac{\varepsilon}{8}, \tag{3.2.17}$$

and we take U_ε large enough so that $m(X \cap U_\varepsilon) > 0$.
Set $X_+ = \{\gamma \in U_\varepsilon : |F(\gamma)| > 0\}$ and $X_- = \{\gamma \in U_\varepsilon : |F(\gamma)| < 0\}$; then $X_\pm$ are measurable and either $mX_+ > 0$ or $mX_- > 0$ (or both).
From measure theory there are compact sets

$$K_\pm \subseteq X_\pm \text{ for which } \{0,1\} \cap K_\pm = \emptyset$$

and

$$m(X_\pm \setminus K_\pm) < \frac{\varepsilon}{16\|F\|_1}. \tag{3.2.18}$$

(3.2.17) and (3.2.18) combine to yield

$$\int_0^1 |F(\gamma)|\,d\gamma - \int_{K_+ \cup K_-} |F(\gamma)|\,d\gamma < \frac{\varepsilon}{4}, \tag{3.2.19}$$

and so

$$\int_{K_+ \cup K_-} |F(\gamma)|\,d\gamma > \frac{3\varepsilon}{4}. \tag{3.2.20}$$

Choose continuous functions $\psi_\pm$ to be ± 1 on $K_\pm$ and 0 on $K_\mp$; further, assume that $0 \leqslant \psi_\pm \leqslant 1$ and let $\psi_\pm(0) = \psi_\pm(1) = 0$. Define $\varphi = \psi_+ + \psi_-$.
Then

$$\int_0^1 F(\gamma)\,\varphi(\gamma)\,d\gamma = \int_{(0,1)\setminus U_\varepsilon} F(\gamma)\,\varphi(\gamma)\,d\gamma + \int_{K_+ \cup K_-} |F(\gamma)|\,d\gamma + r_\varepsilon,$$

where $|r_\varepsilon| < \varepsilon/4$ because of (3.2.18).
Thus, by (3.2.17),

$$\int_0^1 F(\gamma)\,\varphi(\gamma)\,d\gamma = \int_{K_+ \cup K_-} |F(\gamma)|\,d\gamma + r, \qquad |r| < \varepsilon/2. \tag{3.2.21}$$

By our hypothesis and (3.2.21) we conclude that

$$\int_{K_+ \cup K_-} |F(\gamma)|\,d\gamma < \frac{\varepsilon}{2},$$

the desired contradiction.
b) Let

$$X = \{\varphi \in C(\mathbf{T}) : \operatorname{supp} \varphi \subseteq (\lambda, \gamma)\}$$

and

$$Y = \{\psi \in X : \exists\, \theta \in X \text{ such that } \theta' = \psi\},$$

where we take the ordinary derivative in the definition of Y.
Y is a subspace of X and, by hypothesis and the definition of the distributional derivative, we have

$$\forall\, \psi \in Y, \qquad \langle F, \psi \rangle = 0. \tag{3.2.22}$$

Clearly, $X \setminus Y \neq \emptyset$ and we choose $\varphi_0 \in X \setminus Y$ for which $\int_{\mathbf{T}} \varphi_0(\gamma)d\gamma = 1$.

Given $\varphi \in X$ we define

$$\psi = \varphi - \varphi_0 \int \varphi(\gamma)\,d\gamma$$

and

$$\theta(s) = \int_0^s \psi(r)\,dr, \qquad s \in [0, 2\pi).$$

It is easy to check that $\psi, \theta \in X$ and, of course, $\theta' = \psi$.
Consequently, each $\varphi \in X$ has the representation

$$(3.2.23) \qquad \varphi = c\varphi_0 + \psi, \qquad c \in \mathbf{C} \text{ and } \psi \in Y.$$

If $(c_1 - c_2)\varphi_0 + (\psi_1 - \psi_2) = 0$, where $c_i \in \mathbf{C}$ and $\psi_i \in Y$, we see that $c_1 = c_2$ and $\psi_1 = \psi_2$ since $\varphi_0 \in X \setminus Y$.
Thus, the representation (3.2.23) is unique. Therefore, because of (3.2.22) and (3.2.23), we have

$$(3.2.24) \qquad \forall\, \varphi \in X, \qquad \langle F, \varphi \rangle = c \int F(\gamma)\,\varphi_0(\gamma)\,d\gamma = \int k\varphi(\gamma)\,d\gamma$$

since $c = \int \varphi(\gamma)d\gamma$ (here, $k = \int F(\gamma)\varphi_0(\gamma)d\gamma$).
Now, X is dense in the set W of continuous functions on $[\lambda, \gamma]$ which vanish at λ and γ. Therefore, from (3.2.24), $F - k \in L^1[\lambda, \gamma]$ has the property that

$$\forall\, \varphi \in W, \qquad \int_{\mathbf{T}} (F(\gamma) - k)\,\varphi(\gamma)\,d\gamma = 0;$$

and so $F = k$ a.e. by part a).

q.e.d.

Let $E \subseteq \mathbf{T}$ be a closed set and write the complement as

$$E^{\sim} = \bigcup_1^{\infty} I_j \subseteq \mathbf{T},$$

where $\{I_j : j = 1, \ldots\} \subseteq \mathfrak{R}$ is a disjoint family of open intervals. We write $I_j = (\lambda_j, \gamma_j)$. Haar measure m on $\mathbf{T}$ is such that $\varepsilon_j = mI_j = \dfrac{1}{2\pi}(\gamma_j - \lambda_j)$ and so

$$m(E^{\sim}) = \sum mI_j = \sum \varepsilon_j \leqslant 1.$$

Because of Proposition 3.2.5b, if $F \sim T \in A' \cap A'(E)$ then

$$(3.2.25) \qquad F = \sum_1^{\infty} k_j \chi_{I_j} \qquad \text{on } E^{\sim}.$$

3.2.7 Katznelson's theorem: $\varphi \in A(\mathbf{T}) \cap BV(\mathbf{T})$ is synthesizable. The following theorem is due to Katznelson.

Theorem 3.2.4. *If $\varphi \in A(\mathbf{T}) \cap BV(\mathbf{T})$ then φ is synthesizable.*

Proof. i) It is sufficient to prove that $\langle T, \varphi \rangle = 0$, where $F \sim T \in A'$ and $T \in A'(Z\varphi)$. We use the notation of (3.2.7) and Theorem 3.2.3.
Therefore, setting

$$F_\lambda = F * \theta_\lambda,$$

we compute that

$$F'_\lambda = T_\lambda = T * \theta_\lambda,$$

pointwise and distributionally (cf. [Schwartz, 5, Théorème III, p. 54]).
Thus, since $\varphi \in BV(\mathbf{T})$, we integrate by parts and obtain

$$\text{(3.2.26)} \qquad \langle T_\lambda, \varphi \rangle = \int_{\mathbf{T}} T_\lambda(\gamma)\, \varphi(\gamma)\, \mathrm{d}\gamma = -\int_{\mathbf{T}} F_\lambda(\gamma)\, \mathrm{d}\varphi(\gamma).$$

ii) Suppose $F \in C(\mathbf{T})$. Then $\lim_{\lambda \to 0} \|F_\lambda - F\|_\infty = 0$, $\lambda > 0$. Because $\lim_{\lambda \to 0} \langle T_\lambda - T, \varphi \rangle = 0$, $\lambda > 0$ (as we noted in the proof of Theorem 3.2.3), (3.2.26) yields

$$\text{(3.2.27)} \qquad \langle T, \varphi \rangle = -\int F(\gamma)\, \mathrm{d}\varphi(\gamma).$$

From Proposition 3.2.5b and the definition of the Stieltjes integral, we see that the integral in (3.2.27) vanishes since F takes constant values on the intervals contiguous to $Z\varphi$.
iii) If F is not necessarily continuous we use the notation of Section 3.2.6 and write

$$\text{(3.2.28)} \qquad -\langle T_\lambda, \varphi \rangle = \sum_{j=1}^{\infty} \int_{I_j} F_\lambda(\gamma)\, \mathrm{d}\varphi(\gamma)$$

(instead of (3.2.26)), where $(Z\varphi)^\sim = \bigcup_1^\infty I_j$.

Let $m_j(\lambda) = F_\lambda((\gamma_j + \lambda_j)/2)$ and define the functions

$$\forall\, j, \qquad H_{\lambda, j} = F_\lambda - m_j(\lambda) \text{ on } I_j.$$

Clearly, for each j,

$$\text{(3.2.29)} \qquad \int_{I_j} F_\lambda(\gamma)\, \mathrm{d}\varphi(\gamma) = \int_{I_j} H_{\lambda, j}(\gamma)\, \mathrm{d}\varphi(\gamma).$$

In part iv) we shall prove that

(3.2.30) $$\sup_{j,\lambda}\sup_{\gamma\in I_j}|H_{\lambda,j}(\gamma)| = K < \infty.$$

Because of (3.2.29), (3.2.30), and the fact that $\varphi \in BV(\mathbf{T})$, we can use the Lebesgue dominated convergence theorem and take the limit (as $\lambda \to 0$) under the summation sign in (3.2.28).

Since F is constant on I_j we proceed as in part ii) to see that

$$\forall\, j, \qquad \lim_{\lambda\to 0}\int_{I_j} F_\lambda(\gamma)\,\mathrm{d}\varphi(\gamma) = 0, \qquad \lambda > 0.$$

Consequently, $\langle T, \varphi\rangle = 0$.

iv) It remains to prove (3.2.30).

An easy calculation (e.g. Exercise 1.2.6) shows that

$$\exists\, M > 0 \text{ such that } \forall\, \lambda > 0, \qquad \|T_\lambda\|_\infty \leqslant \frac{M}{\lambda}.$$

If $mI_j \leqslant 2\lambda$, then, by the mean-value theorem,

$$\forall\, \gamma \in I_j, \qquad |H_{\lambda,j}(\gamma)| = |H_{\lambda,j}(\gamma) - H_{\lambda,j}((\gamma_j + \lambda_j)/2)| \leqslant 2M.$$

If $mI_j > 2\lambda$ we compute that $H_{\lambda,j} = 0$ on $[\lambda_j + \lambda, \gamma_j - \lambda]$; and we again obtain that $|H_{\lambda,j}(\gamma)| \leqslant 2M$ for $\gamma \in I_j$ by using the mean-value theorem for the difference $H_{\lambda,j}(\gamma) - H_{\lambda,j}(\lambda_j + \lambda)$ (resp. $H_{\lambda,j}(\gamma) - H_{\lambda,j}(\gamma_j - \lambda)$) if $\gamma \in (\lambda_j, \lambda_j + \lambda)$ (resp. $\gamma \in (\gamma_j - \lambda, \gamma_j)$).

q.e.d.

3.2.8 Spectral synthesis and the Kempisty–Denjoy integral. We now return more directly to the integration problem introduced in Section 3.2.1 (cf. (3.2.27)). As a warning to the reader, we note that this section and Section 3.2.9 are more speculative than not; of course, the integration problem itself is important.

Let $\alpha \in (0, \frac{1}{2})$, take $F \in L^1(\mathbf{T})$ to be real-valued, and let $I \subseteq \mathbf{T}$ be a closed interval in $\mathfrak{R}$. Define

$$m(I, \alpha) = \sup\{y : m\{\gamma : F(\gamma) < y\} \leqslant \alpha mI\}$$

and

$$M(I, \alpha) = \inf\{y : m\{\gamma : F(\gamma) > y\} \leqslant \alpha mI\}.$$

Note that if $y_1 < y_2$ then

$$m\{\gamma : F(\gamma) < y_1\} \leqslant m\{\gamma : F(\gamma) < y_2\}$$

and

$$m\{\gamma : F(\gamma) > y_1\} \geqslant m\{\gamma : F(\gamma) > y_2\};$$

and so

$$\alpha mI \geqslant m\{\gamma : F(\gamma) < y_2\} \quad \Rightarrow \quad \alpha mI \geqslant m\{\gamma : F(\gamma) < y_1\}$$

and

$$\alpha mI \geqslant m\{\gamma : F(\gamma) > y_1\} \quad \Rightarrow \quad \alpha mI \geqslant m\{\gamma : F(\gamma) > y_2\}.$$

Proposition 3.2.6. *Let $\alpha \in (0, \frac{1}{2})$, take $F \in L^1(\mathbf{T})$ to be real-valued, and let $I \subseteq \mathbf{T}$ be a closed interval in* Я. *Then*

a) $m(I, \alpha) \leqslant M(I, \alpha)$.

b) $-\infty \leqslant \inf_{\gamma \in I} F(\gamma) \leqslant m(I, \alpha)$ and $-\infty < m(I, \alpha)$.

c) $M(I, \alpha) \leqslant \sup_{\gamma \in I} F(\gamma) \leqslant \infty$ and $M(I, \alpha) < \infty$.

d) *If $F = k$ on I then $m(I, \alpha) = M(I, \alpha) = k$.*

Proof. a) Note that $\operatorname{card}\{u : m\{\gamma : F(\gamma) = u\} > 0\} \leqslant \aleph_0$, and thus $\{y : m\{\gamma : F(\gamma) = y\} = 0\}$ is dense. Take such a y which satisfies

$$(3.2.31) \qquad \alpha mI \geqslant m\{\gamma : F(\gamma) < y\}.$$

Also let z satisfy

$$\alpha mI \geqslant m\{\gamma : F(\gamma) > z\}.$$

If $y > z$ then

$$m\{\gamma : F(\gamma) > y\} \leqslant m\{\gamma : F(\gamma) > z\} \leqslant \alpha mI.$$

Combining this with (3.2.31) gives

$$2\alpha mI \geqslant m\{\gamma : F(\gamma) < y\} + m\{\gamma : F(\gamma) > y\} = mI,$$

a contradiction, since $\alpha < 1/2$.

b) Let $y = \inf\{F(\gamma) : \gamma \in I\}$ (y could be $\pm\infty$) so that $y \leqslant m(I, \alpha)$, since $m(\gamma \in I : F(\gamma) < y\} = 0$. To show that $-\infty < m(I, \alpha)$.

If there is no z for which $m\{\gamma : F(\gamma) < z\} \leqslant \alpha mI$ then for all z, $m\{\gamma : F(\gamma) < z\} > \alpha mI$.

Now, as $z \to -\infty$, $\{\gamma : F(\gamma) < z\}$ decreases, and so

$$mX = \lim_{z \to -\infty} m\{\gamma : F(\gamma) < z\} \geqslant \alpha mI,$$

where $X = \bigcap_z \{\gamma : F(\gamma) < z\}$.

Consequently, $m\{\gamma : F(\gamma) = -\infty\} > 0$ since $F(\gamma) = -\infty$ if $\gamma \in X$.

This is the required contradiction.

c) is proved in a fashion analogous to b).

d) If $y \leqslant k$, $m\{\gamma \in I : F(\gamma) < y\} = 0$; and if $y > k$, $m\{\gamma : F(\gamma) < y\} = mI > \alpha mI$.

Thus $k = m(I, \alpha)$.
A similar argument works for $M(I, \alpha)$.

q.e.d.

Given $\varphi \in C(\mathbf{T})$, and define the upper and lower sums,

$$S(P, F, \varphi) = S(P) = \sum_{1}^{n} M(J_j, \alpha)(\varphi(\gamma_j) - \varphi(\gamma_{j-1}))$$

and

$$s(P, F, \varphi) = s(P) = \sum_{1}^{n} m(J_j, \alpha)(\varphi(\gamma_j) - \varphi(\gamma_{j-1})),$$

for any partition $P: \gamma_0 < \gamma_1 < \ldots < \gamma_n$ and $J_j = [\gamma_{j-1}, \gamma_j]$. *F is integrable with respect to* φ with integral

$$L = (K) \int F \mathrm{d}\varphi$$

if for each $\varepsilon > 0$ there is a partition P_ε such that

$$|S(P) - L| < \varepsilon \tag{3.2.32}$$

for each partition P finer than P_ε. Conditions equivalent to (3.2.32) are also valid in terms of $s(P)$ and in terms of a Cauchy criterion. Further, if $(K) \int F \mathrm{d}\varphi$ exists for some $\alpha \in (0, 1/2)$ then it exists with the same value for all $\alpha \in (0, 1/2)$.

Remark. We use the notation "$(K) \int F \mathrm{d}\varphi$" because of Kempisty's theory of integral (for the case $\varphi(\gamma) = \gamma$) in 1925. Actually, [Denjoy, 1] proved the equivalence of the Kempisty integral and one of his own (vintage 1919) in 1931.

Setting $e_n(\gamma) = e^{in\gamma}$ as we did in Proposition 2.5.3, it is easily checked that if $F \sim T \in A'$ then

$$(K) \int F \mathrm{d}e_n = -\hat{T}(n). \tag{3.2.33}$$

Also, if $F \in BV(\mathbf{T})$ then

$$\forall\, \varphi \in C(\mathbf{T}), \qquad (K) \int F \mathrm{d}\varphi = \int F \mathrm{d}\varphi, \tag{3.2.34}$$

by Proposition 3.2.6 and the definition of the Stieltjes integral; in this case

$$\{s(P, F, e_n), S(P, F, e_n) : n \in \mathbf{Z} \text{ and } P\}$$

is uniformly bounded because of (3.2.34).

Proposition 3.2.7. *Given $\varphi \in A(\mathbf{T})$ and $F \sim T \in A'$.*
a) *If $\{s(P,F,e_n): n \in \mathbf{Z}$ and $P\}$ is uniformly bounded then*

$$(3.2.35) \qquad (K)\int F\,\mathrm{d}\varphi = -\langle T, \varphi\rangle.$$

b) *If (3.2.35) is valid and $T \in A'(Z\varphi)$ then $\langle T, \varphi\rangle = 0$.*

Proof. a) follows by hypothesis and (3.2.33).
b) Choose a partition P_n containing all of the endpoints of the first n contiguous intervals of $Z\varphi$.

We further demand that P_n should satisfy the following property: if

$$\gamma \in P_n \cap Z\varphi,$$

then there are endpoints θ_j and θ_k from the first n contiguous intervals of $Z\varphi$, such that

$$P_n \cap (\theta_j, \theta_k) \subseteq Z\varphi.$$

Because of Proposition 3.2.6, $s(P_n, F, \varphi) = 0$.

q.e.d.

3.2.9 The general problem of integration and synthesis. The situation in Section 3.2.8 appears quite technical, but we shall try to clear things up a bit by the following explanation in terms of determining $A'_S(\Gamma)$.

Given a closed set $E \subseteq \Gamma$ and $T \in A'(E)$. Assume that there is $\mu \in M(E)$ and

$$K(\gamma, \cdot): E \rightarrow A'_S(E)$$

such that

$$T = \int_E K(\gamma, \lambda)\,\mathrm{d}\mu(\lambda)$$

(in the sense of the Bochner integral). Then

$$\forall\, \varphi \in k(E), \qquad \langle T, \varphi\rangle = 0.$$

In Proposition 3.2.7a we started with the specific and simple measure $\mu = \delta$ (assuming for the discussion that $0 \in E$); and as a result were forced into the situation of imposing rigid boundedness conditions on $s(P, F, \varphi)$, whose analogue in the above scheme is $K(\ ,\ )$. By allowing greater flexibility in the choice of μ the type of conditions required on K can be weakened so that the resulting "Proposition 3.2.7a" can have more useful hypotheses. To say anything more definite would require more specific information on the given $T \in A'(E)$. In any case, from this point of view, the integration problem becomes one of constructing useful kernels K and choosing appropriate measures μ.

3.2.10 A characterization of pseudo-measures in terms of L^p. We now consider a variation on the Beurling–Pollard theme which allows us to determine information about F when $F \sim T \in A'$. We shall use the Hausdorff–Young theorem, e.g. [Katznelson, 5, Chapter IV]:

Theorem 3.2.5. *Given $p \in [1,2]$ and define q by the formula $\frac{1}{p} + \frac{1}{q} = 1$.*

a) *If $F \in L^p(\mathbf{T}) (\subseteq L^1(\mathbf{T}))$ then $\hat{F} = \Phi \in L^q(\mathbf{Z})$ and*

$$\|\Phi\|_q \leqslant \|F\|_p.$$

b) *If $\Phi \in L^p(\mathbf{Z})$ then $\sum \Phi(n) e^{in\gamma}$ converges in $L^q(\mathbf{T})$ to a function $F \in L^q(\mathbf{T})$ for which*

$$\|F\|_q \leqslant \|\Phi\|_p.$$

Theorem 3.2.6. *Given $F \sim T \in A'$ where $T \sim \sum' c_n e^{in\gamma}$.*

a) *Then $F \in \bigcap L^p(\mathbf{T})$ and*

$$\forall q < \infty, \|F\|_q \leqslant \left(\sum{}' \left|\frac{c_n}{n}\right|^p\right)^{1/p},$$

where $\frac{1}{p} + \frac{1}{q} = 1$.

b) *If $E = \operatorname{supp} T$ and $mE = 0$ then (using the notation of (3.2.25))*

$$T = \sum_1^{\infty} k_j(\delta_{\lambda_j} - \delta_{\gamma_j}) \quad \text{on} \quad C^1(\mathbf{T}), \tag{3.2.36}$$

where

$$\forall r < 1/(2e\|T\|_{A'}), \qquad \sum_{j=1}^{\infty} e^{r|k_j|} \varepsilon_j < \infty. \tag{3.2.37}$$

Proof. a) If $p > 1$ then $\sum'_n |c_n/n|^p < \infty$, and so we apply Theorem 3.2.5b to obtain $F \in \bigcap \{L^q(\mathbf{T}) : q < \infty\}$ and

$$\forall q < \infty, \qquad \|F\|_q \leqslant \left(\sum_n{}' |c_n/n|^p\right)^{1/p}. \tag{3.2.38}$$

b.i) Since $mE = 0$ we have

$$F = \sum_1^{\infty} k_j \chi_{I_j} \quad \text{a.e.}$$

(3.2.36) follows by the definition of the distributional derivative.

ii) We compute, for $q < \infty$, that

$$(3.2.39)\qquad \|F\|_q = \left(\frac{1}{2\pi}\int_0^{2\pi} |\sum k_j \chi_{I_j}(\gamma)|^q \, d\gamma\right)^{1/q} = \left(\sum_1^\infty \frac{1}{2\pi}\int_{I_j} |k_j|^q \, d\gamma\right)^{1/q},$$

since we can use the Lebesgue dominated convergence theorem and since

$$\forall\, m, \left|\sum_{j=1}^m k_j \chi_{I_j}(\gamma)\right|^q = \sum_{j=1}^m \chi_{I_j}(\gamma)|k_j|^q.$$

Combining (3.2.38) and (3.2.39) we obtain

$$\forall\, q < \infty, \sum_1^\infty \varepsilon_j |k_j|^q \leqslant \left(\|T\|_{A'}\left(\sum_n{}' \frac{1}{|n|^p}\right)^{1/p}\right)^q = K_p^q.$$

iii) We now estimate $\left(\sum_n{}' 1/|n|^p\right)^{q/p}$.
Clearly, if $p > 1$,

$$\sum_2^\infty 1/n^p \leqslant \int_1^\infty \frac{dt}{t^p} = \frac{1}{p-1}.$$

Thus,

$$\left(\sum_n{}' 1/|n|^p\right)^{q/p} \leqslant \left(\frac{2}{p-1} + 2\right)^{q/p} = 2^{q/p}\left(\frac{p}{p-1}\right)^{q/p} = (2q)^{q/p}.$$

iv) Choose

$$r < 1/(2e\|T\|_{A'}).$$

Then

$$\sum_{q=0}^\infty \frac{(rK_p)^q}{q!} \leqslant \sum_{q=0}^\infty (2r\|T\|_{A'})^q \frac{q^{q-1}}{q!}.$$

By the ratio test

$$\overline{\lim_{q\to\infty}} \frac{(2r\|T\|_{A'})^{q+1}(q+1)^q}{(q+1)!} \frac{q!}{(2r\|T\|_{A'})^q q^{q-1}} = 2r\|T\|_{A'} \overline{\lim_{q\to\infty}} \left(\frac{q+1}{q}\right)^{q-1}$$

$$= 2r\|T\|_{A'}\, e < 1.$$

Consequently, $\sum_q (rK_p)^q/q!$ converges, and by the usual properties of interchanging sums in absolutely convergent series, we have

$$\sum_{j=1}^{\infty} e^{r|k_j|}\varepsilon_j = \sum_{j=1}^{\infty}\sum_{q=0}^{\infty} (r|k_j|)^q \varepsilon_j/q!$$
$$= \sum_{q=0}^{\infty}\sum_{j=1}^{\infty} (r|k_j|)^q \varepsilon_j/q! \leqslant \sum_{q=0}^{\infty} (rK_p)^q/q!.$$

q.e.d.

The estimate in (3.2.37) combines with Theorem 3.2.1 to yield an assortment of expected corollaries.

There are easy examples and well-known methods (e.g. [Bary, 2, II, pp. 243 ff.]) to show that $(\cap L^p(\mathbf{T}))\setminus L^\infty(\mathbf{T}) \neq \emptyset$ (cf. the interesting potpourri composed of [Banach, 1, pp. 203–204; Grothendieck, 1; Handelsman and Lew, 1; Rickert, 1; Shapiro, 1]).

On the other hand, if $F \sim T \in A'$, it is not transparent that F is ever unbounded (cf. (E1.3.4)). For convenience, let

$$A_b' = \{T \in A' : F \in L^\infty(\mathbf{T}) \text{ when } F \sim T \in A'\}.$$

In Exercise 3.2.3d we show that there are elements $T \in A'\setminus A_b'$ for which card supp $T = \aleph_0$. In the opposite direction we have

Proposition 3.2.8. *Given*

$$(3.2.40)\qquad T \sim \sum_{n=1}^{\infty} c_n e^{ik_n\gamma} \in A'$$

where $\{k : n = 1, \ldots\}$ *is a lacunary sequence (defined in Exercise* 1.1.4). *Then* $F \in A(\mathbf{T})$, *where* $F \sim T$, *and so* $T \in A_c'(\mathbf{T})$ *(as we show in Exercise* 3.2.5).

Proof. If $k_{n+1}/k_n \geqslant \lambda > 1$ for each n, then

$$\frac{1}{\lambda^{n-1}} \geqslant \frac{k_1}{k_2}\cdot\frac{k_2}{k_3}\cdot\ldots\cdot\frac{k_{n-1}}{k_n} = \frac{k_1}{k_n}.$$

Thus,

$$k_1 \sum_{n=1}^{\infty} \left|\frac{c_n}{k_n}\right| \leqslant \|T\|_{A'} \sum_{n=1}^{\infty} \frac{1}{\lambda^{n-1}} < \infty.$$

q.e.d.

Naturally, given (3.2.40) we can verify directly that $T \in A_c'(\mathbf{T})$.

Remark. $\hat{f} = \varphi \in H^1(\mathbf{T})$ if $f(n) = 0$ for each $n < 0$ and $f \in L^1(\mathbf{Z})$. Clearly, $H^1(\mathbf{T})$ is a closed ideal in $A(\mathbf{T})$, and there is a fundamental theorem by Hardy, e.g. [Katznelson, 5, p. 91], which says:

$$(3.2.41)\qquad \hat{f} = \varphi \in H^1(\mathbf{T}) \;\Rightarrow\; \sum_{1}^{\infty} |f(n)|/n < \infty.$$

In 1961, John and Nirenberg made a study of functions $F \in L^1(\mathbf{T})$ for which

$$(3.2.42) \qquad \sup_I \frac{1}{mI} \int_I |F(\gamma) - F_I| \, d\gamma = \|F\|_* < \infty,$$

where $I \subseteq \mathbf{T}$ is a closed interval in $\mathbb{R}$ and

$$F_I = \frac{1}{mI} \int_I F(\gamma) \, d\gamma;$$

such functions are said to have *bounded mean oscillation* and if (3.2.42) is satisfied we write $F \in BMO$. Fefferman has proved that BMO is the dual of $H^1(\mathbf{T})$. The major recent work on this topic is [Fefferman and Stein, 1].

We mention BMO since

$$L^\infty(\mathbf{T}) \subsetneq BMO \subsetneq \bigcap L^p(\mathbf{T}),$$

and because, by Theorem 3.2.6a and (3.2.41),

$$F \sim T \in A' \quad \Rightarrow \quad F \in BMO.$$

There are functions $F \in BMO$ for which $\overline{\lim}_{|n|} |n\hat{F}(n)| = \infty$ (cf. Exercise 3.2.3).

3.2.11 Measure theoretic criteria for a pseudo-measure to be a measure. With regard to Example 2.5.3c we now prove

Proposition 3.2.9. *Let $E \subseteq \mathbf{T}$ be a closed set for which $mE = 0$. Then*

$$\overline{C^\infty(\mathbf{T}) \cap A_E(\mathbf{T})} = A(\mathbf{T})$$

($A_E(\mathbf{T})$ was defined in (2.5.9)).

Proof. Set $C_E^\infty = C^\infty(\mathbf{T}) \cap A_E(\mathbf{T})$ and let $T \in A'(\mathbf{T})$ have the property that

$$\forall \, \varphi \in C_E^\infty, \qquad \langle T, \varphi \rangle = 0.$$

Since $\varphi = 1 \in C_E^\infty$ we see that $T \in A'$ and so

$$(3.2.43) \qquad \forall \, \varphi \in C_E^\infty, \qquad \int F(\gamma) \, \varphi'(\gamma) \, d\gamma = 0,$$

where $F \sim T$.

The function φ used in the proof of Proposition 3.2.5 can easily be adapted to have the form ψ', $\operatorname{supp} \psi \subseteq (0,1)$.

Thus, if $I \subseteq E^\sim$ is an open interval in $\mathbb{R}$ we use Proposition 3.2.5a in conjunction with (3.2.43) to conclude that $F = 0$ a.e. on I.

Since $mE = 0$ we see that $F = 0$ a.e. and so $T = 0$. The result follows by the Hahn–Banach theorem.

q.e.d.

Proposition 3.2.10. *Let $E \subseteq \mathbf{T}$ be a closed set for which $mE = 0$ and let $T \in A'(E)$. Then $T \in \mathscr{V}$ (defined before Proposition 2.5.6) if and only if $T \in M(E)$.*

Proof. The result follows by Theorem 2.5.5 and Proposition 3.2.9.

q.e.d.

With regard to Proposition 3.2.10, it is shown in [Benedetto, 6, Theorem 2.6] that

Proposition 3.2.11. *Let $E \subseteq \mathbf{T}$ be a closed set for which $mE = 0$ and let $F' = T$ distributionally, where $F \in L^1(\mathbf{T})$, $\hat{F}(0) = 0$, and $\operatorname{supp} T \subseteq E$. Then $T \in \mathscr{V}$ if and only if $T \in M(E)$.*

Proposition 3.2.11 can't be extended to any first order distribution T on $\mathbf{T}$ (the order of a distribution was discussed in Exercise 2.5.5g) as is seen by the example

$$T = \sum_1^{\infty} a_n \delta'_{1/n}, \qquad \sum |a_n| < \infty.$$

A condition for membership in $\mathscr{V}$ is given in [Benedetto, 11]. Also, by standard measure theory,

Proposition 3.2.12. *Given a closed totally disconnected set $E \subseteq \mathbf{T}$ and $T \in A'(E)$. Assume that*

$$\sum T(F_j) \in \mathbf{C}$$

for every disjoint family $\{F_j : j = 1, \ldots\} \subseteq \mathscr{F}$(using the notation of Section 2.5.9). Then $T \in \mathscr{V}$.

There are closed totally disconnected sets $E \subseteq \mathbf{T}$, $mE > 0$, and $T \in A'(E) \setminus M(E)$ such that $T \in \mathscr{V}$ [Katznelson and Rudin, 1].

3.2.12 Kronecker's theorem and the condition that $A'(E) = M(E)$. As we promised in Section 1.3.13 we shall construct uncountable sets $E \subseteq \mathbf{T}$ of strong spectral resolution. Such a set necessarily has the property that $mE = 0$ (cf. Sections 3.2.10 and 3.2.11). The construction is due to [Kahane and Salem, 1] and uses Kronecker's theorem. Of course, in light of Example 2.5.3, it is really only necessary to exhibit a perfect Kronecker set (references were given after Theorem 3.1.6′); and that would be a slightly easier task than the Kahane–Salem construction, although also depending ultimately on Kronecker's theorem.

We made use of a form of Kronecker's theorem in Section 2.4.8, and we shall now prove a stronger version in Theorem 3.2.7 assuming the facts about the Bohr compactification that we listed in Section 3.1.16. Kronecker's theorem is a result in Diophantine approximation and we refer to [Benedetto, 6] for references and a discussion in classical terms. The proofs of Theorem 3.2.7 and Theorem 3.2.8 are obviously true in a more general setting.

Theorem 3.2.7. *Let $E \subseteq \mathbf{T}$ be a finite set and let $\varepsilon \in (0,1)$.*
a) *There is $N > 0$ such that for every homomorphism*

$$\varphi : \mathbf{T} \to \mathbf{T}$$

there is an integer $n_\varphi \in [-N, N]$ such that

$$\forall\, \gamma \in E, \qquad |\varphi(\gamma) - (\gamma, n_\varphi)| < \varepsilon.$$

b) *If E is independent over $\mathbf{Q}$ then there is $N > 0$ such that for every function*

$$\varphi : \mathbf{T} \to \mathbf{T},$$

with the property

$$\forall\, \gamma \in E, \qquad |\varphi(\gamma)| = 1,$$

there is an integer $n_\varphi \in [-N, N]$ for which

$$\forall\, \gamma \in E, \qquad |\varphi(\gamma) - (\gamma, n_\varphi)| < \varepsilon.$$

c) *If E is independent over $\mathbf{Q}$ then there is $N > 0$ such that*

$$\forall\, \mu \in M(E), \qquad \sup_{m \in [-N, N]} |\hat{\mu}(m)| \geqslant (1 - \varepsilon) \|\mu\|_1.$$

Proof. a) Recall that $\beta\mathbf{Z}$ is the Bohr compactification of $\mathbf{Z}$.
If $\psi \in \beta\mathbf{Z}$ then by the definition of the topology on $\beta\mathbf{Z}$,

$$N_\varepsilon(\psi) = \left\{\theta \in \beta\mathbf{Z} : \forall\, \gamma \in E, \, |\psi(\gamma) - \theta(\gamma)| < \frac{\varepsilon}{2}\right\}$$

is an open neighborhood of ψ.
Thus $\{N_\varepsilon(\psi) : \psi \in \beta\mathbf{Z}\}$ is an open cover of the compact space $\beta\mathbf{Z}$ and we let $\{N_\varepsilon(\psi_j) : j = 1, \ldots, k\}$ be a finite subcover.
By the definition of $\beta\mathbf{Z}$, it is possible to choose $n_j \in N_\varepsilon(\psi_j) \cap \mathbf{Z}$, and we let $N = \max\{|n_j| : j = 1, \ldots, k\}$.
b) Let $\langle E\rangle \subseteq \mathbf{T}$ be the group generated by E. Since E is independent each $\gamma \in \langle E\rangle$ has a unique decomposition, $\gamma = \sum_{\lambda \in E} k_\lambda \lambda$, $k_\lambda \in \mathbf{Z}$.
Thus the function $\psi_\varphi : \langle E\rangle \to \mathbf{T}$, where

$$\forall\, \gamma \in \langle E\rangle, \qquad \psi_\varphi(\gamma) = \prod_{\lambda \in E} \varphi(\lambda)^{k_\lambda},$$

is a well-defined homomorphism.

There is a standard procedure, paralleling the proof of the Hahn–Banach theorem, to show that ψ_φ extends to a homomorphism $\varphi_e:\mathbf{T}\to\mathbf{T}$.
The result now follows from part a).
c) Take N from part b). Assume, without loss of generality, that if $\mu\in M(E)$ then $\operatorname{supp}\mu=E$.
If $\mu\in M(E)$ then $\mu=\sum_{\gamma\in E}a_\gamma\delta_\gamma$ and we define $\varphi:E\to\mathbf{T}$ as $\varphi(\gamma)=a_\gamma/|a_\gamma|$. The conditions of part b) are satisfied and for the appropriate $n_\varphi\in[-N,N]$ we have

$$\forall\,\gamma\in E,\qquad \varepsilon>|\varphi(\gamma)-(\gamma,n_\varphi)|=|\varphi(\gamma)\overline{(\gamma,n_\varphi)}-1|\geqslant\big|1-\operatorname{Re}[\varphi(\gamma)\overline{(\gamma,n_\varphi)}\big|.$$

Thus,

$$\forall\,\gamma\in E,\qquad \operatorname{Re}[\varphi(\gamma)\overline{(\gamma,n_\varphi)}]>1-\varepsilon,$$

and so

$$(1-\varepsilon)\sum_{\gamma\in E}|a_\gamma|\leqslant\sum_{\gamma\in E}|a_\gamma|\operatorname{Re}[\varphi(\gamma)\overline{(\gamma,n_\varphi)}]\leqslant|\hat\mu(-n_\varphi)|.$$

This gives the result.

q.e.d.

The format for the following construction is to construct a perfect set $E\subseteq\mathbf{T}$ such that for each $T\in A'(E)$ there is a canonical procedure to write down a sequence $\{\mu_k:k=1,\ldots\}\subseteq M_f(E)$ for which

$$\text{(3.2.44)}\qquad \sup\|\mu_k\|_1<\infty$$

and

$$\text{(3.2.45)}\qquad \hat\mu_k\;\to\;\hat T\qquad\text{in the }\kappa\text{ topology.}$$

We conclude that $T\in M(E)$ by the Alaoglu theorem. Recall that if the norm in (3.2.44) were replaced by $\|\;\;\|_{A'}$ then (3.2.44) and (3.2.45) are equivalent to the statement that the sequence $\{\mu_k:k=1,\ldots\}$ converges to T in the β topology. We mention this now because of the discussion in Section 3.2.13 and Section 3.2.14.

Theorem 3.2.8. *There are non-empty perfect sets, $E\subseteq\mathbf{T}$ for which $M(E)=A'(E)$.*

Proof. i) We shall write $E=\bigcap E_n$, where $\{E_n:n=1,\ldots\}$ is determined inductively and each E_n is a disjoint finite union,

$$E_n=\bigcup_j E_j^n,$$

of compact neighborhoods E_j^n.
Take $E_1\subseteq\mathbf{T}$ to be any compact neighborhood and take $F_1\subseteq\operatorname{int}E_1$ to be a non-empty finite independent subset.

Assume we have $E_1,\ldots, E_{n-1}$ and $F_1,\ldots, F_{n-1}$. Let $F_n = \{\gamma_{j,n}: j = 1,\ldots\} \subseteq \operatorname{int} E_{n-1}$ have the properties that $F_{n-1} \subseteq F_n$, F_n is finite and independent, and $\operatorname{card}(F_n \cap E_j^{n-1}) = 2$ for each j.

We define E_n.

Choose $\varphi_{j,n} \in A(\mathbf{T})$ such that

$$\|\varphi_{j,n}\|_A < 2,$$

$$\varphi_{j,n} = 1 \text{ on a compact neighborhood } V_{j,n} \text{ and } \gamma_{j,n} \in \operatorname{int} V_{j,n},$$

$$\forall\, j \neq k, \qquad \operatorname{supp}\varphi_{j,n} \cap \operatorname{supp}\varphi_{k,n} = \emptyset,$$

and

$$\operatorname{supp}\varphi_{j,n} \subseteq \operatorname{int} E_{n-1}.$$

Take $\varepsilon_n > 0$ with the property that

$$\varepsilon_n(1 + 2\operatorname{card} F_n) < 1/n.$$

Let $N_n > n$ be the integer corresponding to F_n and $\varepsilon = 1/2$ in Theorem 3.2.7.

For each integer $m \in [-N_n, N_n]$ we choose $\psi_m \in A(\mathbf{T})$, as in Exercise 2.5.1b, such that

$$\psi_m = (\gamma_{j,n}, m) \text{ on a compact neighborhood } W_{j,n}$$

and

$$\|\psi_m - (\cdot, m)\|_A < \varepsilon_n.$$

Define

$$E_j^n = V_{j,n} \cap \left(\bigcap_{m=-N_n}^{N_n} W_{j,m} \right)$$

and

$$E_n = \bigcup_j E_j^n.$$

From our definition, E is a non-empty perfect set.

ii) Take $T \in A'(E)$. Preserving the notation of part i) we define

$$\mu_n = \sum_j \langle T, \varphi_{j,n} \rangle \delta_{\gamma_{j,n}}.$$

Thus, $\mu_n \in M(F_n)$ and

(3.2.46) $\quad \|\mu_n\|_1 \leqslant 2(\operatorname{card} F_n)\|T\|_{A'}.$

Set $T_n = T - \mu_n$.

iii) We prove that

(3.2.47) $\quad \forall\, m \in [-N_n, N_n], \qquad \langle T_n, \psi_m \rangle = 0.$

By definition

$$\forall\, m \in [-N_n, N_n] \text{ and } \forall\, \gamma \in E_n, \qquad \psi_m(\gamma) = \sum_j (\gamma_{j,n}, m)\,\varphi_{j,n}$$

and

$$\operatorname{supp} T_n \subseteq \operatorname{int} E_n.$$

Thus,

$$\langle T_n, \psi_m \rangle = \sum_j (\gamma_{j,n}, m) \langle T, \varphi_{j,n} \rangle - \sum_j \sum_k (\gamma_{j,n}, m) \langle T, \varphi_{k,n} \rangle\, \varphi_{j,n}(\gamma_{k,n}) = 0.$$

iv) We next prove that

(3.2.48) $\qquad \forall\, k \in \mathbf{Z}, \qquad \lim_n \hat{\mu}_n(k) = \hat{T}(k).$

From (3.2.47) we have

$$\forall\, m \in [-N_n, N_n], \qquad \hat{T}_n(m) = \hat{T}_n(m) - \langle T_n, \psi_m \rangle,$$

and so, from (3.2.46),

(3.2.49) $\qquad \forall\, m \in [-N_n, N_n], \qquad |\hat{T}_n(m)| = |\langle T_n, (\cdot, m) - \psi_m \rangle|\, \varepsilon_n \|T\|_{A'}$
$$\leqslant \varepsilon_n \|T\|_{A'} (1 + 2 \operatorname{card} F_n).$$

Fix $k \in \mathbf{Z}$. For any $n \geqslant |k|$ we have

$$|\hat{T}_n(k)| \leqslant \|T\|_{A'} / n$$

by (3.2.49) and since $N_n > n$.

v) Because F_n is independent we use Theorem 3.2.7 to estimate

$$\tfrac{1}{2} \|\mu_n\|_1 \leqslant \sup \{ |\hat{\mu}_n(m)| : m \in [-N_n, N_n] \}$$
$$\leqslant \|T\|_{A'} + \sup \{ |\hat{T}_n(m)| : m \in [-N_n, N_n] \}$$
$$\leqslant \|T\|_{A'} + \varepsilon_n \|T\|_{A'} (1 + 2 \operatorname{card} F_n),$$

where the last inequality follows by (3.2.49).

vi) The estimate of part v) yields the $\| \;\; \|_1$-norm boundedness of $\{\mu_n : n = 1, \ldots\}$. Consequently, $T \in M(E)$ by (3.2.48) and the Alaoglu theorem.

q.e.d.

Theorem 3.2.8 should be compared with [McGehee, 1, Section 7].

The above method leads to a very natural question: for any non-empty perfect set $E \subseteq \mathbf{T}$, does there exist a non-empty perfect subset $F \subseteq E$ for which $A'(F) = M(F)$?

Kaufman's method, referred to after Theorem 3.1.6′, yields a positive solution to the question, and, in fact, F can be chosen as a Kronecker set.

3.2.13 Bounded spectral synthesis. A closed set $E \subseteq \mathbf{T}$ is a set of *bounded synthesis*, a *bounded S-set*, if

$$\forall\, T \in A'(E)\ \exists\{\mu_n : n=1,\ldots\} \subseteq M_f(E) \qquad \text{such that}$$

$$\lim_n \mu_n = T \qquad \text{in} \qquad \sigma(A'(\mathbf{T}), A(\mathbf{T})).$$

Bounded S-sets are important from the point of view of actually constructing the process to approximate a synthesizable phenomenon, as well as bounding the $\|\ \|_{A'}$-norms of the approximating measures. If E is an S-set we can always approximate $T \in A'(E)$ by a directed system from $M_f(E)$ but we have no control over the $\|\ \|_{A'}$-norms of the elements from this system. Consequently, it is an interesting problem to determine whether or not a given S-set is a bounded S-set. In light of Theorem 2.2.1 and Theorem 3.2.9 it is worthwhile to note that this problem is "dual" to the "bounded spectral analysis" problem discussed in Section 2.2.7.

Theorem 3.2.9. *Let $E \subseteq \mathbf{T}$ be a compact set. E is a bounded S-set* $\Leftrightarrow$

$$\forall\, T \in A'(\mathbf{T})\ \exists\{\mu_n : n=1,\ldots\} \subseteq M_f(E) \qquad \textit{such that}$$

$$\lim \hat{\mu}_n = \hat{T} \textit{ in the } \beta \textit{ topology}$$

(*the β topology was defined in Section* 2.2.1).

Proof. ($\Rightarrow$) The fact, $\sup_n \|\mu_n\|_{A'} < \infty$, is a consequence of the Banach–Steinhaus theorem. By our hypotheses and properties of $\mathbf{Z}$, it is immediate that $\hat{\mu} \to \hat{T}$ in the κ topology.

($\Leftarrow$) This direction is clear by the Lebesgue dominated convergence theorem.

q.e.d.

The above result is true much more generally; the "$\Leftarrow$" direction is always trivial whereas the "$\Rightarrow$" direction can require a little more effort.

3.2.14 A characterization of bounded-S-sets. It turns out that there are closed sets $E \subseteq \mathbf{T}$ which are S-sets which are not bounded S-sets. The basic examples are given in [Varopoulos, 11, Theorem 3; Lindahl and Poulsen, 1, Chapters XI and XII by Varopoulos; Katznelson and McGehee, 2, Theorem VI].

In order to describe a closely related problem we define

$$\tilde{A}(E) = \left\{\varphi \in C(E) : \|\varphi\|_{\tilde{A}(E)} = \sup \frac{|\langle \mu, \varphi\rangle|}{\|\mu\|_{A'}} < \infty\right\},$$

where the supremum is taken over $M(E) \setminus \{0\}$ and where $E \subseteq \mathbf{T}$ is closed. $\tilde{A}(E)$ is a Banach space normed by $\|\ \|_{\tilde{A}(E)}$ and $A(E) \subseteq \tilde{A}(E)$.

Further,

Proposition 3.2.13. *Let $E \subseteq \mathbf{T}$ be closed and take $\varphi \in C(E)$. $\varphi \in \tilde{A}(E) \Leftrightarrow$ there is a sequence $\{\varphi_n : n = 1, \ldots\} \subseteq A(E)$ for which*

(3.2.50) $\quad \lim \varphi_n = \varphi, \quad$ *uniformly on E*

and

(3.2.51) $\quad \sup_n \|\varphi_n\|_{A(E)} = K < \infty.$

Proof. ($\Rightarrow$) Let Y be $A(E)$ along with all uniform limits of sequences $\{\varphi_n : n = 1, \ldots\} \subseteq A(E)$ for which (3.2.51) holds. Y is closed in $C(E)$.
Take $\varphi \in \tilde{A}(E)$. If $\varphi \in C(E) \setminus Y$ then there is $\mu \in M(E) \setminus \{0\}$ for which $|\langle \mu, \varphi \rangle| > 0$ and $\langle \mu, Y \rangle = 0$.
Since $e_n(\gamma) = e^{in\gamma}$ is an element of $A(E)$ we contradict the hypothesis that $\varphi \in \tilde{A}(E)$.
($\Leftarrow$) Let φ satisfy (3.2.50) and (3.2.51).
If $\varepsilon > 0$ and $\mu \in M(E)$ then for large n

$$|\langle \mu, \varphi \rangle| - \varepsilon \leqslant \left| \int \varphi_n \, d\mu \right| = |\langle \mu, \varphi_n \rangle| \leqslant \|\mu\|_{A'} \|\varphi_n\|_{A(E)} \leqslant K \|\mu\|_{A'}.$$

Thus, $|\langle \mu, \varphi \rangle| \leqslant K \|\mu\|_{A'}$ and K is independent of μ; hence $\varphi \in \tilde{A}(E)$.

q.e.d.

It is clear that $A(\mathbf{T}) = \tilde{A}(\mathbf{T})$. On the other hand, examples for which $A(E) \neq \tilde{A}(E)$ have been given by [Katznelson and McGehee, 1; Varopoulos, 10].

The relation of this topic to the problem of finding bounded S-sets is given by—

Theorem 3.2.10. *Let $E \subseteq \mathbf{T}$ be closed. $A(E)$ is closed in $\tilde{A}(E) \Leftrightarrow E$ is a bounded S-set.*

This result is a theorem in functional analysis and follows from a classical theorem in [Banach, 1, p. 213] (cf. [Dixmier, 1]).

3.2.15 The structure problem. Given $T \in A'(\mathbf{T})$. It can be quite difficult to determine whether or not $T \in A'_S(\mathbf{T})$. Until now any weakening of this problem has been on $\mathbf{T}$, e.g. by considering topologies weaker than the weak $*$ topology in which to approximate $\mathbf{T}$ with elements from $M(\operatorname{supp} T)$. Another way to simplify the problem and at the same time to pose the "structure" problem mentioned in the *Introduction* is to look more closely at the behavior of $\hat{T}$. The structure problem, then, is to investigate the relation between the behavior of $\hat{T}$ and the support properties of T (among other things); and, from our point of view, to gain information about the synthesizability of T in terms of the behavior of $\hat{T}$.

3.2.16 The balayage problem and spectral synthesis. In order to pose some specific problems in terms of our remark in Section 3.2.15 it is convenient for perspective to comment on Beurling's balayage problem.

Let $X \subseteq G$ and $E \subseteq \Gamma$ be closed. *Balayage is possible for X and E* if

$$\forall\, \mu \in M(G)\ \exists\, \nu \in M(X) \qquad \text{such that } \hat{\nu} = \hat{\mu} \text{ on } E.$$

The notion of balayage stems from Poincaré's balayage process in potential theory; and historically E was a collection of potential theoretic kernels. Assume that $E \subseteq \Gamma$ is a compact S-set and that

$$\forall\, \gamma \in E \qquad \text{and} \qquad \forall\, V \subseteq \Gamma,$$

a compact neighborhood of γ,

$$\exists\, \mu \in M_0(V \cap E);$$

then Beurling (1959–60) proved that *balayage is possible for such an E and any closed X if and only if*

$$\exists\, K(X,E) > 0 \textit{ such that } \forall\, \hat{T} = \Phi \in C_b(G) \textit{ for which } \operatorname{supp} T \subseteq E,$$

we can conclude that

$$\sup_{x \in G} |\Phi(x)| \leqslant K(X,E) \sup_{x \in X} |\Phi(x)|$$

(cf. [Beurling, 14; H. Landau, 1]). This led to the natural dual problem; find closed sets $X \subseteq G$ and $E \subseteq \Gamma$ for which it is possible that

(3.2.52) $\quad \forall\, \Phi \in C_b(G)\ \exists\, \hat{S} = \Psi \in C_b(G)$ for which $\operatorname{supp} S \subseteq E$, such that

$$\forall\, x \in X, \qquad \Phi(x) = \Psi(x).$$

We shall refer to (3.2.52) as the *Beurling interpolation problem* (cf. [Helson and Kahane, 2; Kahane, 13, Chapter 10.6–Chapter 10.11]).

3.2.17 The structure problem and spectral synthesis. Let $B(X,E)$ be the restriction to X of $\widehat{M(E)}$ and let $B(X) = B(X, \mathbf{T})$. Given $T \in A'(\mathbf{T})$.

In light of our remarks in Section 3.2.15 it is interesting to know where (in $\mathbf{Z}$) $\hat{T}$ might look like an element of $B(\mathbf{Z})$.

The main problem is to find those sets $X \subseteq \mathbf{Z}$ for which $\hat{T}$ is in the $\sigma(L^\infty(X), L^1(X))$ closure of $B(X, \operatorname{supp} T)$. This is the ordinary synthesis problem for $X = \mathbf{Z}$.

A simpler problem is the following. Given $T \in A'(\mathbf{T})$. Find the sets $X \subseteq \mathbf{Z}$ and the strongest possible topology T for which $\hat{T}$ is in the T closure of $B(X)$. Of course, if T is the $\sigma(L^\infty(X), L^1(X))$ topology then the approximation can be made for any $X \subseteq \mathbf{Z}$ since $\mathbf{T}$ is an S-set. It turns out that *$\hat{T}$ is in the uniform closure of $B(X)$ if and only if whenever $\{f_n : n = 1, \ldots\} \subseteq L^1(X)$ has the properties that*

(3.2.53) $\quad \sup_n \|f_n\|_1 < \infty$

and

(3.2.54) $$\lim_n \|\hat{f}_n\|_\infty = 0$$

then

$$\lim_n \sum_{m \in X} f_n(m)\hat{T}(m) = 0$$

[Hartman, 1] (cf. Exercise 2.2.5).

3.2.18 Refinements of Bochner's theorem and the structure problem. $\hat{T} \in L^\infty(\mathbf{Z})$ may not only be close to an element of $B(\mathbf{Z})$ on certain sets $X \subseteq \mathbf{Z}$ but it might happen that $\hat{T} \in B(X, E)$. Given $T \in A'(\mathbf{T})$. The problem is to find conditions on T, X, and E for which we can conclude that $\hat{T} \in B(X, E)$. Of course if $X = \mathbf{Z}$ and $E = \mathbf{T}$ then we have Bochner's theorem (cf. [Lumer, 1]).

A generalization of Bochner's theorem in one direction is the following, e.g. [Rosenthal, 2; Doss, 1]:

Theorem 3.2.11. *Given $X \subseteq \mathbf{Z}$. If there is $K > 0$ and a sequence $\{\mu_n : n = 1, \ldots\} \subseteq M_f(\mathbf{T})$ such that*

$$\sup_n \|\mu_n\|_1 \leqslant K$$

and

$$\forall\, m \in X, \qquad \lim_n \hat{\mu}_n(m) = \hat{T}(m)$$

then $\hat{T} \in B(X)$ and the corresponding measure μ (for which $\hat{\mu} = \hat{T}$ on X) has the property that $\|\mu\|_1 \leqslant K$.

In the other direction we have

Theorem 3.2.12. *Given a closed set $E \subseteq \mathbf{T}$ and $T \in A'(\mathbf{T})$. $\hat{T} \in B(\mathbf{Z}, E) \Leftrightarrow$ whenever $\{f_n : n = 1, \ldots\} \subseteq L^1(\mathbf{Z})$ has the properties that*

(3.2.55) $$\sup_n \|\hat{f}_n\|_\infty < \infty$$

and

(3.2.56) $$\forall\, \gamma \in E, \qquad \lim_{n \to \infty} \hat{f}_n(\gamma) = 0,$$

then

(3.2.57) $$\lim_n \sum_m f_n(m)\hat{T}(m) = 0.$$

Proof. ($\Rightarrow$) Let $\hat{T} \in B(\mathbf{Z}, E)$ and take $\{f_n : n = 1, \ldots\} \subseteq L^1(\mathbf{Z})$ which satisfies (3.2.55) and (3.2.56).

Since

$$\langle T, \hat{f}_n \rangle = \sum_m f_n(m)\hat{T}(m)$$

and $T \in M(E)$ we obtain (3.2.57) from (3.2.55) and (3.2.56).

($\Leftarrow$) i) We first prove that $\hat{T} \in B(\mathbf{Z})$.

By Bochner's theorem it is sufficient to prove that if the sequence of trigonometric polynomials,

$$\varphi_n(\gamma) = \sum_{j=1}^{k(n)} a_{j,n}(\gamma, m_{j,n}),\ n = 1, \ldots,$$

has the property that

$$\lim_{n \to \infty} \|\varphi_n\|_\infty = 0$$

then

(3.2.58) $$\lim_{n \to \infty} \sum_{k=1}^{k(n)} a_{j,n}\hat{T}(m_{j,n}) = 0.$$

We set

$$f_n = \sum_{j=1}^{k(n)} a_{j,n}\chi_{\{m_{j,n}\}},$$

so that $\hat{f}_n = \varphi_n$.

Thus, $\{f_n : n = 1, \ldots\}$ satisfies (3.2.55) and (3.2.56) and we conclude with (3.2.57); but the left-hand side of (3.2.57) is precisely the left-hand side of (3.2.58), and so $\hat{T} \in B(\mathbf{Z})$.

ii) Suppose $\operatorname{supp} T \nsubseteq E$ and let $\varphi \in A(\mathbf{T})$ have the property that $F \cap E = \emptyset$ and $\langle T, \varphi \rangle \neq 0$, where $F = \operatorname{supp} \varphi$.

Since $T \in M(\mathbf{T})$ we have $T(F) \neq 0$, where T is considered as a set function.

Let $\{U_n : n = 1, \ldots\}$ be a sequence of compact neighborhoods of F for which $U_n \cap E = \emptyset$ and let $\{\hat{f}_n = \varphi_n : n = 1, \ldots\} \subseteq A(\mathbf{T})$ be a sequence of functions for which $\operatorname{supp} \varphi_n \subseteq U_n$, $0 \leqslant \varphi_n \leqslant 1$, and $\varphi_n = 1$ on F.

Then, since $T \in M(\mathbf{T})$,

(3.2.59) $$\lim_{n \to \infty} \langle T, \varphi_n \rangle = T(F).$$

(3.2.55) and (3.2.56) are satisfied for $\{f_n : n = 1, \ldots\}$ and so, by (3.2.57),

$$\lim_{n \to \infty} \langle T, \varphi_n \rangle = 0.$$

This contradicts (3.2.59).

q.e.d.

There are more complicated proofs of Theorem 3.2.12; and the above proof, generalized to any G and Γ is in [Frieberg, 1].

The condition (3.2.56) can obviously be replaced by (3.2.54); and Theorem 3.2.12 should be compared with Hartman's result in Section 3.2.17.

3.2.19 Helson sets and the structure problem. The considerations of Section 3.2.18 lead to an important result concerning Helson sets in $\mathbf{Z}$. For perspective we begin by stating the well-known fact (e.g. [Rudin, 5, Section 5.73])

Proposition 3.2.14. *$X \subseteq \mathbf{Z}$ is an Helson set $\Leftrightarrow$*

$$\forall\, T \in A'(\mathbf{T}), \qquad \hat{T} \in B(X).$$

One direction of Proposition 3.2.14 does not require G to be discrete.

Proposition 3.2.15. *Let $X \subseteq G$ be closed and let $B(X)$ be the space of restrictions of $B(G)$ to X. If $C_b(X) = B(X)$ then X is an Helson set.*

Proof. We let $B(X)$ have the canonical quotient norm $\| \quad \|_{B(X)}$ from $B(G)$ (which in turn has the total variation norm from $M(\Gamma)$).

The identity map

$$B(X) \quad \to \quad C_b(X)$$

is a continuous bijection when $C_b(X)$ is given the sup norm topology.

Thus, by the open mapping theorem, there is $K > 0$ such that

$$(3.2.60) \qquad \forall\, \Phi \in C_b(X), \qquad \|\Phi\|_{B(X)} \leqslant K \|\Phi\|_\infty.$$

Take $\mu \in M(X) = C_0(X)'$. Then $C_0(X) \subseteq B(X)$, by hypothesis, and

$$(3.2.61) \qquad \|\mu\|_1 = \sup\left\{\left|\int_\Gamma \hat{\mu}(\gamma)\,\mathrm{d}\nu_\Phi(\gamma)\right| : \Phi \in C_0(X) \text{ and } \|\Phi\|_\infty \leqslant 1\right\},$$

where $\nu_\Phi \in M(\Gamma)$ has the property that

$$\forall\, x \in X, \qquad \hat{\nu}_\Phi(x) = \Phi(x).$$

If we choose ν_Φ so that $\|\nu_\Phi\|_1 \leqslant \|\Phi\|_{B(X)} + 1$, then (3.2.60) and (3.2.61) combine to give

$$\|\mu\|_1 \leqslant (K+1)\|\hat{\mu}\|_\infty.$$

q.e.d.

[Déchamps–Gondim, 1] has proved that *if $X \subseteq \mathbf{Z}$ is an Helson set and $E \subseteq \mathbf{T}$ is a closed set for which* $\operatorname{int} E \neq \emptyset$ *then*

$$\forall\, T \in A'(\mathbf{T}), \hat{T} \in B(X, E).$$

Exercises 3.2

3.2.1 A characterization of A'

Given the forms $T \sim \sum' c_n e^{in\gamma}$ and $F \sim \sum' \frac{c_n}{in} e^{in\gamma}$. Prove that $F \sim T \in A'$ (resp. $T \in M(\mathbf{T})$) if and only if

$$\text{(E3.2.1)} \qquad \sup \left\| \sum_{k=1}^{m} (\tau_{\gamma_k} F - \tau_{\lambda_k} F) \right\|_{2(\text{resp.}\,\infty)} < \infty,$$

where the supremum is taken over all finite sequences $\{[\lambda_k, \gamma_k] : k = 1, \ldots, m\}$ of non-overlapping intervals in $\mathbf{T}$. The details are carefully written out in [Edwards, 5, Section 16.5.3].

3.2.2 Hadamard measures

$F \in L^2(\mathbf{T})$, $\hat{F}(0) = 0$, is a function of *bounded deviation* if

$$\text{(E3.2.2)} \qquad \forall\, (a,b) \subseteq [0, 2\pi), \qquad \int_a^b F(\gamma)\, e^{in\gamma} d\gamma = O\left(\frac{1}{|n|}\right), \quad |n| \to \infty.$$

This notion was introduced by Hadamard (1892) to generalize the notion of bounded variation.

a) Fix $\alpha > 0$ and prove that

$$F(\gamma) = \gamma^\alpha \sin(1/\gamma^\alpha), \qquad \text{where } \gamma \in [0, 2\pi),$$

satisfies (E3.2.2) but that it does not have bounded variation. References and discussion of functions of bounded deviation are found in [Benedetto, 10].

b) Prove that if F is a function of bounded deviation then $F \in L^\infty(\mathbf{T})$. Note that there are $F \sim T \in A'$ for which $\operatorname{card} \operatorname{supp} T = \aleph_0$ but for which $F \notin L^\infty(\mathbf{T})$ (e.g. Exercise 3.2.3d).

$T \in A'$ is a *Hadamard measure* if F is a function of bounded deviation for $F \sim T \in A'$. What are the spectral synthesis properties of Hadamard measures?

3.2.3 Arithmetic progressions in the construction of pseudo-measures

a) Let $\{c_n : n = 1, \ldots\}$ be a decreasing sequence of non-negative numbers and assume that there is a constant K such that

$$\forall\, n \geqslant 1,\ nc_n \leqslant K.$$

Prove that

$$\forall\, \gamma \in \mathbb{R} \text{ and } \forall\, N, \qquad \left| \sum_{n=1}^{N} c_n \sin n\gamma \right| \leqslant K(\pi + 1).$$

This fact was used in Exercise 1.3.2. The bound K can be refined in some cases, e.g. [Katznelson, 5, p. 22].

b) Given $T \sim \sum_1^\infty c_n e^{in\gamma} \in A'(\mathbf{T})$, where $c_n \geqslant 0$.

Prove that if

$$\sum_1^\infty c_n/n$$

diverges then $T \in A'(\mathbf{T}) \setminus M(\mathbf{T})$. This fact is used in [Benedetto, 9].

c) Let $\lambda, \gamma \in \mathbf{T}$, where $\gamma \neq 0$. Define

$$\mu_N = \sum_{0<|n|\leqslant N} \frac{1}{n} \delta_{\lambda+n\gamma}.$$

Using part a) prove that

$$\|\mu_N\|_{A'} \leqslant 2(\pi+1).$$

Thus, *if a closed set* $E \subseteq \mathbf{T}$ *contains a sequence of finite arithmetic progressions* A_k, card $A_k = k$, *then* E *is not an Helson set*. Substantial refinements of this fact and the construction of elements in $A'(\mathbf{T}) \setminus M(\mathbf{T})$ are found in [Kahane and Salem, 2; 3] (e.g. see [Benedetto, 6, Section 4.3 and Section 7.2] for exposition and references). In fact, Kahane and Salem have shown that the Cantor set $C \subseteq \mathbf{T}$ is not an Helson set; and [Katznelson, 3] has proved, even further, that $C \subseteq \mathbf{T}$ is a set of analyticity (cf. our remarks on the Katznelson conjecture in Exercise 2.4.2).

d) Using c), construct $F \sim T \in A'$ where supp$T \subseteq E = \{0, \frac{1}{n} : n = 1, \ldots\} \subseteq [0, 2\pi)$ and $F \notin L^\infty(\mathbf{T})$. In particular, prove that E is not an Helson set. Note that $\{0, \frac{1}{2^n} : n = 1, \ldots\}$ is an Helson set, e.g. [Benedetto, 6, p. 112]. (*Hint*: A finite arithmetic progression in E can have the form $\{j/K! : j = 1, \ldots, k\}$, where K is fixed and $k \leqslant K$. Let $E_n \subseteq E$ be a finite arithmetic progression with $2M_n + 1$ terms such that if $\gamma \in E_{n+1}$ then $\gamma < \lambda$ for each $\lambda \in E_n$. Choose M_n so that

$$\sum_1^{M_n} \frac{1}{j} > n^3,$$

and let $\mu_n \in M(E_n)$ have mass 0 at the center of E_n and mass $1/j$ (resp. $-1/j$) at the jth point (of E_n) to the right (resp. to the left) of the center. By part c), $\|\mu_n\|_{A'} \leqslant 2(\pi + 1)$. If $G_n \sim \mu_n$ then $|G_n(\gamma)| = \sum_1^{M_n} 1/j$ for each γ in the two intervals contiguous to the center of E_n. Set

$$\nu_n = \sum_{j=1}^{n} \frac{1}{j^2}\mu_j \quad \text{and} \quad F_n = \sum_{j=1}^{n} \frac{1}{j^2} G_j.$$

Thus, $\|\nu_n\|_{A'} \leqslant \frac{\pi}{3}(\pi + 1)$ and for each $1 \leqslant j \leqslant n$,

$$\text{(E3.2.3)} \qquad |F_n(\gamma)| = |G_j(\gamma)|/j^2 \geqslant j$$

on the two intervals contiguous to the center of E_j. A subsequence of $\{\nu_n : n = 1, \ldots\}$ converges to $T \in A'(E)$ by the Alaoglu theorem and $\hat{T}(0) = 0$. It is easy to choose $\varphi_n \in A_E(\mathbf{T})$ so that $\overline{\lim}\,|\langle T, \varphi_n\rangle| = \infty$ and $0 \leqslant \varphi_n \leqslant 1$; thus $T \notin M(E)$. If $F \sim T$ then from (E3.2.3) we see that $F \notin L^\infty(\mathbf{T})$; and this, of course, also tells us that $T \notin M(E)$). With regard to the multiplication defined in part e) we see that if $T \in A' \setminus A_b$ and $\operatorname{supp} T$ is countable then $T^2 \notin A'(\mathbf{T})$; and it is not difficult to construct $T \in A_b$ for which $T^2 \notin A'(\mathbf{T})$.

e) Given a closed set $E \subseteq \mathbf{T}$ for which $mE = 0$. If $F \sim T \in A'$ and $G \sim S \in A'$ we define the distribution

$$TS = (FG)',$$

where the derivative is taken in the distributional sense (cf. Exercise 1.4.5). With this multiplication we know that *if $A' \cap A'(E)$ is a Banach algebra then E is an Helson set.* Verify this statement for the case $E = \{0, \frac{1}{n} : n = 1, \ldots\}$. (*Hint*: Adapt the example of part d). In this regard Ricci observed that *if $A' \cap A'(E)$ is a Banach algebra then $A' \cap A'(E) = A'_b \cap A'(E)$*; further, if the maximal ideal space of $A'(E)$ satisfies certain natural conditions then $A'(E)$ is very close to being $M(E)$ [Benedetto, 4]. Of course, Körner's example, mentioned in Section 1.3.13, tells us that there are Helson sets which are not sets of strong spectral resolution. [Benedetto, 4] has been extended in an interesting way by [Ricci, 1].

3.2.4 Non-pseudo-measures with regular primitives

Given a closed infinite set $E \subseteq \mathbf{T}$ for which $mE = 0$. Find

$$F = \sum_{1}^{\infty} k_j \chi_{I_j}$$

such that $\sum_{1}^{\infty} k_j$ converges and such that the distributional derivative $T = F'$ of F is not in $A'(E)$, e.g. [Benedetto, 5, Theorem 3.1]. Salmons has proved this result using the fact that the union of two countable Helson sets is an Helson set.

3.2.5 $A'_c(\mathbf{T})$ and derivatives of continuous functions

In Exercise 2.1.2 we gave some properties of the space $A'_c(\mathbf{T})$. We now continue this investigation (cf. Exercise 3.2.8).

a) If $\varphi \in A(\mathbf{T})$ then it is obviously not necessarily true that the distributional derivative,

φ', is in $A'(\mathbf{T})$. Prove that if $\varphi \in A(\mathbf{T})$ and $\varphi' = T \in A'(\mathbf{T})$ then $T \in A'_c(\mathbf{T})$. (*Hint*: Use only elementary facts about Cesàro summable series). The Hardy–Littlewood function (defined in Exercise 2.5.2) provides an example of a continuous function φ whose distributional derivative is an element of $A'(\mathbf{T}) \setminus A'_c(\mathbf{T})$.

b) Analogous to our treatment in $A'(\mathbf{R})$ we define $\operatorname{supp}_W T$, for $T \in A'(\mathbf{T})$, to be ZI where $\hat{f} = \varphi \in I$ if and only if

$$\lim_{N \to \infty} \frac{1}{2N+1} \sum_{-N}^{N} |\hat{T} * f(n)| = 0.$$

Prove that $T \in A'(\mathbf{T})$ is an element of $A'_c(\mathbf{T})$ if and only if $\operatorname{supp}_W T = \emptyset$. This fact is not true for $\Gamma = \mathbf{R}$ as we showed in Example 2.1.1b.

c) In the opposite direction to part a) note that (E1.3.4) defines a discontinuous function F, such that $F \sim T \in A'$ and $T \in A'_c(\mathbf{T})$. Find a non-trivial subclass $X \subseteq BV(\mathbf{T})$ such that if $\varphi' = T$ is the distributional derivative of φ then

$$\forall\, \varphi \in X, \qquad \varphi \in C(\mathbf{T}) \;\Leftrightarrow\; T \in A'_c(\mathbf{T}).$$

d) Let X be the class of Fourier series, $\varphi \sim \sum_1^{\infty} f(n) \sin n\gamma$, where $\{f(n) : n = 1, \ldots\}$ decreases to 0. Prove that if $\varphi' = T$ is the distributional derivative of φ then

$$\forall\, \varphi \in X, \qquad \varphi \in C(\mathbf{T}) \;\Leftrightarrow\; T \in A'_0(\mathbf{T}).$$

3.2.6 $A(\mathbf{T}) = L^1(\mathbf{T}) * A(\mathbf{T})$

Let $L^1_+(\mathbf{T}) = \{\varphi \in L^1(\mathbf{T}) : \varphi \geqslant 0 \text{ a.e.}\}$. Prove that

$$A(\mathbf{T}) = L^1_+(\mathbf{T}) * A(\mathbf{T}). \tag{E3.2.4}$$

(*Hint*: Use Cohen's factorization theorem for compact spaces. (E3.2.4) can also be proved classically using properties of convex functions).

3.2.7 An element of $A'(\mathbf{T}) \setminus M(\mathbf{T})$ and its Wiener support

Define

$$\forall\, \varphi \in C^1(\mathbf{T}), \qquad \langle T, \varphi \rangle = \int_{\mathbf{T}} \frac{\varphi(\gamma) - \varphi(0)}{\sin \gamma}\, d\gamma.$$

a) Prove that $T \in A'(\mathbf{T}) \setminus M(\mathbf{T})$. (*Hint*: First calculate $\hat{T}(n)$ in terms of

$$\frac{1}{2\pi} \int_{-\pi}^{\pi} \frac{\sin n\gamma}{\gamma}\, d\gamma$$

plus a term which tends to 0 as $|n| \to \infty$; thus $T \in A'(\mathbf{T})$. If $T \in M(\mathbf{T})$ we have that

$$\exists\, K > 0 \text{ such that } \forall\, (a, b) \subseteq (0, \pi), \qquad \int_a^b \frac{d\gamma}{\sin \gamma} \leqslant K. \tag{E3.2.5}$$

On the other hand

$$\int_a^b \frac{d\gamma}{\sin\gamma} \sim 2\log\frac{b}{a}, \qquad \text{as } a, b \to 0.$$

Letting $b \to 0$ exponentially and $a \to 0$ polynomially we obtain a contradiction to (E3.2.5)).

b) Prove that $\operatorname{supp} T = \mathbf{T}$ whereas $\operatorname{supp}_W T = \{0\}$ (cf. [Benedetto, 11], where general criteria for this phenomena are given).

3.2.8 Sequence spaces and a strong form of uniqueness

Let Ω be the set of non-zero even functions, $\omega: \mathbf{Z} \to \mathbf{R}^+$ such that $\{\omega(|n|): n = 0, 1, \ldots\}$ is decreasing and $\sum \omega(n) < \infty$; if $\omega \in \Omega$ and $\sum \omega(n) = 1$ we write $\omega \in \Omega_0$. Define

$$\forall\, \omega \in \Omega_0 \qquad L^1_\omega = \{\Phi: \mathbf{Z} \to \mathbf{C}: \sum |\Phi(n)|\omega(n) < \infty\}$$

(cf. [Beurling, 12] and the weighted spaces defined in Section 1.4.3),

$$\forall\, \Phi: \mathbf{Z} \to \mathbf{C}, \qquad \|\Phi\|^\Omega = \sup_{\omega \in \Omega_0} \sum |\Phi(n)|\omega(n),$$

and the Banach space,

$$B = \{\Phi: \mathbf{Z} \to \mathbf{C}: \|\Phi\|^\Omega < \infty\}.$$

Recall the definition of $\|T\|^\infty$ in Exercise 2.1.2; because of this, define

$$\forall\, \Phi: \mathbf{Z} \to \mathbf{C}, \qquad \|\Phi\|^\infty = \sup_N \frac{1}{2N+1} \sum_{-N}^{N} |\Phi(n)|.$$

a) Prove that $\|\Phi\|^\Omega < \infty$ if and only if $\|\Phi\|^\infty < \infty$.

b) Prove that

$$B = \cap \{L^1_\omega: \omega \in \Omega_0\} \supseteq L^\infty(\mathbf{Z}).$$

(*Hint*: Ω is a complete metric space with metric $\rho(\omega_1, \omega_2) = \sum |\omega_1(n) - \omega_2(n)|$. Let $\Phi \in \cap \{L^1_\omega: \omega \in \Omega_0\}$.

To prove $\Phi \in B$. Define

$$\Phi_n(m) = \begin{cases} \Phi(m) & \text{for } |\Phi(m)| \leqslant n \\ n & \text{for } |\Phi(m)| > n. \end{cases}$$

Set

$$L_n(\omega) = \sum_m |\Phi_n(m)|\omega(m)$$

and

$$L(\omega) = \sup_n L_n(\omega).$$

Note that $L(\omega) = \sum |\Phi|\omega$ so that L is lower semicontinuous. By Osgood's theorem there is a ball $B(\omega_0, r) \subseteq \Omega$ such that

$$K = \sup_{\omega \in B(\omega_0, r)} \sum |\Phi|\omega < \infty.$$

The real vector space generated by Ω is complete and $B(\omega_0, r) - \omega_0$ is a neighborhood of 0. Thus,

$$\|\Phi\|^{\Omega} \leqslant 4K/r).$$

c) Prove that $\Phi \in B \setminus L^{\infty}(\mathbf{Z})$ where

$$\Phi(n) = \begin{cases} 0 & \text{for } n \leqslant 0 \\ 0 & \text{for } n \neq 2^m \\ 2^m & \text{for } n = 2^m. \end{cases}$$

d) Prove that $L^{\infty}(\mathbf{Z})$, normed by $\| \quad \|^{\infty}$, is not complete. (*Hint*: If the result were true then by the open mapping theorem, $\| \quad \|^{\infty}$ and $\| \quad \|_{A'}$ are equivalent norms on $A'(\mathbf{T})$; and thus by Exercise 2.1.2 $A'_c(\mathbf{T}) = A'_0(\mathbf{T})$, a contradiction).

e) Prove that if $A'(E)$, normed by $\| \quad \|^{\infty}$, is complete then E is a U-set, e.g. [Benedetto, 10]. If E is finite then $A'(E)$ normed by $\| \quad \|^{\infty}$ is complete; are there infinite sets E for which $A'(E)$ normed by $\| \quad \|^{\infty}$ is complete?

f) Show that $\overline{L^{\infty}(\mathbf{Z})} \neq B$, taken with the $\| \quad \|^{\infty}$-norm. (*Hint*: Use the example of part c), assume that

$$\lim_{n \to \infty} \|\Phi - \Phi_n\|^{\infty} = 0$$

for some $\{\Phi_n : n = 1, \ldots\} \subseteq L^{\infty}(\mathbf{Z})$, and obtain a contradiction).

Norms such as $\| \quad \|^{\infty}$ have been used to study spectral synthesis problems by [Benedetto, 11] and [Beurling, 12].

3.2.9 A class of synthesizable pseudo-measures

Given $T \in A'(\mathbb{R})$ and assume that there is $x > 0$ such that $\hat{T}$ has bounded variation on $(-\infty, -x) \cup (x, \infty)$. Prove that $T \in A'_S(\mathbb{R})$, e.g. [Atzmon, 4].

3.2.10 A definition of integral for spectral synthesis

Let $F \in C(\mathbf{T})$ be real-valued. Define

$$F_{\varepsilon}(\gamma) = \sup\{ \inf_{\alpha \in (0, \frac{1}{2})} m(I, \alpha)\}$$

where the supremum is taken over all closed intervals $I \subseteq \mathbf{T}$ in $\mathfrak{R}$ for which $\gamma \in I$ and $mE = \varepsilon$ ($m(I, \alpha)$ was defined in Section 3.2.8). Since $m(I, \alpha_1) \leqslant m(I, \alpha_2)$ when $\alpha_1 < \alpha_2$ we have

$$F_\varepsilon(\gamma) = \sup \{\lim_{\alpha \to 0} m(I, \alpha)\}.$$

a) If $F = k \in \mathbf{R}$ on an open interval I prove that $F_\varepsilon = k_\varepsilon \leqslant k$ on I.
b) Given $F \sim T \in A_b'$. If $F_\varepsilon \sim T_\varepsilon$ prove that supp $T_\varepsilon \subseteq$ supp$T$ and

$$\lim_{\varepsilon \to 0} \|F - F_\varepsilon\|_\infty = 0.$$

3.2.11 A property of $A(\mathbf{T})$
a) Let $\varphi \in A(\mathbf{T})$ have the property that

$$\forall\ F \in A(\mathbf{T}), \qquad F \circ \varphi \in A(\mathbf{T})$$

(cf. Exercise 2.4.4). Prove that

$$\sup_n \|e^{in\varphi}\|_A < \infty.$$

(*Hint*: Define the map

$$\begin{array}{ccc} A(\mathbf{T}) & \to & A(\mathbf{T}) \\ F & \mapsto & F \circ \varphi \end{array}$$

and use the closed graph theorem).
b) On the other hand, prove that the following property can not be valid for a given closed set $E \subseteq \mathbf{T}$, e.g. [Benedetto, 8]: there is $K > 0$ such that for each $n \in \mathbf{Z}$ we can find a real-valued element $\varphi_n \in A(\mathbf{T})$ and a finite disjoint union N_n of closed intervals covering E for which

$$\forall\ n, \qquad \|\varphi_n\|_A < K$$

and

$$\forall\ \gamma \in N_n, \qquad e^{in\gamma} = e^{i\varphi_n(\gamma)}.$$

3.2.12 Non-synthesizable convergence criteria in $A(\mathbf{T})$
a) Let $\hat{f} = \varphi \in C(\mathbf{T})$ be absolutely continuous and assume that $\varphi' \in L^2(\mathbf{T})$. Prove that $\varphi \in A(\mathbf{T})$ and

$$\|\varphi\|_A \leqslant \|\varphi\|_1 + \sqrt{\frac{\pi}{3}}\, \|\varphi'\|_2.$$

(*Hint*: Note that $|f(0)| \leqslant \int |\varphi|$ and

$$\|\varphi'\|_2 = (\sum |n f(n)|^2)^{1/2};$$

apply Hölder's inequality).

b) Find $T \in A'(\mathbf{T})$ and $\{\varphi_j : j = 1, \ldots\} \subseteq C^\infty(\mathbf{T})$ such that

$$\lim \|\varphi_j\|_\infty = 0,$$

$$\forall\, j, \varphi_j' = 0 \text{ on a neighborhood of } \operatorname{supp} T,$$

$$\lim \|\varphi_j'\|_1 = 0, \tag{E.3.2.6}$$

and

$$\varliminf_{j \to \infty} |\langle T, \varphi_j \rangle| > 0.$$

In light of part a), if we used the L^2-norm in (E3.2.6) then we would have to conclude that $\lim \langle T, \varphi_j \rangle = 0$. (*Hint*: Use Exercise 3.2.3d to form $F \sim T \in A' \setminus A_b'$, where $F = \sum k_j \chi_{I_j}$, and choose φ_j to have the value 0 or $1/k_{n_j}$ except for two small intervals, one of which is an open set in $\left(\frac{1}{n_j + 1}, \frac{1}{n_j}\right)$).

c) Prove that for each $\varphi \in A(\mathbf{T})$ there is a sequence $\{\varphi_j : j = 1, \ldots\} \subseteq C^\infty(\mathbf{T})$ for which

$$\lim \|\varphi - \varphi_j\|_A = 0$$

and

$$\lim \varphi_j' = 0 \text{ a.e.}$$

References

[1] Agmon, S.: Complex variable Tauberians. TAMS **74** (1953) 444–481.

[1] Agmon, S. and Mandelbrojt, S.: Une généralization du théorème Tauberien de Wiener. Acta Sci. Math. Szeged **12** (1950) Part B, 167–176.

[1] Akutowicz, E. J. (editor): L'analyse harmonique dans le domaine complexe. Berlin–Heidelberg–New York 1972. Lecture notes, Vol. 336.

[1] Amemiya, I. and Itô, T.: A simple proof of a theorem of P. J. Cohen. BAMS **70** (1964) 774–776.

[1] Arens, R. and Calderón, A. P.: Analytic functions of several Banach algebra elements. Ann. of Math. **62** (1955) 204–216.

[1] Argabright, L. N.: On the mean value of a weakly almost periodic function. PAMS **36** (1972) 315–316.

[1] Arsac, J.: Fourier transforms and the theory of distributions. New York 1966.

[1] Atzmon, A.: Spectral synthesis in regular Banach algebras. Israel J. Math. **8** (1970) 197–212.

[2] —: Non-finitely generated closed ideals in group algebras. J. Fun. Anal. **11** (1972) 231–249.

[3] —: Translation invariant subspaces of $L^p(G)$.

[4] —: Spectral synthesis of functions of bounded variation.

[1] Banach, S.: Opérations linéaires. New York 1955.

[1] Bary, N.: The uniqueness problem of the representation of functions by trigonometric series. AMS Translations Series 1, **3** (1962) 107–195 (the article was originally published in 1949).

[2] —: Trigonometric series. New York 1964.

[1] Benedetto, J.: The Laplace transform of generalized functions. Can. J. Math. **18** (1966) 357–374.

[2] —: Harmonic synthesis and pseudo-measures. U. of Maryland mathematics lecture notes, **5** (1968) 1–316.

[3] —: A strong form of spectral resolution. Ann. di Mat. **86** (1970) 313–324.

[4] —: Support preserving measure algebras and spectral synthesis. Math. Z. **118** (1970) 271–280.

[5] —: Trigonometric sums associated with pseudo-measures. Ann. SNS, Pisa **25** (1971) 230–248.

[6] —: Harmonic analysis on totally disconnected sets. Berlin–Heidelberg–New York 1971. Lecture notes, Vol. 202.

[7] —: Construction de fonctionnelles multiplicatives discontinues sur des algèbres métriques. CRAS, Paris **274** (1972) 254–256.

[8] —: A support preserving Hahn–Banach property to determine Helson *S*-sets. Inv. Math. **16** (1972) 214–228.

[9] —: Idele characters in spectral synthesis on $\mathbf{R}/2\pi\mathbf{Z}$. Ann. Inst. Fourier **23** (1973) 45–64.
[10] —: Pseudo-measure energy and spectral synthesis. Can. J. Math. **26** (1974) 985–1001.
[11] —: The Wiener spectrum in spectral synthesis. Stu. in Applied Math. (MIT) 54 (1975).
[12] —: Real variable and integration. Stuttgart 1975.
[1] Bernard, A. and Varopoulos, N. Th.: Groupes de fonctions continues sur un compact. Applications à l'étude des ensembles de Kronecker. Studia Math. **35** (1970) 199–205.
[1] Berry, A. C.: Necessary and sufficient conditions in the theory of Fourier transforms. Ann. of Math. **32** (1931) 830–838.
[1] Beurling, A.: Sur les intégrales de Fourier absolument convergentes et leur applications à une transformation fonctionnelle. 9th Congrès Math. Scand. (1938) 345–366.
[2] —: Un théorème sur les fonctions bornées et uniformement continues sur l'axe réal. Acta Math. **77** (1945) 127–136.
[3] —: Sur les spectres des fonctions. Colloque sur l'analyse harmonique. Nancy (1947) 9–29.
[4] —: Sur la composition d'une fonction sommable et d'une fonction bornée. CRAS, Paris 255 (1947) 274–275.
[5] —: Sur une classe de fonctions presque-périodiques. CRAS, Paris **225** (1947) 326–328.
[6] —: On the spectral synthesis of bounded functions. Acta Math. **81** (1949) 225–238.
[7] —: On two problems concerning linear transformations in Hilbert space. Acta Math. **81** (1949) 239–255.
[8] —: On a closure problem. Ark. Mat. **1** (1950) 301–303.
[9] —: A theorem on functions defined on a semi-group. Math. Scand. **1** (1953) 127–130.
[10] —: A closure problem related to the Riemann zeta function. Proc. Nat. Acad. Sci. **41** (1955) 312–314.
[11] —: Analyse spectrale de pseudo mesures. CRAS, Paris **258** (1964) 406–409, 782–785, 1380–1382, 1984–1987, 2959–2962, 3423–3425.
[12] —: Construction and analysis of some convolution algebras. Ann. Inst. Fourier **14** (1964) 1–32.
[13] —: A critical topology in harmonic analysis on semigroups. Acta Math. **112** (1964) 215–228.
[14] —: Local harmonic analysis with some applications to differential operators. Belfer Graduate School of Science. Annual Science, Conference Proceedings (1966) 109–125.
[1] Beurling, A. and Helson, H.: Fourier–Stieltjes transforms with bounded powers. Math. Scand. **1** (1953) 120–126.
[1] Beurling, A. and Malliavin, P.: On the closure of characters and zeros of entire functions. Acta Math. **118** (1967) 79–93.
[1] Blanchard, A.: Théorie analytique des nombres premiers. Paris 1969.
[1] Bochner, S.: Lectures on Fourier integrals. Ann. of Math. Studies, Princeton, New York 1959.
[1] Bochner, S. and Chandrasekharan, K.: Fourier transforms. Princeton, New York 1941.
[1] Bochner, S. and Hardy, G. H.: Notes on two theorems of Norbert Wiener. J. Lond. Math. Soc. **1** (1926) 240–244.
[1] Bombieri, E.: Sulle formule di A. Selberg generalizzate per classi di funzioni aritmetiche e le applicazioni al problema del resto nel "Primzahlsatz". Riv. Mat. Univ. Parma **3** (1962) 393–440.
[1] Bourbaki, N.: Intégration. Chapitres 1–4. 2nd edition Paris 1965.

[1] Brainerd, B. and Edwards, R.E.: Linear operators which commute with translation: Part I, Representation theorems. J. Australian Math. Soc. **6** (1966) 289–327.
[1] Buck, R. C.: Operator algebras and dual spaces. PAMS **3** (1952) 681–687.
[2] —: Bounded continuous functions on locally compact spaces. Mich. Math. J. **5** (1958) 95–104.
[1] Burckel, R. B.: Weakly almost periodic functions on semigroups. New York 1970.
[1] Calderón, A. P.: Ideals in group algebras. Symposium on Harmonic Analysis and Related Integral Transforms. Cornell U. 1956.
[1] Cameron, R. H. and Wiener, N.: Convergence properties of analytic functions of Fourier–Stieltjes transforms. TAMS **46** (1939) 97–109.
[1] Carleman, T.: L'intégrale de Fourier. Institut Mittag–Leffler, Uppsala 1944.
[1] Carleson, L.: Sets of uniqueness for functions regular in the unit circle. Acta Math. **871** (1952) 325–345.
[2] —: On convergence and growth of partial sums of Fourier series. Acta Math. **116** (1966) 135–157.
[1] Chandrasekharan, K.: Analytic number theory. Berlin–Heidelberg–New York 1968.
[1] Chauve, M. Sur les séries de Taylor absolument convergentes. CRAS, Paris **268** (1969) 1384–1386.
[1] Cohen, P. J.: Factorization in group algebras. Duke Math. J. **26** (1959) 199–205.
[2] —: On a conjecture of Littlewood and idempotent measures. Amer. J. Math. **82** (1960) 191–212.
[3] —: On homomorphisms of group algebras. Amer. J. Math. **82** (1960) 213–226.
[4] —: A note on constructive methods in Banach algebras. PAMS **12** (1961) 159–163.
[5] —: Idempotent measures and homomorphisms of group algebras. Proc. Int. Cong. Math. (1962) 331–336.
[1] Collins, H. S.: On the space $l^\infty(S)$, with the strict topology. Math. Zeit. **106** (1968) 361–373.
[1] Conway, J. B.: The strict topology and compactness in the space of measures. TAMS **126** (1967) 474–486.
[1] Cotlar, M.: On a theorem of Beurling and Kaplansky. Pac. J. Math. **4** (1954) 459–465.
[1] Cramér, H.: On the representation of a function by certain Fourier integrals. TAMS **46** (1939) 191–201.
[1] Cuccia, C. L.: Harmonics, sidebands, and transients in communication engineering. New York 1952.
[1] Curtis, P. C. and Figá-Talamanca, A.: Factorization theorems for Banach algebras. Function Algebras. Glenview, Ill. 1966. 169–185.
[1] Davenport, H.: On a theorem of P. J. Cohen. Mathematika **7** (1960) 93–97.
[1] Déchamps-Gondim, M.: Ensembles de Sidon topologiques. Ann. Inst. Fourier **22** (1972) 51–79.
[1] Delange, R. The converse of Abel's theorem on power series. Ann. of Math. **50** (1949) 94–109.
[1] De Leeuw, K. and Herz, C.: An invariance property of spectral synthesis. Illinois J. Math. **9** (1965) 220–229.
[1] Denjoy, A.: Sur la definition riemannienne de l'intégrale de Lebesgue. CRAS, Paris **193** (1931) 695–698.
[1] De Vito, C.: Characterization of those ideals in $L^1(\mathbf{R})$ which can be synthesized. Math. Ann. **203** (1973) 171–173.

[1] Dieudonné, J.: Sur le produit de composition (I). J. Pures App. Math. **39** (1960) 275–292.
[1] Dietrich, Jr., W. E.: On the ideal structure of Banach algebras. TAMS **169** (1972) 59–74.
[1] Ditkin, V.: On the structure of ideals in certain normed rings. Uchen. Zap. Mosk. Gos. Univ. Matem. **30** (1939) 83–130. (Math. Reviews **1** (1940) 336.)
[1] Dixmier, J.: Sur un théorème de Banach. Duke Math. J. **15** (1948) 1057–1071.
[2] —: Quelques exemples concernant la synthèse spectrale. CRAS, Paris **247** (1958) 24–26.
[1] Domar, Y.: Harmonic analysis based on certain commutative Banach algebras. Acta Math. **96** (1956) 1–66.
[2] —: On spectral analysis in the narrow topology. Math. Scand. **4** (1956) 328–332.
[3] —: On the uniqueness of minimal extrapolations. Ark. Mat. **4** (1960) 19–29.
[4] —: Some results on narrow spectral analysis. Math. Scand. **20** (1967) 5–18.
[1] Donoghue, Jr., W. F.: Distributions and Fourier transforms. New York 1969.
[1] Doss, R.: Approximations and representations for Fourier transforms. TAMS **153** (1971) 211–221.
[1] Drury, S. W.: Sur la synthèse harmonique. CRAS, Paris **271** (1970) 42–44.
[2] —: Sur les ensembles de Sidon. CRAS, Paris **271** (1971) 162–163.
[1] Dunford, N. and Schwartz, J.: Linear operators. New York.
[1] Dunkl, C. and Ramirez, D.: Topics in harmonic analysis. New York 1971.
[1] Dyson, F. J.: Fourier transforms of distribution functions. Can. J. Math. **5** (1953) 554–558.
[1] Eberlein, W.: Spectral theory and harmonic analysis. Proc. Symposium on Spectral Theory and Differential Problems. Oklahoma A and M, 1955.
[2] —: Characterization of Fourier–Stieltjes transforms. Duke Math. J. **22** (1955) 465–468.
[3] —: The point spectrum of weakly almost periodic functions. Mich. Math. J. **3** (1955–1956) 137–139.
[1] Edwards, R. E.: On factor functions. Pac. J. Math. **5** (1955) 367–378.
[2] —: Comments on Wiener's Tauberian theorems. J. Lond. Math. Soc. **33** (1958) 462–466.
[3] —: Supports and singular supports of pseudo-measures. J. Australian Math. Soc. **6** (1966) 65–75.
[4] —: Uniform approximation on non-compact spaces. TAMS **122** (1966) 249–276.
[5] —: Fourier series. New York 1967.
[1] Eggleston, H. G.: The Bohr spectrum of a bounded function. PAMS **9** (1958) 328–332.
[1] Ehrenpreis, L.: Mean-periodic functions I. Amer. J. Math. **77** (1955) 293–328.
[2] —: Appendix to the paper "Mean-periodic functions I." Amer. J. Math. **77** (1955) 731–733.
[1] Eisen, M. and Gindler, H.: On the spectrum of a Laurent form. Amer. Math. Monthly. Papers in Analysis **731** (1966) 79–90.
[1] Elliott, R. J.: Some results in spectral synthesis. Proc. Camb. Phil. Soc. **61** (1965) 395–423.
[2] —: Two notes on spectral synthesis for discrete abelian groups. Proc. Camb. Phil. Soc. **61** (1965) 617–620.
[1] Eymard, P.: Spectral synthesis for multipliers. Harmonic Analysis Symposium, Warwick 1968.
[2] —: Algèbras A_p et convoluteurs de L^p. Sém. Bourbaki, Exp. 367. Berlin–Heidelberg–New York 1971. Lecture notes, Vol. 180.
[1] Fefferman, C.: Pointwise convergence of Fourier series. Ann. of Math. **98** (1973) 551–571.

[1] Fefferman, C. and Stein, E.: H^p spaces of several variables. Acta Math. **129** (1972) 137–193.
[1] Feldman, G. M.: Spectral theory of representations of locally compact abelian groups. Fun. Anal. and Appl. **6** (1972) 82–83.
[1] Figà-Talamanca, A.: Translation invariant operators in L^p. Duke Math. J. **32** (1965) 459–502.
[1] Figà-Talamanca, A. and Gaudry, G.: Multipliers of L^p which vanish at infinity. J. Fun. Anal. **7** (1971) 475–486.
[1] Filippi, M.: Variété de non-synthèse spectrale sur un groupe abélien localement compact. Israel J. Math. **3** (1965) 43–60.
[1] Frieberg, S. H.: Functions which are Fourier–Stieltjes transforms. PAMS **28** (1971) 451–542.
[1] Fukamiya, M.: Topological method for Tauberian theorem. Tôhoku Math. J. **1** (1949) 77–87.
[1] Ganelius, T.: General and special Tauberian remainder theorems. 13th Congrès Math. Scand (1957) 102–103.
[2] —: Tauberian remainder theorems. Berlin–Heidelberg–New York 1971. Lecture notes, Vol. 232.
[1] Garsia, A. M.: A new proof of Beurling's theorem on the spectrum of bounded functions on the real axis. J. Math. Anal. and Appl. **7** (1963) 436–439.
[1] Gelfand, I.: On absolutely convergent trigonometrical series and integrals. Dokl. Akad. Nauk SSSR **25** (1939) 570–572.
[2] —: Normierte Ringe. Mat. Sbornik (N.S.) **9** (1941) 3–24.
[1] Gelfand, I., Raikov, D. and Shilov, G.: Commutative normed rings. New York 1964.
[1] Gelfand, I. and alii: Generalized functions. New York.
[1] Gilbert, J. E.: Spectral synthesis problems for invariant subspaces on groups II. Function Algebras. Glenview, Ill. 1966, 257–264.
[2] —: On a strong form of spectral synthesis. Ark. Mat. **7** (1969) 571–575.
[1] Glaeser, G.: Synthèse spectrale des idéaux de fonctions lipschitziennes. CRAS, Paris **260** (1965) 1539–1542.
[1] Glasser, M. L.: A note on the Littlewood–Tauber theorem. PAMS **20** (1969) 34–40.
[1] Glicksberg, I.: When is $\mu * L^1$ closed? TAMS **160** (1971) 419–425.
[1] Glicksberg, I. and Wik, I.: The range of Fourier–Stieltjes transforms of parts of measures. Berlin–Heidelberg–New York 1972. Lecture notes, Vol. 266.
[1] Godement, R.: Extension à un groupe abélien quelconque des théorèmes taubériens de N. Wiener et d'un théorème de A. Beurling. CRAS, Paris **223** (1946) 16–18.
[2] —: Théorèmes taubériens et théorie spectrale. Ann. Sci. École Norm. Sup. **64** (1947) 119–138.
[1] Godin, G. The analysis of tides. Toronto 1972.
[1] Goldberg, R. R. and Simon, A. B.: Characterization of some classes of measures. Acta Sci. Math. Szeged. **27** (1966) 157–161.
[1] Golding, J.: Cubism: a history and analysis 1907–1914. New York 1959.
[1] Goldman, S.: Information theory. New York 1953.
[1] Graham, C.: Symbolic calculus for algebras of Fourier–Stieltjes transforms. PAMS **23** (1969) 311–314.
[2] —: Compact independent sets and Haar measure. PAMS **36** (1972) 578–582.
[3] —: Sur une théorème de Katznelson et McGehee. CRAS, Paris **276** (1973) 37–40.

[4] —: The Fourier transform is onto only when the group is finite. PAMS **38** (1973) 365–366.
[5] —: Interpolation sets for convolution measure algebras. PAMS **38** (1973) 512–522.
[1] Grothendieck, A.: Sur certains sous-espaces vectoriels de L^p. Can. J. Math. **6** (1954) 158–160.
[1] Halmos, P.: Measure theory. New York 1950.
[1] Handelsman, R. A. and Lew, J. S.: On the convergence of the L^p norm to the L^∞ norm. Amer. Math. Monthly **79** (1972) 618–622.
[1] Hankinson, M.: The unusual classical synthesizer, ABC Records 1972.
[1] Hardy, G. H.: Divergent series. Oxford 1949.
[1] Hardy, G. H. and Littlewood, J. E.: Tauberian theorems concerning power series and Dirichlet series whose coefficients are positive. Proc. Lond. Math. Soc. **13** (1913) 174–191.
[1] Hartman, S.: The method of Grothendieck–Ramirez and weak topologies in $C(\mathbf{T})$. Studia Math. **44** (1972) 181–197.
[1] Hedberg, T.: Distributions with bounded potentials and absolutely convergent Fourier series. Ark. Mat. **10** (1972) 49–57.
[1] Helson, H.: Spectral synthesis of bounded functions. Ark. Mat. **1** (1951) 497–502.
[2] —: On the ideal structure of group algebras. Ark. Mat. **2** (1952) 83–86.
[3] —: Isomorphisms of abelian group algebras. Ark. Mat. **2** (1953) 475–487.
[4] —: Lectures on invariant subspaces. New York 1964.
[1] Helson, H. and Kahane, J.-P.: Sur les fonctions opérant dans les algèbres de transformées de Fourier de suites ou de fonctions sommables. CRAS, Paris **247** (1958) 626–628.
[2] —: A Fourier method in diophantine problems. J. d'Analyse **15** (1965) 245–262.
[1] Helson, H., Kahane, J.-P., Katznelson, Y. and Rudin, W.: The functions which operate on Fourier transforms. Acta Math. **102** (1959) 135–157.
[1] Herz, C. S.: Spectral synthesis for the Cantor set. Proc. Nat. Acad. Sci. **42** (1956) 42–43.
[2] —: A note on summability methods and spectral analysis. TAMS **86** (1957) 506–510.
[3] —: A note on the span of translations in L^p. PAMS **8** (1957) 724–727.
[4] —: Spectral synthesis for circles. Ann. of Math. **68** (1958) 709–712.
[5] —: The spectral theory of bounded functions. TAMS **94** (1960) 181–232.
[6] —: Synthèse harmonique de distribution dans le plan. CRAS Paris **260** (1965) 4887–4890.
[7] —: Remarques sur la note précédente de M. Varopoulos. CRAS Paris **260** (1965) 6001–6004.
[8] —: Math. Reviews, 2567, **31** (1966) 462–463.
[9] —: The ideal theorem in certain Banach algebras of functions satisfying smoothness conditions. Irvine Symposium on Functional Analysis. Washington, D.C. 1967, 222–234.
[10] —: The theory of p-spaces with an application to convolution operators. TAMS **154** (1971) 69–82.
[11] —: Drury's lemma and Helson sets. Studia Math. **42** (1972) 205–219.
[1] Hewitt, E.: The asymmetry of certain algebras of Fourier–Stieltjes transforms. Mich. Math. J. **5** (1958) 149–158.
[2] —: Harmonic analysis. Some aspects of analysis and probability. New York 1958, 107–168.
[1] Hewitt, E. and Ross, K.: Abstract harmonic analysis. Berlin–Heidelberg–New York.
[1] Hewitt, E. and Stromberg, K. Real and abstract analysis. Berlin–Heidelberg–New York 1965.

[1] Hille, E. and Phillips, R. S.: Functional analysis and semi-groups. AMS Colloquium Publications 1957.
[1] Hoffman, K.: Banach spaces of analytic functions. New York 1962.
[2] —: Lectures on sup norm algebras. NATO Summer School on Topological Algebra, Bruges (1966) 1–74.
[1] Hoffmann-Jørgensen, J.: A generalization of the strict topology. Math. Scand. **30** (1972) 313–323.
[1] Hörmander, L.: Estimates for translation-invariant operators in L^p spaces. Acta Math. **104** (1960) 93–140.
[2] —: Supports and singular supports of convolutions. Acta Math. **110** (1963) 279–302.
[1] Horváth, J.: Topological vector spaces and distributions. Reading, Mass. 1966.
[1] Ikehara, S.: An extension of Landau's theorem in the analytical theory of numbers. J. Math. and Physics **10** (1931) 1–12.
[1] Ingham, A. E.: On Tauberian theorems. Proc. Lond. Math. Soc. **14A** (1965) 157–173.
[1] Jerison, M., Siegel, J. and Weingram, S.: Distinctive properties of Stone–Čech compactifications. Topology **8** (1969) 195–201.
[1] Josephs, J. J.: The physics of musical sound. New York 1967.
[1] Kac, M.: Can one hear the shape of a drum? Amer. Math. Monthly. Papers in Analysis **73** (1966) 1–23.
[1] Kahane, J.-P.: Sur les fonctions sommes de séries trigonometriques absolument convergentes. CRAS, Paris **240** (1955) 36–37.
[2] —: Sur certaines classes de séries de Fourier absolument convergentes. J. Pures App. Math. **35** (1956) 249–259.
[3] —: Sur un théorème de Wiener–Lévy. CRAS, Paris **246** (1958) 1949–1951.
[4] —: Sur un théorème de Paul Malliavin. CRAS, Paris **248** (1959) 2943–2944.
[5] —: Lectures on mean-periodic functions. Tata Institute, Bombay 1959.
[6] —: Sur la synthèse harmonique dans l^∞. An. Acad. Brasil. Ci. **32** (1960) 179–189.
[7] —: Transformées de Fourier des fonctions sommables. Proc. Int. Cong. Math. (1962) 114–131.
[8] —: Algèbres tensorielles et analyse harmonique. Sém. Bourbaki, Exp. 291, 1964–1965.
[9] —: On the construction of certain bounded continuous functions. Pac. J. Math. **16** (1966) 129–132.
[10] —: Sur le théorème de Beurling–Pollard. Math. Scand. **21** (1967) 71–79.
[11] —: Sur les séries de Fourier à coefficients dans ℓ^p. Orthogonal expansions and their continuous analogues. S. Illinois U. Press (1968) 257–272.
[12] —: A metric condition for a closed circular set to be a set of uniqueness. J. Approx. Theory **2** (1969) 233–236.
[13] —: Séries de Fourier absolument convergentes. Berlin–Heidelberg–New York 1970.
[1] Kahane, J.-P. and Katznelson, Y.: Sur la réciproque du théorème de Wiener–Lévy. CRAS, Paris **248** (1959) 1279–1281.
[2] —: Contribution à deux problèmes, concernant les fonctions de la class A. Israel J. Math. **1** (1963) 110–131.
[3] —: Lignes de niveau et séries de Fourier absolument convergentes. Israel J. Math. **6** (1969) 346–353.
[1] Kahane, J.-P. and Salem, R.: Sur les ensembles linéaires ne portant pas de pseudomesures. CRAS, Paris **243** (1956) 1185–1187.
[2] —: Sur les ensembles de Carleson et de Helson. CRAS, Paris **243** (1956) 1706–1708.

[3] —: Construction de pseudomesures sur les ensembles parfaits symétriques. CRAS, Paris **243** (1986) 1986–1988.
[4] —: Ensembles parfaits et séries trigonometriques. Paris 1963.
[1] Kaijser, S.: Representations of tensor algebras as quotients of group algebras. Ark. Mat. **10** (1972) 107–141.
[1] Kaplansky, I.: Primary ideals in group algebras. Proc. Nat. Acad. Sci. **35** (1949) 133–136.
[2] —: Commutative rings. Boston, Mass. 1970.
[1] Karamata, J.: Über die Hardy–Littlewood Umkehrungen des Abelschen Stetigkeilssatzes. Math. Zeit. **32** (1930) 319–320.
[1] Katznelson, Y.: Sur les fonctions opérant sur l'algèbre des séries de Fourier absolument convergentes. CRAS, Paris **247** (1958) 404–406.
[2] —: Sur le calcul symbolique dans quelques algèbres de Banach. Ann. Sci. École Norm. Sup. **76** (1959) 83–127.
[3] —: Sur un théorème de Kahane et Salem concernant les ensembles parfaits symétriques. CRAS, Paris **254** (1962) 2514–2515.
[4] —: Calcul symbolique dans les algèbres homogènes. CRAS, Paris **254** (1962) 2700–2702.
[5] —: An introduction to harmonic analysis. New York 1968.
[1] Katznelson, Y. and Körner, T.: An algebra dense in its tilde algebra. Israel J. Math. **17** (1974) 248–260.
[1] Katznelson, Y. and Malliavin, P.: Un critère d'analyticité pour les algèbres de restriction. CRAS, Paris **261** (1965) 4964–4967.
[2] —: Vérification statistique de la conjecture de la dichotomie sur une class d'algèbres de restriction. CRAS, Paris **262** (1966) 490–492.
[3] —: Analyse harmonique dans quelques algèbres homogènes. Israel J. Math. **5** (1967) 107–117.
[1] Katznelson, Y. and McGehee, C.: Measures and pseudo-measures on compact subsets of the line. Math. Scand. **23** (1968) 57–68.
[2] —: Some Banach algebras associated with quotients of $L^1(\mathbf{R})$. Indiana J. Math. **21** (1971) 419–436.
[1] Katznelson, Y. and Rudin, W.: The Stone–Weierstrass property in Banach algebras. Pac. J. Math. **11** (1961) 253–265.
[1] Kaufman, R.: Gap series and an example to Malliavin's theorem. Pac. J. Math. **28** (1969) 117–119.
[2] —: Pseudofunctions and Helson sets. Astérisque **5** (1973).
[1] Keogh, F. R.: Riesz products. Proc. Lond. Math. Soc. **14A** (1965) 174–181.
[1] Kinukawa, M.: Contractions of Fourier coefficients and Fourier integrals. J. d'Analyse **8** (1960–1961) 377–406.
[2] —: On the spectral synthesis of bounded functions. PAMS **14** (1963) 468–471.
[3] —: A note on the closure of translations in L^p. Tôhoku Math. J. **18** (1966) 225–231.
[1] Koosis, P.: Sur un théorème de Paul Cohen. CRAS, Paris **259** (1964) 1380–1382.
[2] —: On the spectral analysis of bounded functions. Pac. J. Math. **16** (1966) 121–128.
[1] Korevaar, J.: Tauberian theorems. Simon Stevin **30** (1955) 129–139.
[2] —: Distribution proof of Wiener's Tauberian theorem. PAMS **16** (1965) 353–355.
[1] Körner, T.: Pseudofunctions and Helson sets. Astérisque **5** (1973).
[1] Landau, E.: Über einen Satz des Herrn Littlewood. Rend. di Palermo **35** (1913) 265–276.

[1] Landau, H. J.: Necessary density conditions for sampling and interpolation of certain entire functions. Acta Math. **117** (1967) 37–52.

[1] Larsen. R.: An introduction to the theory of multipliers. Berlin–Heidelberg–NewYork 1971.

[1] Leibenson, Z. L.: On the ring of functions with absolutely convergent Fourier series. Uspehi Mat. Nauk (N.S.) **9** (1954) 157–162.

[1] Levinson, N.: On closure problems and the zeros of the Riemann zeta function. PAMS **7** (1956) 838–845.

[2] —: Wiener's life. BAMS **72** [no. 1, part II] (1966) 1–32.

[3] —: On the elementary character of Wiener's general Tauberian theorem. J. Math. Anal. and Appl. **42** (1973) 381–396.

[1] Levitan, B.: On an integral equation with an almost periodic solution. BAMS **43** (1937) 677–679.

[1] Lévy, P.: Sur la convergence absolue des séries de Fourier. Compositio Math. **1** (1934) 1–14.

[2] —: Sur quelques problèmes actuellement irrésolus et sans doute insolubles dans les théories des séries et des intégrales de Fourier. J. École Polytech. **145** (1939) 179–194.

[1] Lindahl, L.-Å. and Poulsen, F. (editors): Thin sets in harmonic analysis. New York 1971.

[1] Lindenstrauss, J. and Pelczynski, A.: Absolutely summing operators in L^p-spaces and their applications. Studia Math. **29** (1968) 275–326.

[1] Littlewood, J. E.: The converse of Abel's theorem on power series. Proc. Lond. Math. Soc. **9** (1910) 434–448.

[1] Loewenstein, E. V.: The history and current status of Fourier transform spectroscopy. Applied Optics **5** (1966) 845–854.

[1] Lohoué, N.: Ensembles de non-synthèse uniforme dans les algèbres $A_p(G)$. Studia Math. **36** (1970) 125–129.

[2] —: Nombres de Pisot et synthèse harmonique dans les algèbres $A_p(\mathbf{R})$. CRAS, Paris **270** (1970) 1676–1678.

[1] Loomis, L.: Abstract harmonic analysis. New York 1953.

[2] —: The spectral characterization of a class of almost periodic functions. Ann. of Math. **72** (1960) 362–368.

[1] Lozinskii, S. M.: On a theorem of N. Wiener. Dokl. Akad. Nauk SSSR **49** (1945) 542–545, **53** (1946) 687–690.

[1] Lumer, G.: Bochner's theorem, states, and the Fourier transforms of measures. Studia Math. **46** (1973) 135–140.

[1] Lyttkens, S.: The remainder in Tauberian theorems. Ark. Mat. **2** (1954) 575–588.

[2] —: The remainder in Tauberian theorems, II. Ark. Mat. **3** (1956) 315–349.

[1] Mackey, G. W.: Commutative Banach algebras. Notas de matematica, **17**, Rio de Janeiro 1959.

[1] Macmillan, D. H.: Tides. London 1966.

[1] Malgrange, B.: Existence et approximation des solutions des équations aux dérivées partielles et des équations de convolution. Ann. Inst. Fourier **6** (1955) 271–355.

[1] Malliavin, P.: Sur l'impossibilité de la synthèse spectrale dans une algèbre de fonctions presque périodiques. CRAS, Paris **248** (1959) 1756–1759.

[2] —: Sur l'impossibilité de la synthèse spectrale sur la droite. CRAS, Paris **248** (1959) 2155–2157.

[3] —: Impossibilité de la synthèse spectrale sur les groupes Abéliens. Inst. Hautes Études Sci., **2** (1959) 61–68.

[4] —: Ensembles de résolution spectrale. Proc. Int. Cong. Math. (1962) 368–378.

[1] Mandelbrojt, S.: Les Taubériens généraux de Norbert Wiener. BAMS **72** [no. 1, part II] (1966) 48–51.

[1] Marcinkiewicz, J.: On Riemann's two methods of summation. J. Lond. Math. Soc. **10** (1935) 268–272.

[2] —: Sur la convergence absolue des séries de Fourier. Mathematica, Cluj **16** (1940) 66–73.

[1] McGehee, C.: Certain isomorphisms between quotients of a group algebra. Pac. J. Math. **21** (1967) 133–152.

[2] —: An account of the proof that the union of two Helson sets is a Helson set 1972.

[3] —: Math. Reviews, 5939, **46** (1973) 1020–1021.

[1] Meyer, Y.: Le spectre de Wiener. Studia Math. **27** (1966) 189–201.

[2] —: Idéaux fermés de L^1 dans lesquels une suite approche l'identité. Math. Scand. **19** (1966) 219–222.

[3] —: Synthèse harmonique. Sem. d'Anal. Harm, Orsay 1968–1969.

[4] —: Les nombres de Pisot et al synthèse harmonique. Ann. Sci. École Norm Sup. **3** (1970) 235–246.

[5] —: Algebraic numbers and harmonic analysis. Amsterdam 1972.

[6] —: Trois problèmes sur les sommes trigonométriques. Astérisque **1** (1973).

[1] Meyer, Y. and Rosenthal, H.: Convexité et ensembles de Ditkin forts. CRAS, Paris **262** (1966) 1404–1406.

[1] Michelson, A. A.: Studies in optics. Chicago 1962.

[1] Mirkil, H.: A counterexample to discrete spectral synthesis. Compositio Math. **14** (1960) 269–273.

[2] —: The work of Šilov on commutative semi-simple Banach algebras. Notas de matematica, **20**, Rio de Janeiro, 1966.

[1] Moh, T. T.: On a general Tauberian theorem. PAMS **36** (1972) 167–172.

[1] Muraz, D.: Critères de compacité étroite sur un groupe abélien localement compact. Bull. Soc. Math. France **96** (1972) 263–271.

[1] Newman, D. J.: The closure of translates in l^p. Amer. J. Math. **86** (1964) 651–667.

[2] —: Translates are always dense on the half-line. PAMS **21** (1969) 511–512.

[1] Nyman, B.: On some groups and semi-groups of translations. Thesis at U. of Uppsala 1950.

[1] Osipov, V. F.: Spectral synthesis on a sphere. Vestnik Leningrad Univ. no. **7** (1972) 156–157.

[1] Paley, R. E. A. C. and Wiener, N.: Fourier transforms in the complex plane. AMS Colloquium Publications 1934.

[1] Pitt, H. R.: General Tauberian theorems. Proc. Lond. Math. Soc. **44** (1938) 243–288.

[2] —: Tauberian theorems. Oxford 1958.

[1] Pitt, H. R. and Wiener, N.: On absolutely convergent Fourier–Stieltjes transforms. Duke Math. J. **4** (1938) 420–436.

[1] Pollard, H.: The closure of translations in L^p. PAMS **2** (1951) 100–104.

[2] —: The harmonic analysis of bounded functions. Duke Math. J. **20** (1953) 499–512.

[1] Povzner, A.: On the spectrum of bounded functions. Dokl. Akad. Nauk. SSSR **57** (1947) 755–758.

[1] Pták, V.: Un théorème de factorisation. CRAS, Paris **275** (1972) 1297–1299.

[1] Rainwater, J.: A remark on regular Banach algebras. PAMS **18** (1967) 255–256.

[1] Ramirez, D. E.: Uniform approximation by Fourier–Stieltjes transforms. Proc. Camb. Phil. Soc. **64** (1968) 323–333.

[2] —: A characterization of quotient algebras of $L^1(G)$. Proc. Camb. Phil. Soc. **66** (1969) 547–551.

[1] Read, H. E.: A concise history of modern painting. New York 1959.

[1] Reiter, H.: Investigations in harmonic analysis. TAMS **73** (1952) 401–427.

[2] —: On a certain class of ideals in the L^1 algebra of a locally compact abelian group. TAMS **75** (1953) 505–509.

[3] —: Contributions to harmonic analysis. II. Math. Ann. **133** (1957) 298–302.

[4] —: Contributions to harmonic analysis. III. J. Lond. Math. Soc. **32** (1957) 477–483.

[5] —: Contributions to harmonic analysis. IV. Math. Ann. **135** (1958) 467–476.

[6] —: Contributions to harmonic analysis. VI. Ann. of Math. **77** (1963) 552–562.

[7] —: Subalgebras of $L^1(G)$. Nederl. Akad. Wetensch. Indag. Math. **27** (1965) 691–696.

[8] —: Classical harmonic analysis and locally compact groups. Oxford 1968.

[1] Ricci, F.: Multiplication of pseudomeasures. 1975.

[1] Richards, I.: On the disproof of spectral synthesis. J. Comb. Theory **2** (1967) 61–70.

[1] Rickart, C.: Banach algebras. New York 1960.

[1] Rickert, N.: A note on positive definite functions. Studia Math. **36** (1970) 223–226.

[1] Riesz, F.: Über die Fourierkoeffizienten einer stetigen Funktion von beschränkter Schwankung. Math. Zeit. **2** (1918) 312–315.

[1] Riesz, F. and St.-Nagy, B.: Functional analysis. New York 1955.

[1] Riss, J.: Élements de calcul différentiel et théorie des distributions sur les groupes abéliens localement compacts. Acta Math. **89** (1953) 45–108.

[1] Rosenthal, H.: On the existence of approximate identities in ideals of group algebras. Ark. Mat. **7** (1967) 185–191.

[2] —: A characterization of restrictions of Fourier–Stieltjes transforms. Pac. J. Math. **23** (1967) 403–418.

[3] —: A characterization of the linear sets satisfying Herz's criterion. Pac. J. Math. **28** (1969) 663–668.

[1] Rossi, B.: Optics. Reading, Mass. 1957.

[1] Royden, H. L.: Real analysis (second edition). New York 1968.

[1] Rubel, L. A. and Ryff, J. V.: The bounded weak * topology and the bounded analytic functions. J. Fun. Anal. **5** (1970) 167–183.

[1] Rudin, W.: The automorphisms and the endomorphisms of the group algebra of the unit circle. Acta Math. **95** (1956) 39–56.

[2] —: The closed ideals in an algebra of analytic functions. Can. J. Math. **9** (1957) 426–434.

[3] —: Weakly almost periodic functions and Fourier–Stieltjes transforms. Duke Math. J. **26** (1959) 215–220.

[4] —: Closed ideals in group algebras. BAMS **66** (1960) 81–83.

[5] —: Fourier analysis on groups. New York 1962.

[6] —: Functional analysis. New York 1973.

[1] Ryan, R.: Fourier transforms of certain classes of integrable functions. TAMS **105** (1962) 102–111.

[1] Saeki, S.: An elementary proof of a theorem of Henry Helson. Tôhoku Math. J. **20** (1968) 244–247.

[3] —: On norms of idempotent measures. PAMS **19** (1968) 600–602.
[3] —: On norms of idempotent measures, II. PAMS **19** (1968) 367–371.
[4] —: Spectral synthesis for the Kronecker sets. J. Math. Soc. Japan **21** (1969) 549–563.
[5] —: A characterization of *SH*-sets. PAMS **30** (1971) 497–503.
[6] —: *S*-tensor products of Banach algebras and restriction algebras of Fourier algebras. Tôhoku Math. J. **24** (1972) 281–300.
[7] —: On strong Ditkin sets. Ark. Mat. **10** (1972) 1–7.
[1] Salem, R.: On singular monotonic functions of the Cantor type. J. Math. and Physics **21** (1942) 69–82.
[2] —: On singular monotonic functions whose spectrum has a given Hausdorff dimension. Ark. Mat. **1** (1950) 353–365.
[3] —: Sur une proposition équivalente à l'hypothèse de Riemann. CRAS, Paris **236** (1953) 1127–1128.
[4] —: Algebraic numbers and Fourier series. Boston, Mass. 1963.
[5] —: Oeuvres mathématiques. Paris 1967.
[1] Salinger, D. L. and Varopoulos, N. Th.: Convolutions of measures and sets of analyticity. Math. Scand. **25** (1969) 5–18.
[1] Schiff, L.: Quantum mechanics. New York 1955.
[1] Schoenberg, I. J.: A remark on the preceding note by Bochner. BAMS **40** (1934) 277–278.
[1] Schreiber, B. M.: On the coset ring and strong Ditkin sets. Pac. J. Math. **32** (1970) 805–812.
[1] Schwartz, L.: Théorie générale des fonctions moyennepériodiques. Ann. of Math. **48** (1947) 857–929.
[2] —: Sur une propriété de synthèse spectrale dans les groupes noncompacts. CRAS, Paris **227** (1948) 424–426.
[3] —: Analyse et synthèse harmoniques dans les espaces de distributions. Can. J. Math. **3** (1951) 503–512.
[4] —: Méthodes mathématiques pour les sciences physiques. Paris 1961.
[5] —: Théorie des distributions. Paris 1966.
[1] Segal, I. E.: The group ring of a locally compact group. Proc. Nat. Acad. Sci. **27** (1941) 348–352.
[2] —: The span of the translations of a function in a Lebesgue space. Proc. Nat. Acad. Sci. **30** (1944) 165–169.
[3] —: The group algebra of a locally compact group. TAMS **61** (1947) 69–105.
[4] —: The class of functions which are absolutely convergent Fourier transforms. Acta Sci. Math. Szeged. **12** (1950) 157–160.
[1] Shapiro, H. S.: (proposer) Problem 5199. Amer. Math. Monthly **72** (1965) 435–436.
[1] Shilov, G. E.: On regular normed rings. Trav. Inst. Math. Steklov **21** (1947) 100–118.
[1] Simon, A. B.: Cesàro summability on groups: characterization and inversion of Fourier transforms. Function Algebras. Glenview, Ill. 1966, 208–215.
[1] Spector, R.: Espaces de mesures et de fonctions invariants par les isomorphismes locaux de groupes abéliens localement compacts. Ann. Inst. Fourier **15** (1965) 325–343.
[2] —: Groupes localement isomorphes et transformation de Fourier avec poids. Ann. Inst. Fourier **19** (1969) 195–217.
[1] Stegeman, J. D.: Extension of a theorem of H. Helson. Proc. Int. Cong. Math. Abstracts, Section 5 (1966) 28.

[1] Sunouchi, G.: On the convolution algebra of Beurling. Tôhoku Math. J. **19** (1967) 303–310.
[1] Titchmarsh, E. C.: The theory of the Riemann zeta-function. Oxford 1951.
[1] Tricomi, F. G.: Integral equations. New York 1957.
[1] Varopoulos, N. Th.: Continuité des formes linéaires positives sur une algèbre de Banach avec involution. CRAS, Paris **258** (1964) 1121–1124, 2465–2467.
[2] —: Sur les ensembles parfaits et les séries trigonométriques. CRAS, Paris **260** (1965) 3831–3834.
[3] —: Sur les ensembles parfaits et les séries trigonométriques. CRAS, Paris **260** (1965) 4668–4670.
[4] —: Sur les ensembles parfaits et les séries trigonométriques. CRAS, Paris **260** (1965) 5997–6000.
[5] —: Sur les ensembles parfaits et les séries trigonométriques. CRAS, Paris **262** (1965) 384–387.
[6] —: Sur un théorème de M. Katznelson. CRAS, Paris **263** (1966) 785–787.
[7] —: Spectral synthesis on spheres. Proc. Camb. Phil. Soc. **62** (1966) 379–387.
[8] —: Tensor algebra and harmonic analysis. Acta Math. **119** (1967) 51–111.
[9] —: Algèbres tensorielles et analyse harmonique. Sem. d'Anal. Harm., Orsay 1967–1968.
[10] —: On a problem of A. Beurling. J. Fun. Anal. **2** (1968) 24–30.
[11] —: Groups of continuous functions in harmonic analysis. Acta Math. **125** (1970) 109–154.
[1] Veech, W. A.: Almost automorphy and a theorem of Loomis. Arch. Math. **18** (1967) 267–270.
[2] —: Properties of minimal functions on abelian groups. Amer. J. Math. **91** (1969) 415–440.
[1] Vretblad, A.: Spectral analysis in weighted L^1 spaces on **R**. Ark. Mat. **11** (1973) 109–138.
[1] Wallin, H.: On Bohr's spectrum of a function. Ark. Mat. **4** (1961) 159–162.
[1] Warner, C. R.: A generalization of the Šilov–Wiener Tauberian theorem. J. Fun. Anal. **4** (1969) 329–331.
[2] —: A class of spectral sets 1970.
[1] Wendel, J. G.: On isometric isomorphisms of group algebras. Pac. J. Math. **1** (1951) 305–311.
[2] —: Left centralizers and isomorphisms of group algebras. Pac. J. Math. **2** (1952) 251–261.
[1] Wermer, J.: On a class of normed rings. Ark. Mat. **2** (1954) 537–551.
[1] Weyl, H.: Ramifications, old and new, of the eigenvalue problem. BAMS **56** (1950) 115–139.
[1] Whitney, H.: On ideals of differentiable functions. Amer. J. Math. **70** (1948) 635–658.
[1] Widder, D. V.: The Laplace transform. Princeton 1946.
[2] —: An introduction to transform theory. New York 1971.
[1] Wielandt, H.: Zur Umkehrung des Abelschen Stetigkeitssatzes. Math. Zeit. **56** (1952) 206–207.
[1] Wiener, N.: The quadratic variation of a function and its Fourier coefficients. J. Math. and Physics **3** (1924) 72–94.
[2] —: On the representation of functions by trigonometrical integrals. Math. Zeit. **24** (1925) 575–616.
[3] —: On a theorem of Bochner and Hardy. J. Lond. Math. Soc. **2** (1927) 118–123.
[4] —: A new method in Tauberian theorems. J. Math. and Physics **7** (1928) 161–184.
[5] —: Generalized harmonic analysis. Acta Math. **55** (1930) 117–258.
[6] —: Tauberian theorems. Ann. of Math. **33** (1932) 1–100.

[7] —: The Fourier integral and certain of its applications. Cambridge 1933.
[8] —: I am a mathematician. Cambridge, Mass. 1964.
[9] —: Nonlinear problems in random theory. Cambridge, Mass. 1966.
[10] —: Wiener memorial issue. BAMS [no. 1, part II] 1966.
[11] —: Time series. Cambridge, Mass. 1970.
[1] Wiener, N. and Wintner, A.: Fourier–Stieltjes transforms and singular infinite convolutions. Amer. J. Math. **60** (1938) 513–522.
[2] —: On singular distributions. J. Math. and Physics **17** (1939) 233–246.
[1] Wik, I.: On a strong form of spectral synthesis. Ark. Mat. **6** (1965) 55–64.
[2] —: Extrapolation of absolutely convergent Fourier series by identically zero. Ark. Mat. **6** (1965) 65–76.
[3] —: Criteria for absolute convergence of Fourier series of functions of bounded variation. TAMS **163** (1972) 1–24.
[1] Williamson, J. H.: Banach algebra elements with independent powers, and theorems of Wiener–Pitt type. Function Algebras. Glenview, Ill. 1966, 186–197.
[1] Zygmund, A.: Trigonometric series. (First edition). Warsaw 1935.
[2] —: Trigonometric series. Cambridge 1959.

Index of proper names

The following books are not referenced in the above list of proper names: [Benedetto, 2; 6; 12; Dunford and Schwartz, 1; Edwards, 5; Hewitt and Ross, 1; Hewitt and Stromberg, 1; Horvàth, 1; Kahane, 13; Kahane and Salem, 4; Katznelson, 5; Loomis, 1; Riesz and Sz.-Nagy, 1; Royden, 1; Rudin, 5; 6].

Index of terms